AF096732

MEDIEVAL SCIENCE IN THE NORTH

KNOWLEDGE, SCHOLARSHIP, AND SCIENCE IN THE MIDDLE AGES

VOLUME 2

General Editor
Helen Foxhall Forbes, *Durham University*

Editorial Board
Nicholas Everett, *University of Toronto*
Giles Gasper, *Durham University*
Christina Lee, *Nottingham University*
Immo Warntjes, *Trinity College, Dublin*
Antony Watson, *Archaeological Institute of the Academy of Sciences, Kazakhstan*
Jonas Wellendorf, *University of California, Berkeley*

Medieval Science in the North

Travelling Wisdom, 1000–1500

Edited by
CHRISTIAN ETHERIDGE AND
MICHELE CAMPOPIANO

BREPOLS

British Library Cataloguing in Publication Data.
A catalogue record for this book is available from the British Library.

© 2021, Brepols Publishers n. v., Turnhout, Belgium.

All rights reserved. No part of this publication may be reproduced, stored in a retrieval system, or transmitted, in any form or by any means, electronic, mechanical, photocopying, recording, or otherwise without the prior permission of the publisher.

D/2021/0095/50
ISBN 978-2-503-58804-9
E-ISBN 978-2-503-58805-6
DOI 10.1484/M.KSS-EB.5.119355

Printed in the EU on acid-free paper.

Contents

Acknowledgements   9

Introduction
Christian ETHERIDGE and Michele CAMPOPIANO   11

Roger of Hereford and the Transformation of *Computus* **in Twelfth-Century England**
C. Philipp E. NOTHAFT   27

Travelling Optics
Robert Grosseteste and the Optics behind the Rainbow
Giles E. M. GASPER, Brian K. TANNER, Sigbjørn SØNNESYN and Nader EL-BIZRI   45

Language and Wisdom
Mathematics and Astronomy in Bacon´s Edition of the *Secretum secretorum*
Michele CAMPOPIANO   77

Wisdom's Trips to Denmark
Sten EBBESEN   97

Medieval Scientific Book Fragments Held in Swedish and Finnish Archives
The Tantalizing Remains of a Greater Scientific Corpus
Christian ETHERIDGE   111

Friars of Science
Dominican Transmission and Usage of Scientific Knowledge in Medieval Scandinavia
Johnny Grandjean Gøgsig JAKOBSEN   141

Master Perus of Arabia
An Exemplary Magician in Medieval Iceland
Marteinn Helgi SIGURÐSSON   159

Science in Medieval Fiction
The Case of *Konráðs saga keisarasonar*
Florian SCHRECK   181

Continental Ironmongers, Whalers, Smugglers, and Craftsmen
Immigration and Trade Routes and their Influence on the London Armourers' Industry
Brad KIRKLAND   201

Index   219

Illustrations

Figure 1.1.	Map of locations discussed within the volume.	10
Figure 2.1.	(a) Demagnification with lens further from the object than its focal length; (b) Magnification when lens is closer to the object than its focal length. Note that the demagnified image is inverted.	58
Figure 2.2.	The experiment of the appearing coin. As water is poured into the vessel, the image of the object rises due to refraction at the water-air interface, enabling it to become visible.	58
Figure 2.3	Graphical illustration of Grosseteste's description of image formation and the position of the virtual image of the object.	60
Figure 2.4	(a) Identically sized objects appear larger as the distance to the observer is reduced due to the increasing angle subtended at the observed; (b) Different size objects at different distances appear of similar size due an identical angle subtended at the observer.	60
Figure 2.5	(a) Magnification of an object by formation of a virtual image on the far side (from the eye) of a lens; (b) Demagnification by formation of a real, inverted image on the same side of the lens as the eye.	61
Figure 2.6.	Data points of Ptolemy and Witelo with best fit to Snell's Law (solid line) for either data set and calculated (dotted) line for the air-water interface using the modern value for the refractive index.	64
Figure 2.7.	Refraction from air to glass data with best fit to Snell's Law.	65
Figure 2.8.	Data for water to glass refraction from Ptolemy and Witelo. Solid line is the best fit to the data using the modified Snell's Law.	67
Figure 2.9.	Plot of Witelo's supposed measurements of the angles of incidence and refraction for the water-air interface, where light is now passing from the denser to the less dense medium.	68
Figure 5.1.	Stockholm, Riksarkivet Var 2, Aristotle, *De memoria et reminiscentia* (On Memory and Reminiscence), thirteenth century.	127
Figure 5.2.	Stockholm, Riksarkivet Var 33, Pseudo-Aristotle, *Secretum secretorum* (The Secret of Secrets), fourteenth century.	128

Figure 5.3.	Stockholm, Riksarkivet Var 41, Arzachel, *Canones in motibus celestium corporum* (Rules for the Movements of the Heavenly Bodies), thirteenth century.	129
Figure 5.4.	Stockholm, Riksarkivet SRA Fr 9399, Averroes, *De substantia orbis* (Concerning the Substance of the Celestial Sphere), fourteenth century.	130
Figure 6.1.	Dominican convents in the Baltic Sea region with provincial borders around 1500.	155

Acknowledgements

On a hot September day in Palermo in 2014, Christian Etheridge and Michele Campopiano were walking with Christian Høgel, from the Centre for Medieval Literature, through the old streets of the city discussing various aspects of medieval history. The subject of medieval science came up and as the special atmosphere of old Palermo began to affect us, an idea started to form. We decided that Christian Etheridge and Michele Campopiano would set up a conference on medieval science and that the Centre for Medieval Literature, based in Odense and York, would fund the proceedings. The following year the conference took place and the journey began which ultimately ended up in the book you now have in your hands. We have many people to thank who helped with this volume along the way.

We would first like to thank Shazia Jagot, who was the co-organizer of the Travelling Wisdom conference with us. Without her it would have been impossible to organize the conference. Unfortunately, for personal reasons, she was unable to contribute to this volume, but we hope that the present volume stands to a testimony to all of her hard work. We would especially like to thank the Centre for Medieval Literature for their sponsoring of this conference, especially the organizers: Lars Boje Mortensen, Christian Høgel, and Elizabeth Tyler. We would also like to give special thanks to Charlotte Vegge Thorup Hansen and Maiken Bundgaard Villumsen whose magnificent powers of organization were fundamental to the running of the conference. We would also like to thank Giles Gasper who suggested that we publish this volume through Brepols. At Brepols, we would like to thank Rosie Bonté for all of her tireless help along the way, as well as Helen Foxhall Forbes at Durham University, in her role as series editor, and Maria Whelan for her excellent copyediting. We would also like to thank Jens Eike Schnall, Dale Kedwards, and Rudy Simek who gave papers at the conference but who were unable to contribute to the current volume. Finally, Christian Etheridge would like to thank his family both in Denmark and England for all the help they have given over the years. He would especially like to thank his dear wife Majken and son Arthur. He could not have made this book without them. Michele Campopiano would like to thank his father Ettore, who passed away last year, for inspiring his interest in the history of science.

Figure 1.1. Map of locations discussed within the volume. Map courtesy of Johnny Grandjean Gøgsig Jakobsen.

CHRISTIAN ETHERIDGE
AND MICHELE CAMPOPIANO*

Introduction

Omnis sapientia a Domino Deo est; soli quod desiderant facere possunt sapientes

[All wisdom is from the Lord God, only the wise people are able to do what they desire.][1]

The above passage is found on the pages of the *Hortus deliciarum*, a late twelfth-century manuscript, written under the aegis of Herrad of Landsberg, abbess of Hohenberg Abbey in Alsace. On folio 32^r of the manuscript an illumination depicts a woman as Wisdom incarnate who holds the above quotation in a scroll. During the Middle Ages, and especially from the twelfth century onwards, wisdom was often personified as a woman, who was given the name of *Philosophia* or Lady Philosophy and who was the source of all knowledge. Schematic diagrams, such as those found in the *Hortus deliciarum*, depict *Philosophia* sitting above learned scholars like Plato and Socrates with the branches of wisdom, also depicted as women, surrounding her. Several genres of medieval science such as astronomy and geometry are depicted as these aspects of *Philosophia*. Therefore, scientific knowledge is depicted in these diagrams as being a part of wisdom in general.[2] The accumulation of this scientific knowledge,

---

* Michele Campopiano was responsible for the section *Science in the Middle Ages*; all other sections were written by Christian Etheridge.
1 From the *Hortus Deliciarum* fol. 32^r. The image is partially reproduced on the front cover of the current volume. Photograph taken by Dnalor_01, Wikimedia Commons CC-BY-SA 3.0. Image reproduced in full in Joyner, *Painting the Hortus Deliciarum*, p. 70.
2 For more on *Philosophia* in medieval depictions see Joyner, *Painting the Hortus Deliciarum*, pp. 71–76, for further on this with a focus on twelfth-century depictions see also Cleaver, *Education in Twelfth-Century Art and Architecture*, pp. 11–17.

> **Christian Etheridge** • is a Mads Øvlisen Novo Nordisk Postdoctoral Fellow at the National Museum of Denmark. He recently defended his doctoral thesis, 'The Transmission and Reception of Science in Medieval Scandinavia 1100–1525' at the University of Southern Denmark. Christian.etheridge@natmus.dk
>
> **Michele Campopiano** • is Senior Lecturer (Associate Professor) at the University of York and von Humboldt Fellow (Experienced Researcher) at the Institute for History at the Technical University of Darmstadt. michele.campopiano@york.ac.uk

or wisdom, in the Middle Ages required travel. This could be the physical travelling of a scholar to a library, university, cathedral school, or other repository of knowledge. It could also be the journey of a book from one individual to the next as the volume moved over time and space from person to person. It could also be the movement of a scientific treatise that was taught, copied, and translated over the centuries.

The current volume is derived from an international conference on medieval science; *Travelling Wisdom: Medieval Science in the North c. 1000–1500*, held at the University of Southern Denmark in the city of Odense during May 2015. This conference was a gathering of specialists in the field of medieval science who over the course of a couple of very fruitful days delivered a series of papers and shared discussions on all aspects of science and scientific practitioners active during the Middle Ages, with a more specific focus on the connection between England and Scandinavia, which shared for a large part of the Middle Ages an important history of exchange and contact.[3] Each of the papers at the conference discussed the role of transmission in the development of a range of scientific endeavours in England and Scandinavia during the period 1000–1500. We wanted to look at this from a multi-disciplinary viewpoint and across language boundaries, which is reflected in both the academic background of the speakers and the series of articles derived from their papers.[4] We had a special interest in the development of scientific knowledge as it moved between various places and languages, therefore the essays collected in this volume address the role translation played in the understanding of science and natural philosophy, especially as treatises moved from Arabic and Greek into Latin and then further into the Northern European vernaculars. We were also interested in studies on the transmission of technology and practical knowledge, not just theoretical, thereby looking to a broader body of interaction rather than just that boosted by literate circles.

This volume brings together nine chapters based on the papers given at the Travelling Wisdom conference on science in the Middle Ages. Our conference in 2015 was focused on the broad theme of 'Travelling Wisdom' in that science, in the form of wisdom, was something that moved throughout Africa, Asia, and Europe during the Middle Ages. This wisdom could travel for example in a scientific manuscript, in a story, in teaching or in the form of practical knowledge. This theme of the journey of scientific knowledge as it travelled unites all the chapters in this book. As outlined above we were interested in narrowing our themes to a smaller geographical area and so chose England and Scandinavia. This reflected our own research interests but also the areas had similarities. Both England and Scandinavia were formed of highly centralized kingdoms, that were geographically on the periphery of Europe but were heavily involved in maritime trading networks that firmly linked them to the

---

3 For an informative recent investigation on the connections between Britain (specifically East Anglia) and Scandinavia see the articles in Bates and Liddiard, eds, *East Anglia and its North Sea World in the Middle Ages*.
4 The gender balance of the speakers at the conference was not a deliberate act on the editors' part, but simply reflected those who responded to the call for papers. The second conference in this series, *Travelling Wisdom: Science in the Medieval Religious Orders* had an even gender balance. We hope to publish the papers of this second conference in the foreseeable future.

rest of the Continent. England and Scandinavia also had a shared political, trading, cultural, and linguistic history that stemmed back to at least the seventh century and continued throughout the Middle Ages. We do not wish to perpetuate an idea that these English and Scandinavian scholars were mere receptacles of a learning that they may have received from somewhere like the University of Paris. Instead as the chapters show, this knowledge was taken, commented on, translated, and fashioned anew into something different than before and formed another strand of the rich tapestry of science in the Middle Ages. Another subject that also binds all of the chapters together is that of translation. The movement of texts from one language to another was one of the most important aspects of medieval science, which required a series of editorial choices on which words and terms to translate and which to leave in the original language. The chapters approach the translation of texts from Arabic and Hebrew into Latin and from Latin into the vernacular.

The first six chapters of this book document the transfer and movement of ideas in the Middle Ages. Individual scholars are the focus of Chapter One by C. Philipp E. Nothaft on Roger of Hereford, Chapter Two by Giles Gasper et. al. on Robert Grosseteste and Chapter Three by Michele Campopiano on Roger Bacon. A series of scholars from Denmark educated in universities abroad form the study of Sten Ebbesen in Chapter Four. The fragments of manuscripts of medieval science owned by Swedes and Finns who studied abroad is the focus of Christian Etheridge in Chapter Five. Finally, the Dominican Order in their province of Dacia and the educational travels of learned friars is the subject of Chapter Six by Johnny Jakobsen.

The next two chapters move from the transmission of scientific treatises through the network of universities, cathedral schools and the mendicant orders to another way of approaching medieval science, namely through literature. Often sidelined by scholars on medieval science for more direct scientific treatises, medieval literature contains a surprising amount of material concerning the science of the Middle Ages. The chapters written by Marteinn Sigurðsson and Florian Schreck have many parallels between them, and it is useful to explore these here. Both writers here focus on the lesser-known Icelandic chivalric sagas (*riddarasögur*) for their choice of material. The chivalric sagas are relatively unknown outside of specialist fields and are overshadowed by the famous family sagas (*Íslendingasögur*). Focusing on tales of chivalry and exotic places, the chivalric sagas were more popular in medieval Iceland than the better-known family sagas and contain a wealth of interesting material. With his focus on *Konráðs saga keisarasonar*, Schreck details how the anonymous writer was able to use a wide range of encyclopaedic texts dealing with subjects such as minerology, astronomy, and zoology to help enlighten their audience. The author of *Konráðs saga keisarasonar* takes the reader on a journey through places where the protagonists find knowledge including the Holy Roman Empire, Constantinople, and Baghdad. Marteinn Sigurðsson chooses *Clári saga keisarasonar* as his text to explore the world of chivalric sagas. Like *Konráðs saga keisarasonar* the world of *Clári saga keisarasonar* is also full of characters from distant places such as France, the Holy Roman Empire, Scotland, Arabia, and Ethiopia. The main protagonist in fact is a learned scholar from Arabia, Master Perus, who using his advanced knowledge is always able to outwit his enemies. This knowledge includes astronomy and the ability to construct automata. The author Jón Halldórsson was himself a university-educated Dominican and Marteinn Sigurðsson points out that his learning underlines the text of *Clári saga keisarasonar*. Both chapters show that the chivalric sagas are a rich mine of contemporary attitudes and understanding of medieval science in Iceland

In many ways the last chapter of this volume, that of Brad Kirkland, is atypical when compared to the others. It concerns technological adaption rather than theoretical science. Yet it is also the culmination of the volume. As well as linking Scandinavia and England with the trade in whalebone and iron, the chapter is like the previous ones concerned with the travelling of ideas, of wisdom, and of knowledge. The armorers that are described in Kirkland's chapter are mobile. They come from all over England, from Sandwich to Durham, and also come from as far afield as France, Flanders, and the Holy Roman Empire. In many ways they mirror the mobility of the scholars described in the previous chapters, as well as that of the literary figures in the chapters of Marteinn Sigurðsson and Schreck. The use and development of technological science changed the course of warfare in the Middle Ages and this practical use of science can be seen in some of the previous chapters too. Medicine was also a practical science as can be seen in the chapters of Etheridge, Jakobsen, and Schreck, as was optics as can be seen in the chapters of Nothaft and Gasper et. al. By showing the change brought about by the exchange of technological scientific ideas, Kirkland shows that medieval science was not just restricted to a clerical elite but had much wider repercussions throughout society.

Science in the Middle Ages

The history of medieval science has increasingly attracted scholarly attention over the last decades. Modern research on the topic has cast off the positivistic approach that previously considered medieval science as a simple repetition of bookish knowledge, and instead has moved to understand its peculiar dynamics and its capacity to produce innovation. No one can deny that the lack of knowledge of Greek in the early medieval West as well as the loss of continuity with the ancient education system built up in the Classical world severely restricted the basis of scientific knowledge. However, science in the early medieval period should not be considered as a pure repetition of the fragments of wisdom contained in the limited number of scientific texts available in the West during the Middle Ages. The early Middle Ages saw the writing of possibly the most important encyclopaedic text on which so much of later medieval culture was based on, namely the *Etymologiae* of Isidore of Seville (c. 560–636).[5] Early medieval scholars tried to address new questions concerning the natural world and were able to absorb new empirical information into their works.[6] Particularly in the British Isles, the development of scientific themes was remarkable. Bede (673–735) shows

---

5 Fontaine, *Isidore de Seville*; Courcelle, *Les lettres grecques en Occident*, pp. 74–78; Riché, *Ecoles et enseignement*, pp. 13–16.
6 Caiazzo, 'Filosofia della natura e fisica elementare nell'Alto Medioevo'.

a clear interest in the natural sciences, as can be seen from his *De natura rerum*, a text which, compared to Isidore's book of the same name, tends to avoid allegorical discussions, focusing instead on physical cosmology. But even more remarkable was the development in Ireland of the *computus*, linked to the disputes over the date of Easter. Studies on the *computus*, initially from Ireland and later elsewhere in the British Isles, were essential in medieval Europe in the centuries to come. Once again essential in this area were the works of Bede, which left a profound influence in other areas related to the study of the calendar, such as historical chronology. Complex discussions concerning methods of computing the date of Easter or on the antipodes demonstrate the complex mathematical, astronomical, and geographical debates that early medieval scholars were able to engage with.[7] The centuries of the early Middle Ages also saw an important development of other scientific aspects, such as the study of music, part of the *quadrivium*. Music is deeply linked to the theory of numbers and studies on ratios and proportions, which acquired a central importance in the teaching of the ninth and tenth centuries.[8] This was also facilitated by the attention paid to a wider range of Latin texts from the Carolingian era.[9] Practical science such as the botanical knowledge shown in monastic herbal gardens and or the hydro-technology used in monasteries in this period demonstrates connections between scientific knowledge and its practical application. In the Carolingian period, the works of philosophers such as John Scotus Eriugena, which have their roots in a revival of the study of Latin works such as Pliny's *Naturalis historia*, Calcidius's translation and commentary of the *Timaeus* and Macrobius's *Commentarii in Somnium Scipionis*, testify to the intense and complex cosmological discussions which could find place in the so-called 'Dark Ages'.[10]

It is, however, without doubt that it was in the High Middle Ages that the development of scientific studies found new stimuli and connections. Contact with works of Arabic science and new waves of translation of Greek texts helped to develop these interests and to develop new avenues of thought. Jewish texts and scholars also contributed decisively to update Latin wisdom, both by making Arabic texts available through translation, often with instructive and innovative commentary, or due to scientific developments from scholars from the Jewish learned community.

Arabic science itself had been deeply influenced by translations of Greek philosophical and scientific texts. Central to this development had been the patronage of philosophy and science under the Abbasid caliphs in Iraq from the late eighth century, under rulers such as Hārūn al-Rashīd (d. 809) and al-Ma'mūn (813–33), who founded the *Bayt al-Ḥikma*, the House of Wisdom, a rich library and translation centre.[11] Scholars in the Islamic world did not simply translate and study Greek wisdom: they also absorbed knowledge from other Middle Eastern traditions, such

---

7 Eastwood, *Ordering the Heavens*, pp. 10–13 and Gautier Dalché, 'La Terre dans le cosmos', pp. 193–98.
8 Beaujouan, 'L'enseignement du "quadrivium"'.
9 Eastwood, 'Early-Medieval Cosmology', p. 307.
10 Eastwood, 'Early-Medieval Cosmology', pp. 307–08.
11 Gutas, *Greek Thought, Arabic Culture*, pp. 53–50, and Baffioni, *Storia della filosofia islamica*, pp. 130–31.

as the Zoroastrians and the so-called Nabateans, the pre-Islamic inhabitants of Iraq and from India as well, in particular in mathematics. Also important were the Sabians (possibly to be identified with the Mandaeans or Harranians), a religious group in Iraq and Mesopotamia whose religious philosophy combined Middle Eastern and gnostic traditions and practices with Aristotelic and hermetic sciences and who deeply influenced Islamic astronomy and astrology.[12] Finally, scholars in the Islamic world gave their own independent contributions to the development of natural sciences themselves.

The first figure in the Latin West who showed a deep interested in Arabic mathematics and astronomy was probably the Benedictine Gerbert of Aurillac, later Pope Sylvester II, who studied these subjects in the latter part of the tenth century in Catalonian centres of learning that bordered the Caliphate of Cordoba.[13] However, it is from the end of the eleventh century that Arabic science started to be absorbed with some continuity in Latin Europe. A crucial channel of transmission of both Greek and Arabic science was southern Italy, and in particular the two intellectual centres of Montecassino, with its Benedictine abbey, and Salerno, with its medical school.[14] The figure of Constantine the African, a Tunisian who converted to Christianity and translated the *Kitāb al-Malikī* of al-Majūsī, is perhaps the best known of scholars working with scientific material in southern Italy at this time.[15] Salerno was already in the eleventh century a centre of translation from Greek as well as Arabic, with Bishop Alfanus of Salerno for example translating the treatise *De natura hominis* by Nemesius of Emesa into Latin, and this attention to translating Greek texts continued in the Kingdom of Sicily with scholars such as Henricus Aristippus (died *c.* 1162), who translated among other things Plato's *Phaedo* and *Meno*.[16]

The largest contribution of translations from Arabic came, however, from Spain. The city of Toledo, captured by Alfonso VI (1065–1109) from the Arabs in 1085, became a major centre of translation in the twelfth century. Of these translators the most influential was Gerard of Cremona, to whom over seventy translations are attributed.[17] Other figures worked as intermediaries in the travels of Wisdom. Michael Scot served as a link between Toledo, where he had been a canon of the cathedral, and the court of the Holy Roman Emperor Frederick II (1220–1250) that was renowned for translating works of science from Arabic. In Toledo he translated from Arabic the *De motibus caelorum* of al-Biṭrūjī in 1217, and Aristotle's *De animalibus* shortly before 1220.[18] The translation movement brought European scholars in contact with practices of astrology, necromancy, and construction of talismans for which the Arabs were in debt not just to Greek treatises, but also to a long tradition

---

12 Green, *The City of the Moon God*.
13 Zuccato, 'Gerbert of Aurillac', pp. 742–63; Burnett, 'Translation and Transmission', pp. 341–42.
14 Jacquart and Bagliani, eds, *La Scuola Medica Salernitana*, and Gallo, ed., *Salerno e la sua scuola medica*.
15 Jacquart, and Burnett, eds, *Constantine the African*; Burnett, 'Humanism and Orientalism', pp. 22–31.
16 Alfano I, *Premnon Physicon*, ed. by Chirico; Haskins and Putnam Lockwood, 'The Sicilian Translators of the Twelfth Century'.
17 Burnett, 'The Coherence of the Arabic–Latin Translation Program'.
18 Burnett, 'Michael Scot and the Transmission of Scientific Culture', p. 102.

of Middle-Eastern religious and magic beliefs. In this light we can see John of Seville and Adelard of Bath who independently translated the *De imaginibus* of Thābit, a representative work of the Sabean intellectual tradition.[19]

The intensive engagement with Latin and Greek scientific treatises in the twelfth century also stimulated new interests and directions of thinking. The best known example of is probably that of the School of Chartres with the Neoplatonic tradition, represented in particular by Plato's *Timaeus*, known in the translation of Calcidius, accompanied by the commentary from the same.[20] The way in which classical heritage and Arabic influences interacted is exemplified by Adelard of Bath in his *Quaestiones naturales*, where he affirms that his work is based on *Arabum studia*: what he learned from the newly translated texts (the presence of Arabic sources in the *Quaestiones* is disputable at best) is, however, the methodological approach, the necessity of apprehending the world through conceptual schemes, a trend that is at the same time Platonic and 'Arabic'.[21]

The increasing role played by experience in the construction of scientific knowledge was also related to the availability of these new texts. Authors such as Albertus Magnus, Robert Grosseteste, or Roger Bacon acquired this 'experience' through traditional texts, as well as the new collections of *experimenta* like pharmacological compilations. It also implied a stronger attention to the data acquired through the senses, and to the ways that they needed to be comprehended and evaluated, due to the influence of a 'new' science such as optics.[22]

The new mendicant orders, represented by authors like Albertus Magnus and Roger Bacon, represented a crucial factor in the elaboration of new study trends.[23] Eager to acquire new knowledge to preach, to defend and spread Christianity, the Franciscans and the Dominicans also played a pivotal role in the development of universities. From the thirteenth century, these universities were at the centre of scientific enterprise in Europe. The new wisdom was absorbed through the faculties of arts and of medicine.[24] This also influenced other scholars, such as jurists. One wonderful example of this is shown by the *Tractatus Tyberiadis* by Bartolo da Sassoferrato, who heavily relied on geometry in his study of the juridical implications of the Tiber floods.[25] The connection between practice and science was indeed very lively in the Middle Ages. Medieval science renewal was also stimulated by economic and administrative transformations, as shown by Fibonacci's works such as *Practica Geometriae* and *Liber Abbaci*, which reflect an awareness of the new developments in trade and finances found in the Italian City-States. This reminds us that there is an often-unseen development of technical and practical questions that stimulates new

---

19 Burnett, 'Talismans: Magic as Science?', p. 6.
20 Waszink, *Timaeus a Calcidio translatus*.
21 Burnett, 'Adelard of Bath and the Arabs', pp. 89–107.
22 Panti, *Moti, virtù e motori celesti nella cosmologia di Roberto Grossatesta*; Draelants, 'Expérience et autorité' and Hackett, '*Ego expertus sum*'.
23 Panti, 'Scienza e teologia'.
24 Shank, 'Schools and Universities in Medieval Latin Science'.
25 Frova, 'Le traité de fluminibus', and Cavallar, '*Quod de tibere dicetur*'.

scientific answers, which we tend to see just in the form that is finally committed to written culture. For example, Fibonacci's *Liber Abbaci* proposes mathematic solutions to problems of credit and trade which clearly have their background in the activity of merchants and travellers of the Mediterranean world.[26] These works influenced in return the education of merchants and, more in general, of burghers, also due to the fact that many of these texts were then later translated or adapted into the vernacular.

Science flourished also in other cultural centres. Nunneries played a crucial role too, as it is clearly shown by the personality of the German abbess Hildegard of Bingen (1098–1179), who left a strong inheritance in fields as various as music, medicine, and natural history.[27] Medieval courts, which with figures like Frederick II had played a major role in the support of medieval scientific interests, continued to promote the work of important scholars: for example, the court of King Charles V of France in which context Nicole Oresme wrote his *Livre de divinacions*.[28] It is, however, at universities that we have to look to find the most significant scholars. Jean Buridan, a teacher at the faculty of arts in Paris, developed the theory of *impetus*, while the Masters of Merton College in Oxford developed new mathematical tools, also to be applied to natural sciences.[29] In Padua, it was a medical master, Giovanni Dondi, who built the *Astrarium*, a mechanism that gave the position of the planets.[30]

The fifteenth century saw an increase of interactions between technical developments and scientific discussions, as can be shown for example by the creation of treatises which described and summarized mechanical devices, such as *De machinis et rebus mechanicis*, written for the king of Denmark Eric VII (1396–1439) by the German (from Werden) Konrad Gruter, who spent considerable time also working in Italy.[31] Humanism intensified interest in the observation of the natural world and in technical problems, partially thanks to renewed interest in ancient scientific texts, partially because it stimulated new investigations into the relationship between humankind and cosmos. The interest in philological problems stimulated careful evaluations of the quality of the copies of ancient scientific texts available to scholars.[32] Astronomy was again one of the fields which was boosted by these new approaches to texts. An example can be given by the relationship existing between the Austrian scholar Georg Peuerbach (1423–1461) and the German scholar Johannes Müller (1436–1476), better known as Regiomontanus, and Cardinal Bessarion, papal legate to the Holy Roman Empire. Bessarion persuaded Peuerbach, who had come to be enthusiastic about the literary ideals of humanism, to produce an improved abridgment of the *Almagest*,

---

26 Catastini, Ghione, and Rashed, *Algebra*, pp. 37–63. On this issue, see also: Gautier Dalché, 'D'une technique à une culture'.
27 Panti, 'A Woman's Voice through the Centuries'; Kelby-Fulton, 'Hildegard of Bingen'; Glaze, 'Medical Writings'.
28 Coopland, *Nicole Oresme and the Astrologers*.
29 Sylla, 'The Oxford Calculators in Context'.
30 Giovanni Dondi dall'Orologio, *Tractatus astrarii*.
31 Konrad Gruter von Werden, *De machinis et rebus mechanicis*.
32 Marcacci, 'Umanesimo scientifico', pp. 151–53.

relying on Peuerbach's knowledge of Greek.³³ Peuerbach died in 1461, and the task was completed about two years later by Regiomontanus. This work, printed in 1496, was also used heavily by Copernicus.³⁴ Print was an important factor of innovation for fifteenth-century science, allowing a quick diffusion of both new treatises and ancient texts: Peuerbach's textbook *Theoricae novae planetarum* (1454), which replaced earlier works on planetary theory, was printed by Regiomontanus around 1472, reaching nearly sixty editions by the seventeenth century!³⁵ Mathematics was also a field that experienced a great creative spell in the fifteenth century, to a large extent thanks to the vital connection between theoretical questions and financial and technical problems. The name which more easily springs to mind is of course that of the Franciscan Luca Pacioli (*c.* 1447–1517), who promoted innovative teaching methods in book-keeping and was also a friend of Leonardo da Vinci: the great artist benefitted from the mathematical learning of the friar.³⁶

Scientific Traditions

The scientific treatises in the chapters in this volume can be said to stem from various traditions. Two of the most important are the Islamic/Arabic and the Jewish traditions. Nothaft describes the tradition of translations into Latin of Arabic astronomical treatises that came to England from the first half of the twelfth century onwards and how Roger of Hereford used these newly translated Arabic sources in his works of computus. Nothaft also notes the importance of Jewish scholarship for Roger of Hereford's calculations on the calendar while Gasper et. al. in Chapter Two note that Robert Grosseteste also embraced medieval Arabic scholarship on science and that he was reliant on Arabic thought and translations from Arabic in his pioneering works on optics. Furthermore, Campopiano shows that Roger Bacon's work on the Arabic *Secretum secretorum* is fundamental to his thought. Bacon was also well acquainted with scientific treatises in Arabic as well as works of Jewish philosophy and was extremely interested in the process of translation of these works into Latin. The importance of translated Arabic texts on medieval sciences in the Kingdom of Sweden during the Middle Ages is shown by Etheridge. These are texts that range from astronomical treatises by al-Zarqālī or Ibn Rushd, to texts on medicine by Ibn Sīnā through to Pseudo-Aristotelian works stemming from Arabic sources such as the *Liber de causis* and the aforementioned *Secretum secretorum*. Finally, Marteinn Sigurðsson, in his study of the works of the Dominican Jón Halldórsson shows that the literary figure of Master Perus is a representation of Arabic learning. As well as being the main protagonist of Jón Halldórsson's exempla he is also a revered figure of Arabian learning who outwits various members of the European nobility.

---

33 Overfield, *Humanism*, pp. 61–71.
34 North, "Astronomy and Astrology", pp. 471–72.
35 North, "Astronomy and Astrology", pp. 471–72.
36 Pisano, 'Details of the Mathematical Interplay'; Sangste and Scataglinibelghitar, 'Luca Pacioli'.

The Classical scientific tradition, derived from Greek and Roman antiquity, was wide reaching and included many notable figures such as Claudius Ptolemy who wrote important treatises on astronomy as did Galen for medicine. The most significant and influential of these figures for science in the Middle Ages were Plato and Aristotle. Platonic theory and teaching on vision, for example, were a strong influence on Robert Grosseteste's optical science, as demonstrated by Gasper et. al., who also show that Aristotle's works on physics, meteorology, and the rainbow were a key influence on Grosseteste's thought. Plato and Neoplatonism also had an influence on the *Secreta secretorum* tradition used by Roger Bacon, as shown by Campopiano who acknowledges that the figure of Aristotle is key in the understanding of Bacon's own edition of the *Secretum secretorum*. This text utilized the Pseudo-Aristotelian *Liber de causis* and other hermetic texts of which Aristotle was seen as a central figure. Bacon also commented on many of Aristotle's scientific treatises. Ebbesen notes the importance of Aristotle's teachings of logic and how they formed part of the curriculum in the schools and the University of Paris. For this he uses the example of Gunner of Viborg who engages returning Danish students from the University of Paris with a scholastic *Questio*. Etheridge notes that Aristotle's treatises on natural philosophy are the most common scientific works that survive today from the manuscript fragment collections that were used in the medieval Kingdom of Sweden. Many of these are likely to have their origin in the environment of the University of Paris. These types of Aristotelian texts are shown by Etheridge, by using surviving booklists and teaching curricula, to have been owned by scholars from the Kingdom of Sweden who taught at universities such as Paris and Vienna. Finally, Jakobsen discusses the commentaries of Plato and Aristotle that were to be found in the libraries of the Dominican Order in the province of Dacia.

Cathedral Chapters and Monastic Schools

The study and reception of medieval science in the twelfth century took place in the schools of cathedrals and monasteries. The learned environment of the cathedral and monastic schools of Northern France, namely those of Paris, especially St Victor, and also Chartres Cathedral, are explored by Ebbesen. The reception and creation of scientific texts in twelfth-century cathedral and monastic schools in England and Scandinavia are further explored in several of the chapters. Nothaft investigates the learned environment of Great Malvern Priory in England as well as Hereford Cathedral which is further explored by Gasper et. al. Scandinavian cathedrals and monastic houses in the twelfth century are the subject of Ebbesen. Cathedrals continued to be important centres for the reception and transmission of scientific material in England and Scandinavia, especially before the advent of universities becoming important centres of learning. In England this is around the mid-thirteenth century, while in Scandinavia it is the end of the fifteenth century. Lincoln Cathedral is the focus of Gasper et. al. and also Campopiano. Danish cathedrals such as Lund, Roskilde, Schleswig, and Viborg are dealt with by Ebbesen. Cathedrals in the Kingdom of Sweden, notably Skara, Västerås, Turku, and Uppsala are looked at by Etheridge and

Jakobsen. Hólar and Skálholt cathedrals in Iceland, along with Nidaros Cathedral in Norway, are the subject of Marteinn Sigurðsson. Finally, Etheridge looks at the great Birgittine abbey of Vadstena that had one of the greatest libraries of any contemporary institution in medieval Europe and contained many works of a scientific nature.

The Medieval University

One of the main themes of this volume is the medieval university, which was one of the most important centres of transmission of scientific material in the Middle Ages. The curriculum of the faculty of arts taught astronomy, arithmetic, and geometry, as well as Aristotelian natural philosophy, while the faculty of medicine taught astrology as well as medicine. The one university that runs through and binds many of the chapters of this volume together is the University of Paris. As Campopiano shows, Roger Bacon studied and taught in Paris and was influenced by the intellectual environment there. While Giles Gasper et. al. put forward the possibility that Robert Grosseteste also studied there and was certainly influenced by learned currents emanating from the University of Paris. Ebbesen, in his chapter, follows the educational career of several Danes at the university and on their return to Denmark. Etheridge demonstrates that several of the manuscript fragments from the medieval Kingdom of Sweden must stem from the Parisian scholarly environment and offers up the careers of several Swedish and Finnish scholars at the university as a point of comparison. Jakobsen shows how learned Dominican scholars from the province of Dacia had Paris as the centre of their network of learning, with several learned Dacian friars having studied there. Finally, Marteinn Sigurðsson is able to extract biographical details from the vita of the Norwegian Dominican Jón Halldórsson to illuminate both his learning in Paris in some detail and its influence on his later scholarship.

Other universities are also important in the context of the chapters of this volume. One of the most notable was Oxford which was central to the careers of both Robert Grosseteste and Roger Bacon as shown in the chapters by Gasper et.al. and Campopiano. Oxford and Cambridge took over from Paris as the most important universities for the study of scientific subjects for scholars from England but were also attended by some Scandinavian scholars as shown by Etheridge. The University of Bologna was most famous for canon law but was also an important centre for the study of science. This university was favoured by some Scandinavians throughout the Middle Ages as shown by Sten Ebbesen, Etheridge and with biographical details, again drawn from the vita of Jón Halldórsson, by Marteinn Sigurðsson. The universities of the Holy Roman Empire took over from Paris as a place of study for Scandinavian scholars from the end of the fourteenth century. These universities then formed the most important centres for the dissipation of scientific knowledge to Scandinavia up to the end of the Middle Ages as shown by the cases of the universities of Erfurt, Greifswald, Leipzig, and Vienna by Ebbesen, Etheridge, and Jakobsen. Finally, by the end of the fifteenth century the universities of Copenhagen and Uppsala were founded in Scandinavia where they had a small but notable influence on medieval science at the end of the Middle Ages as shown by Ebbesen, Etheridge, and Jakobsen in their chapters.

Mendicants

A model of the idea of travelling wisdom in the Middle Ages can be found in the intellectual activity of the mendicant orders. The friars belonging to the Dominican and Franciscan orders, which are the focus of the chapters in this volume, were highly mobile. Unlike earlier religious orders, the mendicants were not tied to a monastery and could instead move around between friaries. Their books were also mobile, and libraries were in a constant state of fluidity. The mendicants also had their own parallel education system of various studia that was used in the education of the friars. Some of these, such as the *studia artium* and *studia naturalia*, focused on the teaching of scientific texts. The mendicants and their friaries were found throughout England and Scandinavia and could also move between the kingdoms. One example of this is that of Friar Peter, a Danish Franciscan who was the visitor (an official that conducted visitations to maintain discipline) of the Dublin friary in the late thirteenth century, a city then under English rule.[37] The Franciscan Order is covered in several of the chapters with the Franciscan Roger Bacon being the focus of Campopiano, while the great friend of the Franciscans, Robert Grosseteste, is the focus of Gasper et. al. Both these chapters look at the important role played by the Oxford Franciscan Friary and its connection to the university, while Etheridge looks at Scandinavian Franciscans who studied there. The Franciscan Order in Scandinavia itself is looked at by Etheridge with a focus on the manuscripts and scholars of the friaries at Nyköping and Stockholm in the Kingdom of Sweden and Lund in the Kingdom of Denmark. Jacobsen looks further at the studia of the Lund Friary.

The Dominican Order was also prominent in the distribution of scientific knowledge in the Middle Ages and is the main focus of Jacobsen which follows the careers of notable Scandinavian Dominicans, as well as explaining the system of learning in the order as it related to their province of Dacia. Jakobsen looks at the Dominican friaries of Helsingborg, Lund, and Ribe in Denmark, while also exploring those of Sigtuna, Skänninge, Stockholm, and Visby in the Kingdom of Sweden. Jakobsen further stresses the internationality of the Dominican Order and the mobility of its friars in the pursuit of knowledge by showing Scandinavian connections to the friaries of Greifswald and Lübeck in Germany and Tallinn in Estonia. Etheridge continues this internationality by looking at Scandinavians who were at the Dominican studia in Oxford as well as investigating scientific manuscripts that were held at the friary of Sigtuna. Finally, Marteinn Sigurðsson investigates the career of Jón Halldórsson, the Dominican bishop of Skálholt, and his connection to the friary of Bergen in Norway.

The Role of Scandinavia and England in Medieval Science

Wolfgang Undorf in his magisterial work on early printed volumes in Scandinavia decisively counters the opinion that the North was peripheral in terms of printed

---

37   The information on Friar Peter is found in the *Liber Exemplorum*: see Jones, *Friars' Tales*, p. 138.

works received and used there. Scandinavia was in his opinion a full part of the European book trade, not a peripheral outlier.[38] Undorf furthermore argues for more comparative work with Scandinavia and England.

> Future comparative studies of England, on one hand, and Denmark and/or Sweden, on the other, might prove very fruitful for our understanding of the dependences and individuality of northern European book culture in the early modern period. Both regions were geographically peripheral but often proved to be vital players in the ecclesiastical, academic and intellectual life of Europe.[39]

Likewise, we hope that this volume will show that in matters of medieval science, both Scandinavia and England were not peripheral areas. The examples given in the chapters in this book show that, time and again, Scandinavian and English scholars and authors were part of an international network of ideas that utilized scientific works from Classical Antiquity, the Islamic World, Byzantium, and Jewish scholarship. This synthesis was added to with centuries of independent writing, commentaries, and observations from the Western Christian world. In these matters Scandinavia and England were not on the periphery, but a crucial part of this endeavour.

Works Cited

Alfanus I, *Premnon Physicon*, ed. by Irene Chirico (Rome: Edizioni di Storia e Letteratura, 2011)

Baffioni, Carmela, *Storia della filosofia islamica* (Milan: Mondadori, 1991)

Bates, David, and Robert Liddiard, eds, *East Anglia and its North Sea World in the Middle Ages* (Woodbridge: Boydell, 2015)

Beaujouan, Guy, 'L'enseignement du "quadrivium"', in *La scuola nell'Occidente latino dell'alto medioevo*, vol. 2 (Spoleto: CISAM, 1971), pp. 639–67

Burnett, Charles, 'Adelard of Bath and the Arabs', in *Rencontres de cultures dans la philosophie médiévale: Traductions et traducteurs de l'antiquité tardive au XIVe siècle*, ed. by Jacqueline Hamesse and Marta Fattori (Louvain-La-Neuve: Institut d'Études Médiévales de l'Université Catholique de Louvain, 1990), pp. 89–107

——, 'The Coherence of the Arabic–Latin Translation Program in Toledo in the Twelfth Century', *Science in Context*, 14 (2001), 249–88

——, 'Humanism and Orientalism in the Translations from Arabic into Latin in the Middle Ages', in *Wissen über Grenzen Arabisches Wissen und lateinisches Mittelalter*, ed. by Andreas Speer and Lydia Wegener (Berlin: De Gruyter, 2006), pp. 22–31

——, 'Michael Scot and the Transmission of Scientific Culture from Toledo to Bologna via the Court of Frederick II Hohenstaufen', *Micrologus*, 2 (1994), 101–26

---

38 Undorf, *From Gutenberg to Luther*, pp. 310–14.
39 Undorf, *From Gutenberg to Luther*, p. 313.

—, 'Talismans: Magic as Science? Necromancy among the Seven Liberal Arts', in *Magic and Divination in the Middle Ages: Texts and Techniques in the Islamic and Christian Worlds* (Aldershot: Variorum, 1996), pp. 1–15

—, 'Translation and Transmission of Greek and Islamic Science to Latin Christendom', in *The Cambridge History of Science*, vol. 2, ed. by David C. Lindberg, and Michael H. Shank (Cambridge: Cambridge University Press, 2013), pp. 341–64

Caiazzo, Irene, 'Filosofia della natura e fisica elementare nell'Alto Medioevo', in *La conoscenza scientifica nell'Alto Medioevo* (Spoleto: CISAM, 2020), pp. 1059–86

Catastini, Laura, Franco Ghione, and Roshi Rashed, *Algebra. Origini e sviluppi tra mondo arabo e mondo latino* (Rome: Carocci, 2016)

Cavallar, Osvaldo, '*Quod de Tibere dicetur*: fiumi, incrementi fluviali, mulini ad acqua e giuristi', in *La civiltà delle acque tra Medioevo e Quattrocento* (Florence: Olschki, 2010), II, 91–119

Cleaver, Laura, *Education in Twelfth-Century Art and Architecture: Images of Learning in Europe c. 1100–1220* (Woodbridge: Boydell and Brewer, 2016)

Coopland, George William, *Nicole Oresme and the Astrologers: A Study of his Livre de Divinacions* (Cambridge, MA: Harvard University Press, 1952)

Draelants, Isabelle, 'Expérience et autorité dans la philosophie naturelle d'Albert le Grand', in *Expertus sum. L'expérience par les sens dans la philosophie naturelle médiévale*, ed. by Thomas Bénatouïl and Isabelle Draelants (Florence: SISMEL, 2011), pp. 89–122

Eastwood, Bruce S., 'Early-Medieval Cosmology, Astronomy, and Mathematics', in *The Cambridge History of Science*, vol. 2, ed. by David C. Lindberg, and Michael H. Shank (Cambridge: Cambridge University Press, 2013), pp. 302–22

—, *Ordering the Heavens: Roman Astronomy and Cosmology in the Carolingian Renaissance* (Leiden: Brill, 2007)

Fontaine, Jacques, *Isidore de Seville et la Culture classique dans l'Espagne Wisigothique*, 3 vols (Paris: Etudes Augustiniennes, 1959–1983)

Frova, Carla, 'Le traité de fluminibus de Bartolo da Sassoferrato (1355)', *Médiévales*, 36 (1999), 81–89

Gallo, Italo, ed., *Salerno e la sua Scuola Medica* (Naples: Guida, 2008)

Gautier Dalché, Patrick, 'D'une technique à une culture. Carte nautique et portulan au XIIe et au XIIIe siècle', in *L'uomo e il mare nella civiltà occidentale da Ulisse a Cristoforo Colombo. Atti del convegno, Genova, 1–4 giugno 1992* (Genoa: Società Ligure di Storia Patria, 1992), pp. 284–312

—, 'La Terre dans le cosmos', in *La Terre. Connaissance, représentations, mesure au Moyen Age*, ed. by Patrick Gautier Dalché (Turnhout: Brepols, 2013), pp. 161–257

Giovanni Dondi dall'Orologio, *Tractatus astrarii*, ed. by Emmanuel Poulle (Geneva: Librairie Droz, 2003)

Glaze, Florence Eliza, 'Medical Writings: "Behold the Human Creature"', in *Voice of the Living Light. Hildegard of Bingen and her World*, ed. by Barbara Newman (Berkeley: University of California Press, 1998), pp. 125–48

Green, Tamara M., *The City of the Moon God: Religious Traditions of Harran* (Leiden: Brill, 1992)

Gutas, Dimitri, *Greek Thought, Arabic Culture: The Graeco-Arabic Translation Movement in Baghdad and Early 'Abbasid Society (2nd–4th/5th–10th C.)* (London: Routledge, 1998)

Hackett, Jeremiah, 'Ego expertus sum: Roger Bacon's Science and the Origins of Empiricism', in *Expertus sum. L'expérience par les sens dans la philosophie naturelle médiévale*, ed. by Thomas Bénatouïl and Isabelle Draelants (Florence: SISMEL, 2011)

Haskins, Charles H., and Dean Putnam Lockwood, 'The Sicilian Translators of the Twelfth Century and the First Latin Version of Ptolemy's Almagest', *Harvard Studies in Classical Philology*, 21 (1910), 75–102

Jacquart, Danielle, and Paravicini Bagliani, Agostino, eds, *La Scuola Medica Salernitana: Gli autori e i testi* (Florence: SISMEL, 2007)

Jacquart, Danielle, and Charles Burnett, *Constantine the African and 'Alī Ibn al-'Abbās al-Maǧūsī. The Pantegni and Related Texts* (Leiden: Brill, 2014)

Jones, David, *Friars' Tales: Thirteenth-century exempla from the British Isles*, Manchester Medieval Sources Series (Manchester: Manchester University Press, 2011)

Joyner, Danielle B., *Painting the Hortus Deliciarum: Medieval Women, Wisdom, and Time* (University Park: Penn State University Press, 2016)

Kelby-Fulton, Kathryn, 'Hildegard of Bingen', in *Medieval Holy Women in the Christian Tradition, c. 1100–c. 1500*, ed. by Alastair Minnis and Rosalynn Voaden, Brepols Collected Essays in European Culture, 1 (Turnhout: Brepols, 2010), pp. 343–69

Konrad Gruter von Werden, *De machinis et rebus mechanicis. Ein Maschinenbuch aus Italien für den König von Dänemark. 1393–1424*, ed. by Dietrich Lohrmann, Horst Kranz, and Ulrich Alertz, 2 vols (Vatican City: Biblioteca Apostolica Vaticana, 2006)

Marcacci, Flavia, 'Umanesimo scientifico e formazione dell'Uomo nell'Età Moderna', in *Formare e tras-formare l'uomo. Per una storia della filosofia come paideia*, ed. by Patrizia Manganaro and Emmanuele Vimercati (Pisa: ETS, 2017), pp. 151–66

North, John, 'Astronomy and Astrology', in *The Cambridge History of Science*, vol. 2, ed. by David C. Lindberg, and Michael H. Shank (Cambridge: Cambridge University Press, 2013), pp. 456–84

Overfield, James H., *Humanism and Scholasticism in Late Medieval Germany* (Princeton: Princeton University Press 1984)

Panti, Cecilia, *Moti, virtù e motori celesti nella cosmologia di Roberto Grossatesta. Studio ed edizione dei trattati 'De sphera', 'De cometis', 'De motu supercelestium'* (Florence: SISMEL, 2001)

—, 'Scienza e teologia agli esordi della scuola dei Minori di Oxford: Roberto Grossatesta, Adamo Marsh e Adamo di Exeter', in *I francescani e le scienze* (Spoleto: CISAM, 2012), pp. 311–51

—, 'A Woman's Voice through the Centuries. Hildegard of Bingen's Music Today', in *The Past in the Present. A Multidisciplinary Approach*, ed. by Fabio Mugnaini, Pádraig Ó Héalaí, and Tok Thompson (Catania: Ed.it, 2006), pp. 15–40

Pisano, Raffaele, 'Details of the Mathematical Interplay between Leonardo da Vinci and Luca Pacioli', *BSHM Bulletin: Journal of the British Society for the History of Mathematics*, 31.2 (2016), 104–11

Riché, Pierre, *Ecoles et enseignement dans le Haut Moyen Age* (Paris: Aubier, 1979)

Sangste, Alan and Giovanna Scataglinibelghitar, 'Luca Pacioli: The Father of Accounting Education', *Accounting Education: An International Journal* 19.4 (2010), 423–38

Shank, Michael, 'Schools and Universities in Medieval Latin Science', in *The Cambridge History of Science*, vol. 2, ed. by David C. Lindberg and Michael H. Shank (Cambridge: Cambridge University Press, 2013), pp. 207–39

Sylla, Edith, 'The Oxford Calculators in Context', *Science in Context*, 1 (1987), 257–79

Undorf, Wolfgang, *From Gutenberg to Luther: Transnational Print Cultures in Scandinavia 1450–1525*, Library of the Written Word 37: The Handpress World 28 (Leiden: Brill, 2014)

Waszink, Jan H. *Timaeus a Calcidio translatus commentarioque instructus* (London, Leiden: Brill, 1975)

Zuccato, Marco, 'Gerbert of Aurillac and a Tenth-Century Jewish Channel for the Transmission of Arabic Science to the West', *Speculum* 80 (2005), 742–63

C. PHILIPP E. NOTHAFT

Roger of Hereford and the Transformation of *Computus* in Twelfth-Century England

Between the onset of the Middle Ages and the end of the first millennium of the Christian era, the study of mathematical astronomy in Latin Europe was to a large extent subordinate to the discipline known as *computus*, which employed calendrical cycles to aid the Church in its task of calculating the date of Easter. Significant changes to this overall situation first become visible in the early eleventh century, with the incipient spread of texts devoted to the astrolabe.[1] More significant still, however, are the developments of the twelfth century, which stands out for the way it reconnected Christian Europe with Graeco-Arabic science through a major wave of Arabic-to-Latin translations. While modern historians have invested a considerable amount of energy in retracing the steps of this knowledge transfer, one topic that has so far received little attention is how the 'new' astronomy and astrology made available through translation interacted with the 'old' bodies of knowledge represented by *computus*.[2] The prevailing lack of interest in *computus* as a component of the 'Renaissance of the Twelfth Century' can be traced back to the otherwise pioneering works of Charles Homer Haskins (1870–1937), most notably his 1924 *Studies in the History of Mediaeval Science*, in which he depicted the twelfth century as a period of heightened computistical activity only to add that '[i]n such conservative circles it was natural that Arabic astronomy should penetrate slowly, and we are not surprised that Roger of Hereford should inveigh against the ignorance of the computists as late as 1176'.[3]

The work Haskins had in mind here was a *Compotus* in five books and twenty-eight chapters, which survives in Oxford, Bodleian Library, MS Digby 40, fols 21^r–50^v (s. xii/

---

1 See, for example, Jacquemard, 'La réinvention de l'astrolabe au Moyen Âge'; Juste, 'Hermann der Lahme und das Astrolab'.
2 Among the very few studies to shed any light on this question are Moreton, 'Before Grosseteste'; Moreton, 'The *Compotus* of "Constabularius" (1175)'.
3 Haskins, *Studies in the History of Mediaeval Science*, p. 87.

C. Philipp E. Nothaft • is a Research Fellow at Trinity College Dublin. He has written extensively on medieval chronology, calendars, mathematical astronomy, and intellectual history. cpenothaft@hotmail.com

xiii), and Cambridge, University Library, MS Kk.1.1, fols 222^v–39^r (s. xiii$^{1/2}$).[4] Another manuscript attesting to its diffusion in thirteenth-century England is Cambridge, St John's College, MS F.25, fols 34^r–40^v (s. xiii$^{2/2}$), which offers a stand-alone version of the calendar contained in book IV. The calendar's excerptor, who worked shortly after 1273, identifies his source as the *Compotus* of 'Master Roger of Hereford, who composed this calendar at the behest of Lord Gilbert, bishop of London, in the year 1176 from the Lord's incarnation'.[5] That this information is reliable may be seen from the acrostic hidden in the initial letters of the work's summary of chapters: *Gilleberto Rogerus salutes h<ic> d<icit>*. It is hence clear that the *Compotus* was written for Gilbert Foliot, who served as bishop of Hereford from 1148 to 1163 and as bishop of London from 1163 to 1187. Among Gilbert's preserved letters is one he sent not long after Easter 1173 to his cousin Robert, who would be consecrated bishop of Hereford on 6 October 1174. The letter is witnessed by one 'Rogerus de Herefordia',[6] presumably the same Roger who finished the *Compotus* between 1 January and 5 September 1176.[7] Starting with Haskins in 1915, modern scholars have identified him with the astrologer Roger of Hereford, who was among the first Latin authors to write texts on judicial astrology on the basis of newly translated Arabic sources.[8] The identification seems justified not just because of the chronological proximity — Roger's astronomical tables for the meridian of Hereford date from 1178 — but because the fifth book of

---

4   Some brief excerpts from Book V appear in Oxford, Bodl. Lib., MS Ashmole 1796, fol. 172^v (s. xiv/xv). Another copy was once in the library of the Austin Friars of York. See *The Friars' Libraries*, ed. by Humphreys, p. 109. I have previously dealt with Roger of Hereford's *Compotus* in Nothaft, *Scandalous Error*, pp. 106–15. The present chapter expands on this account.

5   Cambridge, St John's College, MS F.25, fol. 40$^{r–v}$: 'Quoniam per vicium scriptorum multe falsitates in kalendario naturalis compoti continentur […] ad dilucidationem dicti kalendarii quedam philosophie dicta de compoto magistri Rogeri Herefordensis, qui dictum kalendarium ad instantiam domini Gileberti Londonensis episcopi composuit anno ab incarnatione domini M°C°LXX°VI°, excerpere curavi. Et sciendum quod in hoc compoto triplex est materia: vulgaris scilicet compotus, naturalis, astronomica consideratio circa motum solis et lune. Et dividitur iste compotus in 5 libros, ut sint 3 primi de vulgari compoto. Et in primo agitur de tabulis tam immobilium quam mobilium festorum. In secundo de vulgari motu solis et de hiis que ad ipsum pertinent. In tertio de vulgari motu lune et de pertinentibus ad ipsum. In quarto de naturali compoto. In quinto de motu solis et lune secundum astronomiam et convenientia et differentia triplicis materie'. The version of the calendar presented here was later falsely ascribed to Robert Grosseteste, bishop of Lincoln (d. 1253). See Lindhagen, 'Die Neumondtafel des Robertus Lincolniensis'; Nothaft, *Scandalous Error*, pp. 76–79.

6   *The Letters and Charters of Gilbert Foliot*, ed. by Brooke, Morey, and Brooke, pp. 421–22 (no. 375).

7   The date can be narrowed down thanks to a passage in Roger of Hereford, *Computus*, ed. by Lohr, III. 15, ll. 205–06 (p. 178), where it is stated that the Jews now count the 15th year of the 260th lunar cycle since creation. This matches the year 4936 of the Jewish world era, which ran from 18 September 1175 to 5 September 1176.

8   See Haskins, 'The Reception of Arabic Science in England', pp. 65–67, which was reprinted (with revisions) as Haskins, *Studies in the History of Mediaeval Science*, pp. 123–26. Most of the literature subsequent to Haskins is cited in the introduction to Roger of Hereford, *Computus*, ed. by Lohr, pp. xix–xxvi. To the titles mentioned there one may add Reese, Craun, and Mason, 'Twelfth-Century Origins of the 7980-Year Julian Period'; Burnett, 'Mathematics and Astronomy in Hereford', pp. 55–57; Southern, *Robert Grosseteste*, pp. lii–liii.

the *Compotus* shows traces of an advanced understanding of mathematical astronomy that very few computists would have possessed at the time.[9]

Contrary to what Haskins may have meant to imply in his statement quoted above, however, Roger did not use this skill-set to lambast the computists for their ignorance. Instead of passing judgment in a facile manner, his preface contains some revealing remarks about the controversies that had broken out in his environment with regard to the study of celestial motions. It reveals that Roger wrote at a time when a new class of 'astrologers' had appeared on the scene, flaunting their ability to investigate the truth about 'the motions of both the heavens and the stars'. These astrologers faulted *computus* for deviating from this truth and even argued that it should not be considered a part of philosophy.[10] The computists, meanwhile, were embroiled in their own factional dispute between adherents to two different forms of *computus*, 'vulgar' and 'natural'. Roger describes their disagreement as follows:

> Sed et compotistae inter se tamquam intestina proelia commouentes, naturales uulgarem computum a sua subtilitate discrepantem magisque sensuum opinionem quam rationis ueritatem exsequentem abiciunt; econtra uulgares naturalem a sensibus amotum solique rationi patentem uanam inanemque scientiam, quam nec oculus uidit nec auris audiuit, appellant. Sunt et item huius scientiae tractatores, qui sine distinctione naturalis et artificialis computi multa interponunt superflua, alii uero uolentes ecclesiasticae uulgari consuetudini tantum satisfacere multa abicere necessaria inuenti sunt.[11]

> (Yet even among the computists, who are, as it were, stirring up a civil war among themselves, the natural ones cast aside the vulgar *computus* for not conforming to their standard of subtlety and for following the opinion of the senses more than the truth of reason. The vulgar [computists], by contrast, call the natural [*computus*], which is remote from the senses and only accessible to reason, a vain and empty science, which the eye has not seen and the ear has not heard. Besides, there are also some dealing with this science who insert many superfluous things without distinguishing the natural from the artificial *computus*, whereas others, who want to focus only the vulgar usage of the Church, have been found to discard many necessary things.)

---

9   See BL, MS Arundel 377, fols 86ᵛ–87ʳ, which is reproduced in Burnett, *The Introduction of Arabic Learning into England*, pp. 54–55, and Burnett, 'Mathematics and Astronomy in Hereford', p. 56. A fuller version of these tables is in Madrid, Biblioteca nacional de España, MS 10016, fols 73ʳ–83ᵛ, 85ʳ.

10 Roger of Hereford, *Computus*, ed. by Lohr, praefatio, ll. 23–27 (pp. 127–28): 'Hanc tamen tantae excellentiae scientiam astrologi naturae superiorum secreta motuumque tam caeli quam stellarum certitudinem inuestigantes computumque ab illa certitudine multum discrepare reperientes falsam ab omni philosophica disciplina abiciendam arbitrantur'.

11 Roger of Hereford, *Computus*, ed. by Lohr, praefatio, ll. 27–37 (p. 128). Note that Roger here uses the term *computus artificialis* synonymously with *computus vulgaris*, as also seen from Roger of Hereford, *Computus*, ed. by Lohr, I. 1, l. 34 (p. 134).

The *computus naturalis*, which according to Roger's own words was prized for its greater subtlety, had originally been developed in the eleventh century as a response to observed inaccuracies in the calendrical cycles the Church used to calculate the age of the Moon and the dates of the mobile feast days. Its core idea was to make use of small fractions of time in order to locate the precise moment when the Sun and Moon were in conjunction, with observed eclipses serving as reference points for the calculation. In the idealized picture that the natural *computus* painted of celestial motions, each conjunction was separated from any preceding or subsequent one by a multiple of the same value for the synodic lunar month, which the 19-year cycle of the Church implied to last 29 days and $12\frac{174}{235}$ hours.[12] Roger of Hereford acknowledged the affinity between this approach and the use of mean-motion parameters in the context of mathematical astronomy, but not without stressing the difference of perspective that was introduced by recognizing the non-uniformity of the apparent motions of Sun and Moon, which astronomers explained by invoking epicycles and eccentric deferents. The same non-uniformity made it necessary to distinguish between mean and true longitudes and to employ so-called equations in order to derive the latter from the former. In its concern for the true inequality of celestial motions, astronomy differed in a significant way from both the vulgar *computus*, where periods were unequal merely as a matter of convention, because periods of time had to be expressed as multiples of complete days, and the natural *computus*, which sought to divide everything into equal parts.[13]

Roger's work was hence unique among the computistical textbooks of his time for the way it presented and compared three contrasting approaches to lunar reckoning — *computus usualis*, *computus naturalis*, and *astronomia* — with the express goal of investigating 'the convergence and difference' between them and, above all, of explaining 'why the vulgar new moon [*primatio*] has now receded so much from the truth'. In pursuing this critical project, Roger promised to refer not just to the doctrines of the Latins, but also to those of the 'Chaldaeans and Hebrews', his terms for Arabs and Jews, whose lunar calendars had only recently come to the attention of Latin computists.[14] The nuanced and constructive criticism Roger of Hereford's *Compotus* expressed towards the ecclesiastical mode of lunar reckoning can make for an illuminating case study of how scientific knowledge transferred to northern Europe in the course of the twelfth century could impact traditional canons of learning. I shall seek to justify this claim in the following pages, which are intended to shed light on some of the historical threads that connect *computus* and mathematical

---

12 Nothaft, *Scandalous Error*, pp. 64–76.
13 Roger of Hereford, *Computus*, ed. by Lohr, I. 1, ll. 15–37 (pp. 133–34); v. 23, ll. 8–18 (p. 213); v. 24, ll. 98–131 (pp. 221–22); v. 25, ll. 43–65 (pp. 223–24).
14 Roger of Hereford, *Computus*, ed. by Lohr, praefatio, ll. 63–70 (p. 129): 'Primo itaque de omnibus maxime necessariis in uulgari computo eorumque omnium rationes subiungendo tractaui. Deinde naturalem, quantum operis brevitas patiebatur, cum rationibus suis subiunxi. Tertio astronomicas subtilitates circa motum solis et lunae et de conuenientia et differentia horum trium et quare primatio uulgaris iam a ueritate tantum recesserit et de sententia Chaldaeorum, Hebraeorum et Latinorum circa idem apposui'.

astronomy.[15] Put simply, the widespread practice of *computus* in Latin Europe had played a major role in preparing the intellectual ground for a relatively swift reception of Graeco-Arabic astronomy in the course of the twelfth century. This reception in turn caused *computus* to undergo a lasting transformation to which Roger's treatise is one of our most outspoken witnesses.

The idea of computing the course of the Moon 'according to nature', to which Roger devoted the entire fourth book of his *Compotus*,[16] had germinated in the eleventh century in response to observed discrepancies between the computed and the actual age of the Moon. A pioneering figure in this regard was the monk Hermann of Reichenau (1013–1054), who in *c.* 1040 took the crucial step of dividing the $6939\frac{3}{4}$ days in a 19-year lunar cycle by the 235 lunar months that were assumed to be comprised in the same period. This gave Hermann a value for the mean lunation or synodic month, which he hoped would provide the foundation for a more fine-grained and accurate calculation of the times of the new and full moons. A related project of his was to develop a predictively successful theory of lunar and solar eclipses, to which end Hermann equated the same $6939\frac{3}{4}$ days with 254 sidereal months, but the unsatisfactory results of his calculations left him wondering whether the motions of Sun and Moon were really as uniform as his method assumed.[17] A Lotharingian computist named Gerland later modified Hermann's approach, yet still failed to predict the solar eclipse that was seen over large parts of Europe on 23 September 1093. Instead of admitting defeat, however, he went on to use this eclipse as the basis for a revised lunar calendar, which listed the dates and hours of 940 consecutive conjunctions of the Sun and Moon between September 1055 and August 1131.[18]

One significant point Gerland never properly acknowledged was that his improved lunar tables started almost three complete days ahead of the corresponding new moon in the ecclesiastical calendar — on 23 September 1055, at the third hour of the day, instead of the conventional 26 September in year eleven of the 19-year cycle. Roger of Hereford, who used Gerland's work as the basis for his own discussion of the natural type of *computus*, was more outspoken. According to him, the divergence between *computus vulgaris* and *computus naturalis*, or between *computus* and observation, had caused such consternation among his contemporaries that some tried to account for it in rather fanciful ways. One explanation appealed to the two-day gap between the creation of the Moon on the fourth day and the creation of Adam and Eve on the sixth, while others pointed out that the interval between the Moon's last and first visibility varied in accordance with the seasons. Yet others renounced all attempts to downplay the gap as merely a temporary aberration and instead insisted that it had been caused by an error

---

15 In addition to this main goal, my chapter may be read as an update and corrective to the pioneering account of Roger of Hereford's *Compotus* in Moreton, 'Before Grosseteste', pp. 581–86.
16 Roger of Hereford, *Computus*, ed. by Lohr, IV. 20–22 (pp. 185–212).
17 For the pertinent details, see Borst, 'Ein Forschungsbericht Hermanns des Lahmen'; Warntjes, 'Hermann der Lahme und die Zeitrechnung'.
18 *Der Computus Gerlandi*, ed. by Lohr, pp. 204–32, 431–37; Borst, *Der karolingische Reichskalender*, pp. 265–67.

in the received lunar calculation, more specifically by an overestimation of the length of the month.[19]

Although it seems no longer possible to identify holders of all the different opinions Roger recorded in 1176, there is strong evidence that doubts about the 19-year cycle and its ability accurately to represent the Moon's synodic period had been harboured in England since the early years of the twelfth century. A harbinger of later discussions in this regard was Walcher of Malvern (d. 1135), who had been born in Lotharingia, but spent a large part of his later life at the priory of Great Malvern in Worcestershire. One thing Walcher shared with his compatriot Gerland was his interest in tracking the Moon's course 'according to nature', which led him to create his own set of improved lunar tables. The empirical basis for these tables was a total eclipse of the Moon, observed by Walcher on 18 October 1092 and timed with great precision using an astrolabe. With the eclipse as his starting point, Walcher went on to calculate the current date, hour, and quarter-hour of all 940 conjunctions from January 1036 to December 1111.[20]

The accuracy of Walcher's new lunar tables was without precedent in medieval Latin astronomy, but some troubling discrepancies remained to be accounted for. When another lunar eclipse was witnessed close to midnight on 11 January 1107, his tables implied that Sun and Moon had been in opposition six hours earlier, on what was still 10 January. By the time the next eclipse appeared on 31 December 1107, the discrepancy had unexpectedly grown to sixteen hours, forcing Walcher to conclude that the course of the Moon through the zodiac was less uniform than the simple algorithms of his *computus naturalis* presupposed.[21] The lack of an adequate lunar theory prevented Walcher from making any further progress until *c.* 1120, when he had the good fortune of meeting Petrus Alfonsi, an Iberian Jewish convert who had come from Aragón to France and England to introduce Latin students to the secrets of Arabic astronomy. Walcher's conversations with 'Master Petrus' taught him a range of important new insights, for example on the difference between mean and true motions, and did much to further his understanding of the conditions that governed the occurrence of eclipses, yet his ability to recognize the flaws inherent in the 19-year cycle remained limited. When it came to the mean motion of the Moon, for instance, the prior of Great Malvern showed no awareness that the $13 + \frac{1}{3} + \frac{1}{36} + \frac{1}{144}$ zodiacal parts per day he had extracted when working on his lunar tables exceeded the $13;10,24,52°/d$ at which the Moon traversed the zodiac according to Petrus Alfonsi. The difference between the two values was apparently cloaked by the fact that the former had been calculated using Roman duodecimal fractions and a division of the zodiac into 365 parts, whereas the latter value was expressed in the sexagesimal system Walcher adopted from Alfonsi.[22]

---

19 Roger of Hereford, *Computus*, ed. by Lohr, v. 25, ll. 66–108 (pp. 224–25).
20 See Walcher of Malvern, *De lunationibus*, ed. by Nothaft, c. 4 (pp. 114–69).
21 Walcher of Malvern, *De lunationibus*, ed. by Nothaft, c. 6 (pp. 188–91).
22 Walcher of Malvern, *De Dracone*, ed. by Nothaft, c. 2 (p. 198); c. 4 (p. 204).

A more decisive change was ushered in by the translation of works such as the *zīj* (astronomical tables) of al-Khwārizmī (c. 825), which Petrus Alfonsi had used as his basis for instructing Walcher. A translation of these tables effected by Adelard of Bath, known as *Ezic Elkaurezmi*, familiarized Latin readers with a 30-year cycle that had served Arabic astronomers in creating a fixed version of the Islamic lunar calendar. The cycle equated 10,631 days with 360 lunations and, in doing so, gauged the mean synodic month at 10,631 ÷ 360 = 29;31,50d or 29d 12h 44m. Not long after the appearance of the *Ezic* in c. 1126, this value aroused the curiosity of an anonymous West English computist, who demonstrated conclusively that it was shorter than the value Gerland had extracted from the 19-year cycle of the 'Romans', which was $29\frac{1}{2}$ days with a surplus of $\frac{2}{3}$ of an hour + $2\frac{11}{12}$ moments + $\frac{1}{48}$ of a moment + $7\frac{1}{4}$ atoms per lunation (1 hour = 40 moments = 22,560 atoms). The discrepancy amounted to 0;4,10d per 19-year cycle and was enough to make the Roman computation fall behind the Arabic one at an approximate rate of one day every 285 years.[23] Following Adelard's *Ezic*, a host of new parameters and computational techniques entered Latin Europe in connection with the so-called Toledan Tables, which had been constructed in al-Andalus in the second half of the eleventh century. Like other sets of astronomical tables cast in the Ptolemaic tradition, the Toledan Tables offered a convenient shortcut for calculating times of new and full moons in the form of tables for mean syzygy, which were founded on a very accurate mean synodic month length of 29;31,50,8,20d.[24] The same value, rendered differently as 29 days, 12 hours, and 793 *ḥalakim* or 'parts' (1 hour = 1080 parts), provided the computational basis for the medieval Jewish calendar, whose details had started to become available in Latin by the mid-twelfth century.[25]

The fifth and final book of Roger of Hereford's *Compotus* offers us an impressive glimpse of a twelfth-century scholar who had assimilated much of this new knowledge and was able to apply it critically to examine the technical assumptions that underpinned the traditional *computus*. In opposition to the practice accepted among 'natural' computists, who treated the synodic month as perfectly uniform, Roger drew attention to the way the Moon moved not in a simple circular path around the Earth, but on an epicycle whose centre was carried by a deferent circle eccentric to the Earth, the centre of which was in turn carried around the centre of the Earth by another small circle. The complications wrought by this interplay of circles made it mandatory to distinguish between the Moon's mean and true longitude, the difference between which — according to Roger — could reach up to ±8;39° or the equivalent of about fifteen hours.[26] Under ideal circumstances, the natural *computus* operated with the same rates of mean motion as those recorded in astronomical tables, although differences in notation could render a comparison cumbersome. Roger was aware

---

23 Nothaft, 'Roman vs. Arabic Computistics'.
24 *The Toledan Tables*, ed. by Pedersen, pp. 1327–40; Chabás and Goldstein, *A Survey of European Astronomical Tables*, pp. 139–44.
25 Nothaft, *Medieval Latin Christian Texts on the Jewish Calendar*, pp. 43–68. The history of the parameter is discussed in Neugebauer, 'From Assyriology to Renaissance Art'.
26 Roger of Hereford, *Compotus*, ed. by Lohr, v. 24 (pp. 217–22).

of Walcher of Malvern's opinion according to which the Moon travelled $13 + \frac{1}{3} + \frac{1}{36} + \frac{1}{144}$ degrees per day, a value he ascribed to the 'computists' in general, noting that it relied on a division of the zodiac into 365 rather than 360 parts. As a result, one computistical 'degree' was smaller than the type of degree used in astronomical sources, where the Moon's daily rate of motion was given as approximately 13;10,35°.[27]

To Roger's mind, it was nevertheless certain that the computists erred about the Moon's course and that this was the main reason why the vulgar and natural forms of their craft had ended up placing their new and full moons on different dates. In demonstrating this point to his readers, Roger made sure to take into account the estimates of two other nations, Arabs ('Chaldaeans') and Jews ('Hebrews'), whose systems of lunar reckoning he had already described in earlier sections of his work.[28] Neither of these systems employed the same parameter as the *computus* of the Latins, where an average lunar month lasted 29 days, 12 hours, 29 moments, 348 atoms. The Jewish calendar relied on an estimate that was $139\frac{1}{9}$ atoms below the Latin value (29 days, 12 hours, 29 moments, and 208 atoms), whereas the Arabs even subtracted 160 atoms (29 days, 12 hours, 29 moments, and 188 atoms).[29] That these differences in parameters were conducive of strongly divergent results became clear from a look at the new moon of September 1176, which served Roger in comparing six different styles of lunar reckoning. Calculating his hours from the beginning of the preceding night, he provided his readers with the following list of new moon times:[30]

> Vulgar *computus*: 9 September[31]
> Natural *computus*: 6 September, at the 6th hour[32]
> Jewish calendar: 5 September, at the 24th hour (23 hours and 774 parts)[33]
> Arabic calendar: 5 September, at the 17th hour[34]

---

27 Roger of Hereford, *Computus*, ed. by Lohr, v. 25, ll. 28–43 (p. 223). The daily mean motion of 13;10,35° mentioned in this chapter could have been derived by rounding the value in the Toledan Tables (13;10,34,52,48,47°/d) or indeed any other set of tables available in medieval Europe at the time. See Chabás and Goldstein, *A Survey of European Astronomical Tables*, p. 58.
28 Roger of Hereford, *Computus*, ed. by Lohr, II. 5, ll. 21–43 (p. 153); III. 15, ll. 129–214 (pp. 175–78).
29 See the tabular comparison in Roger of Hereford, *Computus*, ed. by Lohr, v. 26 (p. 227), which also displays the respective differences for lunar years and (multiples of) 19-year cycles, albeit with corrupt values.
30 Roger of Hereford, *Computus*, ed. by Lohr, v. 26, ll. 10–22 (p. 225).
31 AD 1176 was the 18th year of the 19-year cycle, which had the lunation of October begin on 9 September. See the lunar cycle calendar edited in Borst, *Der karolingische Reichskalender*, p. 1707.
32 September AD 1176 was the start of the 46th year of Gerland's 76-year cycle, with the letter *H* and the number *VI* on 6 September. See *Der Computus Gerlandi*, ed. by Lohr, p. 221, and the adaptation of the same calendar in Roger of Hereford, *Computus*, ed. by Lohr, IV. 21 (p. 203).
33 This is correct for the conjunction of Tishri in year 4937 of the Jewish world era (= AD 1176/77), which fell on Sunday, 5 September, at 23h 774p. (23:43h) from the previous equinoctial sunset. The precise time is accurately stated in Roger of Hereford, *Computus*, ed. by Lohr, v. 27, ll. 20–21 (p. 230).
34 The lunation starting in September 1176 corresponds to the month of Rabiʾ I in year 572 of the Arabic calendar, whose standard epoch in astronomical tables is the noon of 14 July AD 622 = Julian Day Number (JDN) 1,948,438. Between the beginning of this date and the start of Rabiʾ I in year 572 there are 19 complete 30-year cycles of 19 × 10,631 = 201,989 days and 14 months of 14 × 29d 12h 44m = 413d 10h 16m. The total is 202,402d 10h 16m, which if added to JDN 1,948,438 leads to JDN 2,150,840 =

Astronomical (mean syzygy): 5 September, at the 22nd hour (21 hours and 14 moments)[35]

Astronomical (true syzygy): 5 September, before sunrise

Even though Roger remained silent on the details behind these calculations, it appears likely that his results for the astronomical style of reckoning were ultimately founded on the Toledan Tables, which would have placed the mean conjunction of 5 September 1176 at 02:16h from noon, or 20:16h from the previous equinoctial sunset. Roger's result (21 hours and 14 moments = 21:21h) is higher by 1 hour and 5 minutes, which suggests very strongly that he availed himself of a Christian adaptation of the Toledan Tables made *c.* 1141 by the astrologer Raymond of Marseille, who estimated the time difference between Toledo and Marseille as 1 hours and 6 minutes.[36] In Roger's opinion, the astronomical approach to calculating the time of mean conjunction related to the Arabic calendar in the same way as did the *computus naturalis* to the *computus vulgaris*. His way, however, of equating the 'mean motion of astronomy' with the 'natural *computus* of the Chaldaeans' reflects a misunderstanding, at least insofar as the mean synodic month in the Toledan Tables had the same implicit length as the 'Hebrew' value, being longer than the average month of the Arabic calendar by $20\frac{8}{9}$ atoms.[37]

More important than this confusion on Roger's part was his conclusion according to which the Arabic value of 29 days, 12 hours, 29 moments, and 188 atoms (or 29 days, 12 hours, and 44 minutes) was the best available estimate for the mean synodic month, accurate enough to hold the middle ground between the fluctuations to which the Moon's true course was subject.[38] It accordingly served to expose the error inherent in the 19-year lunar cycle that underpinned the *computus* in both its vulgar and natural variants. The 'natural *computus* of the Chaldaeans', as obtained from astronomical tables, placed the conjunction of September 1176 eight hours later than did Gerland's lunar tables, at the 22nd hour of 5 September rather than the 6th hour of the following day. This discrepancy had arisen despite Gerland's effort to

---

5 September, with a remainder of 10h 16m. For this to support Roger's result (the 17th hour from previous sunset) would require counting the days and hours from midnight rather than noon of 14 July AD 622.

35 The precise time of 21 hours and 14 moments, which are equivalent to 21 hours and 21 sexagesimal minutes, is stated in Roger of Hereford, *Computus*, ed. by Lohr, v. 27, ll. 18–20 (p. 230).

36 See Raymond of Marseille, *Liber cursuum planetarum*, ed. by d'Alverny, Burnett, and Poulle, II. 1 (p. 200). Roger's astronomical tables for the meridian of Hereford, made in 1178 (see n. 9 above), were an adaptation of Raymond's tables.

37 Roger of Hereford, *Computus*, ed. by Lohr, v. 26, ll. 25–31 (p. 226): 'Sed prius notandum est, quod est Chaldaeorum et naturalis et uulgaris computi consideratio. Naturalis quae dicta est et est secundum medium motum astronomiae; uulgaris vero, ubi cyclus idem est, scilicet annorum XXX, sed lunationes, quas menses dicunt, magis a lunae uisione incipiunt et per integros fiunt dies, ut sit alia XXX dierum alia XXIX additis diebus XI in XXX annis'. Roger of Hereford, *Computus*, ed. by Lohr, v. 27, ll. 28–29 (p. 230): '[E]t Chaldaeorum naturalis computi, qui est medius motus astronomiae'.

38 Roger of Hereford, *Computus*, ed. by Lohr, v. 26, ll. 31–37, 56–59, 145–53 (pp. 226, 229).

base his tables on the solar eclipse of 23 September 1093, and hence on an empirically timed conjunction. Given the known difference between the Latin and Chaldaean lunation lengths, which was 160 atoms, the obvious conclusion was that most of the discrepancy had only accrued over the past 83 years, between September 1093 and September 1176.[39]

A yet more dramatic difference emerged between the astronomical reckoning and the time of the new moon according to the vulgar *computus*, which Roger decided to place at the end, rather than the beginning, of 9 September 1176, claiming that it currently deviated from the 'computation of the Chaldaeans' by 4 days and 2 hours.[40] In Roger's calculation, this difference had accumulated over 1124 years, thereby offering a rough notion of when the 19-year cycle, in its present form, had been instituted.[41] Unlike Gerland's variant of the *computus naturalis*, which had only been inaugurated in 1093, the vulgar *computus* was an ancient institution that had remained unchanged for many centuries owing to its role in fixing the ecclesiastical feast days. This held true in particular for the *terminus paschalis* or 14th day of the paschal lunation, whose 19 different dates were widely believed to have been sanctioned by the authority of the Council of Nicaea in 325, or even revealed by an angel to the Egyptian monk St Pachomius.[42] As one would imagine, the nimbus the 19-year cycle had acquired over the centuries could inspire among later computists some rather desperate attempts to uphold its validity in the teeth of the astronomical evidence. A noteworthy example that puts some of Roger's own statements into sharper relief appears in a *Compotus* written in 1175 by an author known as Magister Cunestabulus. To an even greater degree, perhaps, than Roger of Hereford, Cunestabulus was a writer steeped in Graeco-Arabic astronomy, capable of quoting authors such as Ptolemy, Hipparchus, al-Farghānī, Arzachel, and Thābit ibn Qurra. Yet this access

---

39 Roger of Hereford, *Computus*, ed. by Lohr, v. 27, ll. 5–35 (p. 230). The stated result of 7 hours, 9 moments, and 92 atoms (ll. 34–35) was evidently derived from the table in Roger of Hereford, *Computus*, ed. by Lohr, v. 26 (p. 227), by combining the differential between Latins and Chaldaeans for four 19-year cycles (listed as 6h 25m 188a, which is an error for 6h 26m 376a) and seven years (7 × 3m 228a = 23m 486a), while ignoring the two embolismic months contained in these seven years (6h 25m 188a + 23m 486a = 7h 9m 92a). The correct discrepancy for 83 years containing 1026 months would have been 7 hours, 16 moments, and 36 atoms.
40 Roger of Hereford, *Computus*, ed. by Lohr, v. 27, ll. 38–41 (p. 231): 'Vulgaris vero Latinorum computus, qui lunam in uespere uulgaris diei accensam iudicat, discessit iam a computatione Chadaeorum per dies IIII et horas II'. The difference of 4d 2h results if 9 September is included in the interval between that day and the 22nd hour of 5 September, where Roger previously located the mean conjunction 'according to Chaldaeans'.
41 Roger of Hereford, *Computus*, ed. by Lohr, v. 27, ll. 41–52 (p. 231). As before (n. 40), this calculation relied on the garbled values in Roger's comparative table, whereas on a difference of 160 atoms per lunation the stated 4d 2h should have accrued after 13,818 lunations or 1117.2 years.
42 Roger of Hereford, *Computus*, ed. by Lohr, v. 26, ll. 45–47 (p. 226): 'Unde nec ipsum propter terminos paschales, scilicet lunam XIIII^{am}, a Nicaeno concilio confirmatos mutare audemus'. Roger of Hereford, *Computus*, ed. by Lohr, v. 25, ll. 93–94 (p. 224): '[U]nde ab illa antiquorum distinctione nihil in lunationibus uulgaris computi mutare audemus'. On the background, see Jones, 'A Legend of St Pachomius'; McCluskey, 'Changing Contexts and Criteria'; Lejbowicz, 'Les Pâques baptismales d'Augustin d'Hippone'.

to new sources did not automatically translate into a willingness to criticize the received *computus* for its flaws.[43]

Cunestabulus's apologetic attitude towards the discipline is on full display in Chapter 38 of his *Compotus*, which was there to explain why the Moon was 'sometimes seen sooner' than on its first day according to the calendar.[44] The chapter mentioned the same discrepancies between the Latin, Jewish, and Arabic lunations that Roger of Hereford was to discuss in his treatise the following year, stating that the Arabs counted 160 atoms fewer than the Latins, whereas the Jews only subtracted $139\frac{1}{9}$.[45] Rather than coming to the same critical conclusions as Roger, however, Cunestabulus tried to exonerate the inventors of the Church's lunar calendar by appealing to the Arabic theory of the 'access and recess of the eighth sphere', which predicted that the length of the tropical year had changed between antiquity and the present. From the various 19-year lunar cycles ascribed to Greek astronomers such as Meton, Euctemon, and Callippus, but also from certain statements by Hipparchus and Ptolemy recorded in the *Almagest*, one could gather that the ancients had shared a widespread belief that 19 solar years comprised exactly 235 lunations, even as their estimates for the length of this period varied. If this basic assumption was correct, the historical variations of the tropical year wrought by the 'access and recess of the eighth sphere' should have also affected length of the lunar month, which, just like the solar year, could be expected to oscillate between two extremes and eventually go full circle. As Cunestabulus wrote at the end of this chapter:

> Quod si uerum est, reuertetur ueritas totius nostri compoti, quando aequinoctia et solstitia ad datas sequentes reuertentur. Alieni quoque, licet nunc melius dicere uideantur, de multo manifestiori falsitate sunt arguendi.
>
> > (If this is truly the case, the truth of our whole *computus* will be restored once the equinoxes and solstices will return to the following dates. Likewise, the foreigners, although they may now seem to make better pronouncements, deserve to be convicted of a much more manifest error.)[46]

An astronomical apologetic similar, but not quite identical, to the one used by Cunestabulus in 1175 also crossed the mind of Roger of Hereford, who raised the possibility that the 19-year cycle with its new moon dates had been put in order at a time when the Sun moved more swiftly through the zodiac than it did in present times, explaining why the Moon now needed less time to catch up than used to be the case. Roger's riposte to this idea, which would have equally applied to Cunestabulus's argument, was to insist that the spheres of the Moon and the

---

43 See Nothaft, 'A Reluctant Innovator'.
44 Magister Cunestabulus, *Computus*, ed. by Lohr, c. 38, pp. 112–14. The chapter is entitled 'Quare luna videatur interdum citius, quam dicatur prima'.
45 Magister Cunestabulus, *Computus*, ed. by Lohr, c. 38, pp. 113–14, ll. 51–60.
46 Magister Cunestabulus, *Computus*, ed. by Lohr, c. 38, p. 114, ll. 68–71. By 'the foreigners', Cunestabulus appears to have meant Arabs and Jews.

Sun were both contained within the eighth sphere. Since the eighth sphere was the entity that performed the 'access and recess' it was inevitable that the Sun and Moon were affected by this motion in equal measure, in which case their rate of mutual elongation was guaranteed to remain the same.[47] In stark contrast to what Cunestabulus had hoped would be the case, Roger concluded that 235 lunations and 19 solar years were simply not equivalent. From a closer look at his astronomical tables he could infer that the Sun took 19 Julian years, 20 hours, and 58 minutes to perform 19 (sidereal) revolutions of the zodiac,[48] while 254 revolutions by the Moon took 19 years, 16 hours, and 53 minutes.[49] Roger claimed that these two periods 'came together' only after 251 years, 15 days, 11 hours, and 14 minutes, but this was in all likelihood an arithmetical blunder for what should have been a vastly larger sum of years.[50]

While there was no hope for the 19-year cycle as far as integrating mean motions were concerned, Roger discerned a theoretical possibility that the variable true motions of the Sun and Moon converged in this way on very rare occasions. He even surmised that this had happened once or twice during antiquity, giving the ancients an empirical basis for their assumption that 19 solar years equalled 235 lunar months. From one cycle to another, the error caused by this equation was certainly not dramatic, 'but from a large number of cycles, as is now obvious, a great divergence arises'.[51] Roger's initial reaction to this insight had been relatively harsh: 'Sed quam incongruum est Latinos uiros summae discretionis in re tam celebri adeo manifeste errare!' (Yet how unfitting it is that Latin men of the highest distinction should have erred so manifestly about a thing of such importance!).[52] He softened his verdict in the final chapter of his *Compotus*, which presented the Church's 19-year cycle as the product of a carefully deliberated compromise:

---

47 Roger of Hereford, *Computus*, ed. by Lohr, v. 26, ll. 82–95 (p. 228). The access-and-recess model discussed in this passage differs significantly from the one championed by Cunestabulus, which was based on the Toledan Tables. Roger's model involves a zigzag function with an amplitude of 8° and a period of 2 × 900 = 1800, which he apparently drew from the twelfth-century treatise on planetary motions discussed in Nothaft, 'Ptolemaic Orbs in Twelfth-Century England', pp. 158–61.

48 Roger of Hereford, *Computus*, ed. by Lohr, v. 26, ll. 95–102 (p. 228). This value was very probably derived from the Toledan Tables, where the mean solar motion was 0;59,8,11,28,27°/d with a resulting sidereal year of 365;15,23,29d = 365d 6;9,24h (Chabás and Goldstein, *A Survey of European Astronomical Tables*, p. 57). If multiplied by 19, this sidereal year yields 6939d 20;58,28,24h.

49 Roger of Hereford, *Computus*, ed. by Lohr, v. 26, ll. 102–04 (p. 228). The mean lunar motion in the Toledan Tables is 13;10,34,52,48,47°/d (Chabás and Goldstein, *A Survey of European Astronomical Tables*, p. 58). If the resulting sidereal month is multiplied by 254, this yields 6939d 16;52,54,37... h.

50 Roger of Hereford, *Computus*, ed. by Lohr, v. 26, ll. 104–06 (p. 228). One way of explaining this number is to multiply 19y 20h 58m and 19y 16h 53m in the following three steps: (1) 53m × 58m = 51h 14m; (2) 20h × 16h = 13d 8h; (3) 19y × 19y = 361y. It is probably no coincidence that the sum of (1), (2), and (3) is 361y 15d 11h 14m, which Roger's text decreases to 251y 15d 11h 14m.

51 Roger of Hereford, *Computus*, ed. by Lohr, v. 26, ll. 115–16 (p. 228): 'Sed ex multitudine ciclorum, ut nunc patet, magna oritur differentia'.

52 Roger of Hereford, *Computus*, ed. by Lohr, v. 26, ll. 52–53 (p. 226).

> Et quia conati sunt sapientes Latinorum uulgaribus condescendendo huius aliquam saltem tradere notitiam, ne multitudinis annumeratione confunderentur, quam breuissime potuerunt, ne nimis tamen a ueritate discederent, cyclum XIX annorum ordinauerunt. Et hac diligentia compulsi inciderunt in hunc, si fas est dicere, errorem, ut diceretur ab eis lunatio tantae esse quantitatis.
>
>> (And since the Latin sages lowered themselves when they tried to transmit at least some knowledge of this thing to the simple folk, so that [the simple folk] would not be confused by large sums, they put together a 19-year cycle that was as short as they could make it without digressing too much from the truth. And compelled by this attentiveness they fell into the error — if it is appropriate to call it that — of saying that the lunation has this exact length.)[53]

In light of the 19-year cycle's evident success as a compromise between simplicity and accuracy, the appropriate goal was not to discard this cycle, but to make small amendments in order to bring the *computus* back into agreement with observable reality. Roger's specific recommendation was to reset the so-called Golden Numbers that marked the new moons in Latin calendars by four inclusive days, promising that this way 'the calendar will be able to last for many centuries'.[54] For Roger, who closed his treatise by declaring himself as an admirer of the existing system,[55] there was hence little reason to despair. The ecclesiastical lunar calendar was in error, but this error was easily quantified and counteracted, provided that other computists (and, more importantly, the Church) found themselves willing to follow his lead.

That this reformist message did not fall on deaf ears may be seen most clearly from the *Suppletio defectuum*, an encyclopaedic poem written by Alexander Neckam (1216), who borrowed from Roger of Hereford's *Compotus* when he brought up 'the Chaldaeans, whose first day of the Moon now precedes the conjunction, now follows upon it'.[56] Inspired by Roger's words, Neckam openly decried the four-day discrepancy that had appeared between the 'vulgar' reckoning and the actual conjunctions of Sun and Moon, closing with a passionate appeal to remove an error that, in his words,

---

53 Roger of Hereford, *Computus*, ed. by Lohr, v. 28, ll. 17–23 (pp. 231–32).
54 Roger of Hereford, *Computus*, ed. by Lohr, v. 28, ll. 34–35 (p. 232): 'Et sic kalendarium per multa saecula durare poterit'.
55 Roger of Hereford, *Computus*, ed. by Lohr, v. 28, ll. 45–46 (p. 232): 'Ego vero, quotiens haec considerauero, inaestimabili afficior admiratione'. See also Roger's repeated use of the expression 'miro artificio' to describe the construction of the ecclesiastical calendar: Roger of Hereford, *Computus*, ed. by Lohr, I. 1, l. 34 (p. 134); I. 2, l. 163 (p. 139); I. 3, l. 50 (p. 142); v. 28, l. 10 (p. 231).
56 Alexander Neckam, *Suppletio defectuum*, ed. by Hochgürtel, II. 733–34 (p. 139): 'Approbo Caldeos, quorum primacio lune | Nunc preit accensum, nunc comitatur eum'. Alexander here drew inspiration from Roger of Hereford, *Computus*, ed. by Lohr, v. 26, ll. 31–33 (p. 226): 'Chaldaei vero in hac naturali computatione aliquando praeueniunt aliquando subsequuntur accensionem'.

was bound to make 'future generations grieve'.⁵⁷ The requirement to move the dates of the 'new moon of our calendar further up by four places' was also expressed early in the thirteenth century by one Conrad of Strasbourg, whose brief *Compotus* incorporated a table for the 'Latin', 'Hebrew', and 'Chaldaean' lunation lengths that was evidently drawn from Roger's work.⁵⁸

A far more widely diffused reaction to Roger's ideas was subsequently shown by John of Sacrobosco, who is famous for his university textbooks on mathematics and astronomy. Close to 200 manuscripts still attest to the popularity of his work on time reckoning, known as *Computus* or *De anni ratione* (*c.* 1232), whose introductory chapter draws the line between astronomy and *computus* in a way directly inspired by Roger of Hereford: 'Quicquid etiam Astronomus de temporis fractionibus determinat, fit gratia motus astrorum, ipsius enim est motum ipsum considerare, Computistae vero temporum discretiones' (Whatsoever the astronomer determines with regard to the fractions of time, he does for the sake of celestial motions. For his is to consider motion itself, whereas the computist's [remit are] the divisions of time).⁵⁹ In assessing the problem posed by the 19-year cycle, Sacrobosco utilized data drawn from Ptolemy's *Almagest*, which put him in a position to argue that 235 synodic months were $1\frac{1}{3}$ hour short of 19 Julian years (6939d 18h). Over the course of the more than 1230 years that separated the birth of Christ from his own year of writing, this gap had led to an accumulated error of 3 days and 14 hours, with the result that the day designated as *luna* 1 in the calendar was often the fourth day of the Moon when compared to astronomical reality.⁶⁰ Sacrobosco discerned the necessity of resetting the Golden Numbers, such that the start of the January lunation in year 1 of the 19-year cycle appeared next to 20 January instead of 23 January, but warned that such a correction had to be postponed as long as the Church felt bound by the decrees of the General Council of Nicaea.⁶¹

Such qualms about the permissibility of altering the established *computus* go some way towards explaining why the correction of the ecclesiastical calendar, despite being a frequently addressed desideratum in medieval sources, had to wait

---

57 Alexander Neckam, *Suppletio defectuum*, ed. by Hochgürtel, II. 845–48 (p. 144): 'Terminus id paschalis agit, ne forte uacillet | Inmutatus; ob hoc secla futura gement. | Vrgens articulus hunc errorem reuocabit; | O utinam ueri cogeret istud amor'.

58 See the treatise in Trier, Stadtbibliothek, MS 8° 1077/1263, fols 1ʳ–5ᵛ, 9ʳ–10ʳ (fol. 9ᵛ): 'Ex predictis patet quod ad verificandum primationem nostri kalendarii quelibet primatio quarto loco in kalendario est anterioranda'. Another copy appears in Bruges, Openbare Bibliotheek, MS 528, fols 1ʳ–6ᵛ (fol. 6ʳ).

59 John of Sacrobosco, *Computus*, ed. by Melanchthon, sig. Bʳ⁻ᵛ. Compare the corresponding passage in Roger of Hereford, *Computus*, ed. by Lohr, v. 25, ll. 43–46 (p. 223): 'Differunt autem, quod astronomia, quicquid de tempore dicit, gratia motus est. Eius enim est motum ipsum considerare. Computus autem, quicquid dicit de motu, gratia temporis fit. Eius est enim tempora distinguere'. Further borrowings from Roger of Hereford will be documented in a forthcoming critical edition of Sacrobosco's *Computus* by Alfred Lohr.

60 John of Sacrobosco, *Computus*, ed. by Melanchthon, sigs F2ʳ–F3ʳ. See Pedersen, 'In Quest of Sacrobosco', p. 189.

61 John of Sacrobosco, *Computus*, ed. by Melanchthon, sig. F3ʳ: 'Sed quia in Concilio generali aliquid de Calendario transmutare prohibitum est, oportet modernos adhuc sustinere huiusmodi errores'.

until the sixteenth century to become reality. While Roger's *Compotus* can hardly be counted among the immediate sources behind the calendar reform decreed by Pope Gregory XIII in 1582, the influence of his critique on the further development of computistical literature should not be underestimated. By confronting the existing set of rules with new concepts and parameters specific to mathematical astronomy, the astrologer from the West Midlands was able to demonstrate conclusively that 'naturalizing' one's lunar computation the way Gerland and Walcher of Malvern had done was not nearly enough to neutralize the flaws of the ecclesiastical calendar, as both flavours of *computus* still relied on the same 19-year cycle. As a reaction to this insight, thirteenth-century writers such as Robert Grosseteste abandoned the *computus naturalis* and established in its place a new subgenre sometimes known as the *computus philosophicus*, which continued Roger's approach of supplementing traditional calendrical lore with information drawn from astronomical sources in order to offer readers a more accurate understanding of the courses of Sun and Moon. By fostering this important step in the evolution of Latin computistical writing, Roger of Hereford managed to inscribe himself in a momentous intellectual trajectory, the outcomes of which are not limited to the Gregorian calendar.[62]

Works Cited

Manuscripts and Archival Sources

Bruges, Openbare Bibliotheek, MS 528
Cambridge, St John's College, MS F.25
Cambridge, University Library, MS Kk.1.1
London, British Library, MS Arundel 377
Madrid, Biblioteca nacional de España, MS 10016
Oxford, Bodleian Library, MS Ashmole 1796
—, MS Digby 40
Trier, Stadtbibliothek, MS 8° 1077/1263

Primary Sources

Alexander Neckam, *Suppletio defectuum*, ed. by Peter Hochgürtel, Corpus Christianorum Continuatio Mediaevalis, 221 (Turnhout: Brepols, 2008)
Der Computus Gerlandi, ed. by Alfred Lohr (Stuttgart: Steiner, 2013)
The Friars' Libraries, ed. by K. W. Humphreys, Corpus of British Medieval Library Catalogues, 1 (London: The British Library, 1990)

---

62 For further details, see Robert Grosseteste, *Compotus*, ed. by Lohr and Nothaft; Nothaft, *Scandalous Error*, pp. 116–303.

John of Sacrobosco, *Computus*, ed. by Philipp Melanchthon, in *Ioannis de Sacro Busto Libellus de Sphaera; Eiusdem autoris libellus, cuius titulus est Computus, eruditißimam anni & mensium descriptionem continens* (Wittenberg: Clug, 1538), sigs B^r–H3^r

The Letters and Charters of Gilbert Foliot, Abbot of Gloucester (1139–48), Bishop of Hereford (1148–63) and London (1163–87), ed. by Z. N. Brooke, Dom Adrian Morey, and C. N. L. Brooke (Cambridge: Cambridge University Press, 1967)

Magister Cunestabulus, *Computus*, ed. by Alfred Lohr, Corpus Christianorum Continuatio Mediaevalis, 272 (Turnhout: Brepols, 2015)

Raymond of Marseille, *Liber cursuum planetarum*, in *Opera omnia*, vol. 1, ed. by Marie-Thérèse d'Alverny, Charles Burnett, and Emmanuel Poulle (Paris: CNRS Éditions, 2009), pp. 126–341

Robert Grosseteste, *Compotus*, ed. by Alfred Lohr and C. Philipp E. Nothaft (Oxford: Oxford University Press, 2019)

Roger of Hereford, *Computus*, ed. by Alfred Lohr, Corpus Christianorum Continuatio Mediaevalis, 272 (Turnhout: Brepols, 2015)

The Toledan Tables, ed. by Fritz Saaby Pedersen, 4 vols (Copenhagen: Reitzel, 2002)

Walcher of Malvern, *De Dracone*, in C. Philipp E. Nothaft, *Walcher of Malvern:* De lunationibus *and* De Dracone (Turnhout: Brepols, 2017), pp. 193–217

Walcher of Malvern, *De Lunatione*, in C. Philipp E. Nothaft, *Walcher of Malvern:* De lunationibus *and* De Dracone (Turnhout: Brepols, 2017), pp. 89–191

Secondary Studies

Borst, Arno, 'Ein Forschungsbericht Hermanns des Lahmen', *Deutsches Archiv für Erforschung des Mittelalters*, 40 (1984), 379–477

—, *Der karolingische Reichskalender und seine Überlieferung bis ins 12. Jahrhundert*, 3 vols, MGH Libri mem., 2 (Hannover: Hahnsche Buchhandlung, 2001)

Burnett, Charles, *The Introduction of Arabic Learning into England* (London: British Library, 1997)

—, 'Mathematics and Astronomy in Hereford and Its Region in the Twelfth Century', in *Medieval Art, Architecture and Archaeology at Hereford*, ed. by David Whitehead (Leeds: British Archaeological Association, 1995), pp. 50–59

Chabás, José and Bernard R. Goldstein, *A Survey of European Astronomical Tables in the Late Middle Ages* (Leiden: Brill, 2012)

Haskins, Charles Homer, 'The Reception of Arabic Science in England', *English Historical Review*, 30 (1915), 56–69

—, *Studies in the History of Mediaeval Science* (Cambridge, MA: Harvard University Press, 1924)

Jacquemard, Catherine, 'La réinvention de l'astrolabe au Moyen Âge: l'exemple du *De statu mundi*', in *La technologie gréco-romaine: transmission, restitution et médiation*, ed. by Philippe Fleury, Catherine Jacquemard, and Sophie Madeleine (Caen: Presses universitaires de Caen, 2015), pp. 191–211

Jones, Charles W., 'A Legend of St Pachomius', *Speculum*, 18 (1943), 198–210

Juste, David, 'Hermann der Lahme und das Astrolab im Spiegel der neuesten Forschung', in *Hermann der Lahme: Reichenauer Mönch und Universalgelehrter des 11.*

Jahrhunderts, ed. by Felix Heinzer and Thomas Zotz (Stuttgart: Kohlhammer, 2016), pp. 273–84

Lejbowicz, Max, 'Les Pâques baptismales d'Augustin d'Hippone, une étape contournée dans l'unification des partiques computistes latines', in *Computus and Its Cultural Context in the Latin West, AD 300–1200*, ed. by Immo Warntjes and Dáibhí Ó Cróinín (Turnhout: Brepols, 2010), pp. 1–39

Lindhagen, Arvid, 'Die Neumondtafel des Robertus Lincolniensis', *Arkiv för Matematik, Astronomi och Fysik*, 11.2 (1916), 1–41

McCluskey, Stephen C., 'Changing Contexts and Criteria for the Justification of Computistical Knowledge and Practice', *Journal for the History of Astronomy*, 34 (2003), 201–17

Moreton, Jennifer, 'Before Grosseteste: Roger of Hereford and Calendar Reform in Eleventh- and Twelfth-Century England', *Isis*, 86 (1995), 562–86

—, 'The *Compotus* of "Constabularius" (1175): A Preliminary Study', in *Langage, Sciences, Philosophie au XIIe siècle*, ed. by Joël Biard (Paris: Vrin, 1999), pp. 61–82

Neugebauer, Otto, 'From Assyriology to Renaissance Art', *Proceedings of the American Philosophical Society*, 133 (1989), 391–403

Nothaft, C. Philipp E., *Medieval Latin Christian Texts on the Jewish Calendar* (Leiden: Brill, 2014)

—, 'Ptolemaic Orbs in Twelfth-Century England: A Study and Edition of the Anonymous *Liber de motibus planetarum*', *Mediterranea: International Journal for the Transfer of Knowledge*, 3 (2018), 145–210

—, 'A Reluctant Innovator: Graeco-Arabic Astronomy in the *Computus* of Magister Cunestabulus (1175)', *Early Science and Medicine*, 22 (2017), 24–54

—, 'Roman vs. Arabic Computistics in Twelfth-Century England: A Newly Discovered Source (*Collatio Compoti Romani et Arabici*)', *Early Science and Medicine*, 20 (2015), 187–208

—, *Scandalous Error: Calendar Reform and Calendrical Astronomy in Medieval Europe* (Oxford: Oxford University Press, 2018)

Pedersen, Olaf, 'In Quest of Sacrobosco', *Journal for the History of Astronomy*, 16 (1985), 175–221

Reese, Ronald Lane, Edwin D. Craun, and Charles W. Mason, 'Twelfth-Century Origins of the 7980-Year Julian Period', *American Journal of Physics*, 51.1 (January 1983), 73

Southern, Richard W., *Robert Grosseteste: The Growth of an English Mind in Medieval Europe*, 2nd edn (Oxford: Clarendon Press, 1992)

Warntjes, Immo, 'Hermann der Lahme und die Zeitrechnung: Bedeutung seiner Computistica und Forschungsperspektiven', in *Hermann der Lahme: Reichenauer Mönch und Universalgelehrter des 11. Jahrhunderts*, ed. by Felix Heinzer and Thomas Zotz (Stuttgart: Kohlhammer, 2016), pp. 285–321

GILES E. M. GASPER, BRIAN K. TANNER,
SIGBJØRN SØNNESYN, AND NADER EL-BIZRI

Travelling Optics

Robert Grosseteste and the Optics behind the Rainbow[*]

Robert Grosseteste's treatise on the rainbow, *De iride*, has attracted attention on a number of grounds, notably for Grosseteste's explanation of the rainbow itself, taking its place in the longer history of scientific interest in the rainbow as a natural phenomenon, and in the articulation of the laws of refraction and reflection. The argument of this treatise also relies on the Aristotelian notion of subalternation of scientific disciplines, one to another. The study of the rainbow is subalternated to optics, which, being again subalternated to geometry, can provide a causal explanation of the phenomenon in a way that physics, the general study of nature, cannot. Both of these are needed. What the current investigation will focus on is the first half of the

---

[*] The authors would like to acknowledge the input and activities of the Ordered Universe project (www.ordered-universe.com) whose deliberations on the subject of the rainbow they have enjoyed since 2010, and without whom the conjunction of scientific, historical, philosophical, and philological methods would be impossible to replicate. They would like to thank in particular Professor Hannah Smithson (University of Oxford), Professor Tom McLeish (University of York), Professor Neil Lewis (Georgetown University) and Professor Cecilia Panti (Università di Roma 'Tor Vergata') for their wise guidance and comments. Brian and Giles would like to thank Christian Etheridge for the invitation to talk at the University of Southern Denmark in 2015 which provided the spur for this contribution to the volume.

> **Giles E. M. Gasper** • is Professor of High Medieval History at Durham University. He is principal investigator on the inter-disciplinary Ordered Universe project on the writings of Robert Grosseteste. g.e.m.gasper@durham.ac.uk
>
> **Brian K. Tanner** • is Emeritus Professor of Physics at Durham University and Fellow of the Royal Society of Arts. He is a member of the Ordered Universe project. b.k.tanner@durham.ac.uk
>
> **Sigbjørn Sønnesyn** • is Postdoctoral Fellow (High Medieval History) at the University of Oslo. He is editing six and translating ten of Robert Grosseteste's thirteen scientific opuscula for the Ordered Universe project. sigbjorn.o.sonnesyn@durham.ac.uk s.o.sonnesyn@imv.uio.no
>
> **Nader El-Bizri** • is Professor of philosophy and civilization studies at the American University of Beirut, and Leverhulme Visiting Professor at Durham University. NB44@aub.edu.lb.

Medieval Science in the North: Travelling Wisdom, 1000–1500, ed. by Christian Etheridge and Michele Campopiano, KSS 2 (Turnhout: Brepols, 2021), pp. 45-75
BREPOLS PUBLISHERS   10.1484/M.KSS-EB.5.122880

treatise, and the extended *resolutio* in which Grosseteste not only asserts the principle of subalternation but rehearses and explores a series of demonstrations of optical phenomena. These demonstrations will be analysed from contextual and scientific perspectives, with comparison to Grosseteste's successors in specific instances, for example the Silesian scholar Witelo on the measurement of refraction. Following Birkenmajer, Crombie makes the argument that Witelo's *Perspectiva* includes a number of passages which appear to have been taken directly from Grosseteste's writings.[1]

This multi-disciplinary approach allows a fuller assessment of Grosseteste's intellectual methods and choices. The sources at Grosseteste's disposal illustrate the extent to which he had embraced the wisdom of ancient Greek and medieval Arabic thought, and how far such wisdom had travelled within the Latin West. Why the particular demonstrations of optical phenomena were selected by Grosseteste illustrate the independence and intricacy of his cast of mind, as well as his capacious knowledge of past experience.

Robert Grosseteste's (*c.* 1170–1253) Anglo-Norman treatise on the loss and restoration of creation, the *Château d'amour*, incorporates an extended allegorical description of the castle, in which Jesus Christ took refuge, as the Virgin Mary. The poem conducts a tour of the castle, with close attention to architectural and military detail. At the centre of the castle, in the middle of the tallest tower, the description reaches its denouement:

> En cele bele tur et bone | I ad de ivoire une trone | Ki plus ad en sei blancheur | Ke en mi esté le beau jur. | Par engine est compassez; | Al munter i ad set degrez | Ki par order cochez sunt | Ni a si bele chose el mund | Le arc du ciel entur s'estent | Od la colur k'a li apent.
>
> > (Inside this fine and beautiful tower there is an ivory throne which shines more brightly than daylight in midsummer. It is skilfully designed, with seven steps arranged to approach it. Nothing in the world is so beautiful. The rainbow with all its colours extends around it.)[2]

The *Château d'amour* was quite probably composed before Grosseteste became bishop of Lincoln in 1235; half of the medieval records of the poem state the author's title as *Magister*, that is 'Master', rather than bishop.[3] To visit the question of pre-episcopal authorship also raises the question of audience. Mackie, in particular, has advanced the convincing suggestion that the poem was composed for the Franciscans, to the Oxford community of which Grosseteste became lector in 1228/1229.[4] If these suggestions are followed then the composition of Le *Château d'amour* may have

---

1 Crombie, *Robert Grosseteste and the Origins of Experimental Science*, pp. 213–16.
2 Murray, *Le Château d'Amour de Robert Grosseteste, évêque de Lincoln* pp. 106–07. English translation from Mackie, 'Grosseteste's Anglo-Norman Treatise', p. 167.
3 Mackie, 'Grosseteste's Anglo-Norman Treatise', pp. 153–54.
4 Mackie, 'Grosseteste's Anglo-Norman Treatise', pp. 154–56. See also her 'Scribal Intervention and the Question of Audience'.

coincided with the last of Grosseteste's scientific works, including, above all the *De iride – On the Rainbow*. If the image of the rainbow is given place of honour within the poem, the same might be said, analogously, of Grosseteste's discussion of how natural rainbows and their colours are formed within his scientific canon.

On the Rainbow, composed at the end of the 1220s, is, arguably, the most sophisticated of his scientific 'opuscula'.[5] These short works composed from *c.* 1195 until *c.* 1228 explore, for the most part, questions arising from natural phenomena, including those that are cosmological and meteorological.[6] All are related to study of the quadrivium in particular, that is, the mathematical arts of arithmetic, geometry, music, and astronomy. These shorter works sit alongside the fuller, worked-up and polished commentary on Aristotle's *Posterior Analytics* and the unfinished commentary, in note form, on Aristotle's *Physics*. A central aspect to Grosseteste's intellectual exploration in these three decades, as all of these works reveal, was the reception of ancient natural philosophy, Aristotle, Ptolemy, and others, and the thought they had inspired amongst medieval Islamicate philosophers, Ibn Rushd (Averroes), Ibn Sīnā (Avicenna), Al-Kindī (Alkindus), and others. Especially with respect to natural philosophy Grosseteste's generation was among the first in the Latinate European milieu to absorb these materials, newly translated into Latin from Greek and Arabic, in various locations, but principally in the Iberian Peninsula.[7] The new knowledge then travelled north to the schools and early universities of Europe, principally, although not exclusively, Paris and Oxford.

---

5   These exist in a variety of critical editions. That by Baur, *Die Philosophischen Werke des Robert Grosseteste, Bischofs von Lincoln* remains the only complete collection to date. Modern critical editions exist of the treatises *On the Sphere*, *On Comets*, *On the Super-Celestial Motions*, and of *On Light*: see *Moti, virtù e motori celesti nella cosmologia di Roberto Grossatesta*, ed. by Panti; and *Roberto Grossatesta, La Luce*, ed. by Panti. The text of the critical edition of the *De luce*, without critical apparatus is printed as 'Robert Grosseteste's De luce: A Critical Edition', ed. by Panti. In addition to these a critical edition of the treatise *De colore*, ed. by Greti Dinkova-Bruun, has been produced by the Ordered Universe project. A new six-volume series from Oxford University Press, produced by the Ordered Universe project (www.ordered-universe.com), will provide a complete set of modern critical editions: *The Scientific Works of Robert Grosseteste: Editions, English Translations and Interdisciplinary Analysis*, the first volume of which is published as Gasper and others, ed., *Knowing and Speaking*.

6   On the sequence of Grosseteste's works see Panti, 'Robert Grosseteste and Adam of Exeter's Physics of Light', p. 185. For major discussions of the chronology of Grosseteste's works, including the scientific, see Dales, 'Robert Grosseteste's Scientific Works'; McEvoy, 'The Chronology of Robert Grosseteste's Writings'; Southern, *Robert Grosseteste*, pp. 111–40. Many of the issues connected to the differences between McEvoy and Southern reflect different views on Grosseteste's putative career, as outlined in Goering, 'Where and When did Grosseteste Study Theology?' Panti's chronology places the scientific, quadrivial works before the exegetical and speculative theology, pastoral works, and translations from Greek, which is more reflective of early thirteenth-century norms, and the lack of positive evidence to the contrary. She also emphasizes the internal evidence of the individual treatises for indications of changes to reading habits or intellectual interests, see Panti, 'The Evolution of the Idea of Corporeity'.

7   See, for example, Burnett, 'The Institutional context of Arabic-Latin Translations of the Middle Ages', on the role of Jewish interlocutors see Glick, 'Science in Medieval Spain'. On translation from Greek see, Berschin, *Greek Letters and the Latin Middle Ages*.

Grosseteste's *On the Rainbow* has received more attention than most of the other scientific opuscula, including a translation into English.[8] It is important as an example of applied optics, as an expression of the subalternation of sciences (where one science adopts the principles of another), and as a contribution to unlocking the physics of the rainbow.[9] The treatise is the focus of discussions of medieval experimental methods and observations of natural phenomena, and to discussions, older and recent, of theories of vision, perception and description of colour, and geometrical optics, including the theory of refraction.[10] It is part of a process in which the European opticians developed new approaches to the rainbow, influenced by and influential upon changes in the understanding of optics. The treatise is also important as a marker for the relative accessibility of texts in optics and related fields that were translated from Greek and Arabic.[11]

What follows offers a synopsis of Grosseteste's treatise *On the Rainbow* and some further contextual remarks on the circumstances of its composition. A comparative outline of the history of the rainbow amongst authors writing in Arabic in the early to high medieval period, and the classical European tradition as inherited by Grosseteste, emphasizes the somewhat eclectic nature of Grosseteste's intellectual inheritance on this score. Finally, attention is paid to the first half of the treatise and the articulation of the principles which should underpin the study of the rainbow. The application and exposition of optics and refraction form the focus. Comparison with some of Grosseteste's thirteenth-century successors, notably Witelo, helps to highlight Grosseteste's achievements.

Synopsis

On the Rainbow sets out to explain a theory of how rainbows are formed, and how this theoretical account works out in practice. Grosseteste begins by explaining his departure from the prevailing Aristotelian paradigm, based on a distinction between modes of argument propounded in Aristotle's *Posterior Analytics*.[12] Aristotle himself had studied the rainbow, arguing from observed phenomena to their causes, and giving an explanation based on the discipline of physics. Grosseteste, by contrast, sets out to argue from principles taken from the science of optics, to which the study

---

8 Lindberg, 'Robert Grosseteste'.
9 Boyer, 'Robert Grosseteste on the Rainbow'; Boyer, *The Rainbow*, pp. 88–94; Lindberg, 'Roger Bacon's Theory of the Rainbow'; Lee and Fraser, *The Rainbow Bridge*.
10 Smith, *From Sight to Light*, pp. 256–60; Laird, 'Robert Grosseteste on the Subalternate Sciences', 166–67; Crombie *Robert Grosseteste and the Origins of Experimental Science*; Eastwood, 'Robert Grosseteste's Theory of the Rainbow'; Eastwood 'Medieval Empiricism'; Lindberg, *Theories of Vision from Al-Kindi to Kepler*, pp. 94–103; Smithson, 'All the Colours of the Rainbow'; Smithson and others 'A Color Coordinate System from a Thirteenth-Century Account of Rainbows'; Turbayne, 'Grosseteste and an Ancient Optical Principle', and Eastwood, 'Grosseteste's "Quantitative" Law of Refraction'.
11 El-Bizri, 'Grosseteste's Meteorological Optics', p. 23; see also Panti, 'The Theological Use of Science in Robert Grosseteste and Adam Marsh'.
12 Robert Grosseteste, *Commentarius in Posteriorum Analyticorum libros*, ed. by Rossi, I.13.

of the rainbow is subalternated. Having sketched out the trajectory of his argument, Grosseteste proceeds in the first of the two main parts of the treatise, to set out an account of the causes of rainbows as delineated by the science of optics.

He bases his theory on the behaviour of light described in geometrical terms, and by positing a tripartite division of the science of optics, according to the three principal ways in which a light ray may reach the eye, or reach the object of vision from the eye. The first, a direct and straight line between the light source and the eye, belongs to sight; the second, the light being reflected off a reflective surface at an angle to both the light source and the eye, belongs to the science of mirrors; and the third, in which the light ray passes through several diaphanous media during which passage it is refracted and forms a series of angular interconnections, is the basis for Grosseteste's theory of the formation of rainbows.

Having described the principles of this theory in some detail, explaining what happens to a ray of light that passes through several diaphanous media on account of which its straight path is broken, Grosseteste moves to apply this theory to the way in which a light ray passes through several diaphanous media when it hits clouds of a certain form, and at a certain point in the formation of condensed drops of water from its vaporized form. It is the repeated breaking of the light ray as it passes through the cloud and its droplets that ultimately constitute the cause of the appearance of rainbows. Finally, Grosseteste gives an account of the colours of the rainbow, based on his own theory of colour, and by explaining the reasons for why the colours appear to differ under varying meteorological conditions.

Composition

Where Grosseteste composed the treatise *On the Rainbow* is an intriguing question. Relatively little is known about his life until 1225. Grosseteste may first enter the historical record as witness to a charter of Bishop Hugh of Lincoln, after 1189 and some time before April 1192.[13] He is specified as a master, implying that he then was in his early twenties. Nothing certain is known of his earlier life, bar the statement from the historian Nicholas Trevet (c. 1257 × 65 – in or after 1334), writing between 1314 and 1320, that Grosseteste was from Suffolk.[14] His first known post was as a member of Bishop William de Vere's household in Hereford, between c. 1195 and 1198, which

---

13 This was a confirmation charter for possession of the monks of St Andrew in Northampton: British Library, MS Royal 11 B IX fols 24ᵛ–25. See Southern, *Robert Grosseteste*, pp. 63–64.
14 Nicholas Trivet, *Annales sex regum Angliae*, ed. by Hog, p. 242. On the shortcomings of Hog's edition see: Mantello, 'The editions of Nicholas Trevet's *Annales sex regum Angliae*'; on Trevet see: Clark, 'Trevet, Nicholas'. The identification of Stradbroke in Suffolk as the village of Grosseteste's birth, perpetuated in Stevenson's *Robert Grosseteste*, p. 1, is not supported by any medieval evidence, and seems to have been added, on authority unknown, by Archbishop Matthew Parker to his 1570 edition of Matthew of Westminster's *Flores historiarum*, in sections of this work now identified as Matthew Paris — Matthew of Westminster himself is unlikely to have existed; for details see *Flores Historiarum*, ed. by Luard, p. xlvi and Gransden, *Historical Writing in England*, pp. 333–34. The suggestion in

dissolved on the bishop's death. Where he was between then and the next firm date, his appointment as Rector of Abbotsley in Lincoln diocese, is unclear. He was probably in France at the time of the papal interdict in England, 1208–1214, but whether this included study at Paris or elsewhere is unclear. Grosseteste does appear in charters relating to legal cases concerning Hereford diocese in the second decade of the thirteenth century, which might imply a continued association to the region perhaps in connection with Archdeacon Hugh Foliot. Whether this included teaching, or being at Oxford, is also unclear. Some circumstantial evidence might indicate a presence at Paris, such as the lack of specific reference to Aristotelian texts in the treatises *On the Generation of Sounds* and *On Comets*, despite their clear use in these works, might reflect the ban on Aristotle's natural philosophy at Paris in 1210, and repeated in 1214. The treatise on calendar reform, the *Compotus*, probably composed in the early 1220s, makes use of the meridian of Paris, and later in life Grosseteste had a close friendship with William of Auvergne (1180/90–1249) who was active in the Paris schools in the early thirteenth century, before becoming bishop of Paris in 1228.[15] Nicole Schulman has also advanced the intriguing thesis that Grosseteste may have been married, with a family, in Paris, on the basis of a legal dispute over the bequest of the house of one Robertus Grosse Teste to the church of Sainte-Opportune, subject to it belonging to his three children as long as they lived, witnessed by William of Auvergne.[16] Whether this relates to the same Robert Grosseteste is not firmly established.

From 1225, however, Grosseteste can be placed with some confidence in English circles, and at some point, probably in this decade, he embarked on the next stage of his intellectual formation, taking up a magistry in theology. As he himself described his career he was 'first a cleric, then a Master of Theology and priest, and then bishop'. Quite when the magistry began is not easy to identify. As Ginther has pointed out, the documentary evidence, as it stands, indicates a regency that began in or around 1229, and ending with his election to Lincoln as bishop in 1235.[17] At the same time Grosseteste received further ecclesiastical preferment, being appointed as Archdeacon of Leicester in 1229, and, possibly in the same year, or 1230, possibly 1231, as the first lector to the Oxford Franciscans.[18] *On the Rainbow* falls then within a period of intellectual and documentable circumstantial change in Grosseteste's life, and, arguably, a point at which his scientific, quadrivial, penchant was developing alongside his exegetical and theological interests.

Richard of Bardney's *Vita Roberti Grosthed*, that Grosseteste's birthplace was the Lincolnshire village of Stowe, should also be treated with caution owing to its late date (1502) and fantastical presentation of its protagonist's life: the text is printed in Wharton, *Anglia Sacra*, pp. 325–41.

15 See Robert Grosseteste, *The Letters of Robert Grosseteste Bishop of Lincoln*, trans. by Mantello and Goering pp. 270–71, and *Roberti Grosseteste episcopi quondam Lincolniensis epistolae*, ed. by Luard, Letter 78.

16 Schulman, 'Husband, Father, Bishop? Grosseteste in Paris'.

17 Ginther, *Master of the Sacred Page*, p. 5; a variety of dates from 1214 to 1225 have been suggested: Callus 'Robert Grosseteste as Scholar', pp. 6–11; McEvoy, *Robert Grosseteste*, p. 25; Southern, *Robert Grosseteste*, pp. 69–71; although none is conclusive.

18 Southern, *Robert Grosseteste*, pp. 74–75.

Textual Debts and Intellectual Inheritance

The treatise *On the Rainbow* deploys a number of named sources, indicating a wider range of references from previous generations, although the full corpus of Greek and Arabic writing on optics would not be available in Latin translation until the mid-thirteenth century.[19] Aristotle's *Meteorology, Posterior Analytics, On Animals* (the collective title in the Middle Ages for Aristotle's various works on animals) and probably *On the Heavens*, form an important basis for Grosseteste's discussion. These works sit alongside the emission theory of vision deriving, ultimately, from Plato's *Timaeus*, reaching Grosseteste via intermediaries including important texts on geometrical optics. Grosseteste makes reference to (pseudo) Euclid's *Optics* and *Catoptrics*, alongside a familiarity with other sources for the study of optics, such as al-Kindī's *De aspectibus*.[20] These figures presumably belong to the ranks of the *mathematici et physici* (mathematicians and physicists) who study what is beyond nature.[21] Grosseteste inherited a complex series of investigations on the phenomenon of the rainbow, from various cultural perspectives, and is significant for his attempt to utilise the new learning available to him.

Grosseteste's principal source is the account of the rainbow in Book Three of Aristotle's *Meteorology*.[22] This presents a physical, as opposed to a geometric, argument for the rainbow's appearance and shape. According to Aristotle, the rainbow is produced by reflection off a cloud and consists of three colours; the shape of the bow is related to the form of the clouds onto which it is projected.[23] The Aristotelian thesis entails that the rainbow is caused by the reflections of sunlight on a dense dark cloud that is at a limited distance from an observer. When it is at the point of raining, and the air in the cloud is turning into small raindrops, if the sun is opposite the cloud and its light is bright, then the cloud acts like a mirror, and its reflection issues forth as a colouration of the rainbow. This is seen as an accumulation of small mirrors too small to be visible individually, wherein the reflections appear as a single and whole continuous magnitude that is made up of all of them, and under these conditions the rainbow is a reflection of sunlight on the cloud to the sight of an observer.[24] Within Latin Christendom this theory remained essentially unchallenged, and was not discussed in any great detail from Late Antiquity through to the High Middle Ages.[25] Isidore of Seville attributed the shape of the bow to the clouds and the sun,

---

19   Lindberg, *Theories of Vision*, p. 103.
20   Smith, *From Sight to Light*, pp. 259–60, makes the unconvincing suggestion that Grosseteste may have known in part, or superficially, the optical work of Ibn al-Haytham.
21   Grosseteste, *De iride*, § 2. All Latin and English quotations from this treatise in what follows derive from the forthcoming edition produced under the aegis of the Ordered Universe research project: *Colour and the Refraction of Rays: Robert Grosseteste's De colore* 'On Colour' and *De iride* 'On the Rainbow'.
22   See Boyer, *The Rainbow*, pp. 38–55.
23   See Sayili, 'The Aristotelian Explanation of the Rainbow'.
24   Aristotle, *Meteorologica*, ed. and trans. by Lee, Book III, Parts 4–5, esp. 373a35–375a8.
25   Lee and Fraser, *The Rainbow Bridge*, pp. 138–39.

and identified four colours, and related these to the four elements. Bede followed Isidore in the observation that: 'Arcus in aere quadricolor ex sole aduerso nubibusque formatur' (the Rainbow with its four colours is formed in the air from the directly opposed sun and the clouds)'.[26]

Matters were different within the early-medieval Islamicate context. The principal theories regarding the explication of the phenomenon of the rainbow in the classical Arabic scientific milieu of the Islamic civilization and its intellectual history would have been articulated originally as analytic commentaries on Aristotle's *Meteorology*, which itself was translated into Arabic (c. 801) by Yūḥannā ibn al-Biṭrīq under the title: *Kitāb al-āthār al-ʿulwīyya*.[27] A principal analytic commentary on this transmitted Aristotelian thesis is embodied in the theory of Ibn Sīnā in his treatment of *al-āthār al-ʿulwīyya* (*Meteorology*) within his *Kitāb al-Shifāʾ* (*Book of Healing*, Part V). In this, Ibn Sīnā argued that the *qaws quzaḥ* (rainbow) results from the reflection of sunlight on the mist that is formed of small transparent *rashsh* (dewdrops) as they are dispersed in the wet air between the cloud and the sun or the observer. The cloud would act as a dark background for the mist, similar to the quicksilver lining at the rear surface of the glass that forms a mirror. For example, a crystal would not act as a mirror if it did not have some mask behind it as an opaque coloured object, and if a transparent expanse is behind it, then it does not reflect like a mirror.[28] A similar view is encountered in the tenth-century proto-encyclopaedic *Rasāʾil Ikhwān al-Ṣafāʾ* (*Epistles of the Brethren of Purity*), just a few decades before the time of Ibn Sīnā, and also set in the context of commenting on Aristotle's *Meteorology*. The anonymous adepts of the Iraqi coterie of the *Ikhwān al-Ṣafāʾ* (Brethren of Purity) noted in their epistolary compendium that the cause of the occurrence of the rainbow was due to the irradiation of sunlight on a humid and dense vapour that fills the atmosphere.[29] However, there is no evidence that Ibn Sīnā benefited directly from their meteorological explanations despite the wide dissemination of their works in his epoch and the fact that he was already acquainted with their teachings in his youth. Ibn Sīnā's focus on the mist dewdrops, instead of placing his emphasis on the dense dark cloud in acting as a continuous mirror, signalled a theoretical possibility that this meteorological phenomenon might be studied by way of geometric optics and through some form of controlled testing by simulated modelling as it happened later with Kamāl al-Dīn al-Fārisī (d. 1320).[30]

Another principal theory in explicating the phenomenon of the rainbow is articulated in the optical research of Ibn al-Haytham (Alhazen, d. 1041) as primarily set

---

26 Isidore of Seville, *De natura rerum*, ed. and trans. by Fontaine c. 31, English translation in Isidore of Seville, *On the Nature of Things*, trans. by Kendall and Wallis, Bede, *De natura rerum*, ed. Jones, c. 31, English translation in *On the Nature of Things and on Times*, trans. by Kendall, and Wallis.
27 *Kitāb al-āthār al-ʿulwīyya*, ed. by Petraitis. Christian era dating is used throughout the present discussion.
28 Ibn Sīnā, *Al-Maʿādin wa al-āthār al-ʿulwīyya, Kitāb al-Shifāʾ* 5, ed. by Madkour and others, p. 51.
29 *Epistles of the Brethren of Purity* (*Rasāʾil Ikhwān al-Ṣafāʾ*), ed. and trans. by Baffioni, Ep. 18, Ch. 11, p. 209 of the annotated English translation, and pp. 225–26 of the Arabic edition.
30 Rashed, 'Le modèle de la sphère transparente et l'explication de l'arc-en-ciel', pp. 124–25, 131–33, 135–36 and Topdemir, 'Kamal Al-Din Al-Farisi's Explanation of the Rainbow', pp. 76–77.

out in his *Kitāb al-manāẓir* (*Book of Optics; De aspectibus; Perspectiva*).³¹ Even though Ibn al-Haytham was Ibn Sīnā's contemporary, we do not have any evidence that they were aware of each other's works, with the former being based in Egypt and the latter in Iran. Ibn al-Haytham's studies in meteorological optics rested on the findings of his research in catoptrics and dioptrics within an Arabic mathematical tradition that appealed to the legacies of Euclid and Ptolemy, as well as to those of Apollonius of Perga and Archimedes. Ibn al-Haytham built upon the research of his predecessors in this regard, and most notably the dioptrics of Abū al-ʿAlāʾ ibn Sahl (tenth century) who disclosed a principle that is akin to what is more commonly known since the seventeenth century as 'Snell's law of refraction' (namely a principle attributed to Willebrord Snellius that determines the refractive index of a transparent medium in connection with a given geometric shape, which can act as a basis for designing lenses).³² Ibn al-Haytham's studies in catoptrics and dioptrics, as the respective sciences of the reflection and refraction of light with their optical instruments, underpinned his *Maqāla fī al-hāla wa qaws quzaḥ* (*Discourse on the Halo and the Rainbow*).³³

Approximately speaking, Ibn al-Haytham held that every *jism kathīf* (opaque body) is coloured, and its *lawn* (colour) would be a visible property that is intermixed with the secondary accidental light that is emitted from that lit object and propagates with it.³⁴ The *ṣūra*; *eidos* (form) of colour is akin to that of light, and colours always accompany light and will never appear without illumination.³⁵ In the *Qawl fī al-ḍawʾ* (*Discourse on Light*) Ibn al-Haytham distinguished between accidental light that is *ʿaraḍī thānī* (secondary) and emitted from lit surfaces of opaque objects, and substantial light that is *jawharī awwal* (primary) and irradiated from luminous bodies. Both types of light follow the same principles of emission, rectilinear propagation, reflection, and refraction. Consequently, the propagation of colour as what always accompanies secondary light entails that it obeys the same laws governing light. Moreover, the essence of colour and light is not associated with the nature and properties of the transparent body in which they travel, rather the transparent media affect the propagation of colour and light whether they spread rectilinearly or get reflected or refracted. In this way Ibn al-Haytham rejected the Aristotelian doctrine of the *diaphanous* wherein light actualizes the potencies of the transparent medium.³⁶ Ibn al-Haytham explicated the rainbow in terms of catoptrics by taking it

---

31   Ibn al-Haytham, *Kitāb al-manāẓir*, ed. by Sabra; Ibn al-Haytham, *The Optics of Ibn al-Haytham*, trans. by Sabra, hereinafter to be referred to in the body of the text as: *Optics*. See also El-Bizri, 'A Philosophical Perspective on Alhazen's Optics' and Rashed, *Les mathématiques infinitésimales du ixe au xie siècle*.
32   Rashed, *Geometry and Dioptrics in Classical Islam*, pp. 63, 108, 152, 181 and Rashed, 'A Pioneer in Anaclastics'.
33   See Würschmidt, 'Die Theorie des Regenbogens'.
34   Ibn al-Haytham, *Optics*, I.2 [12]; the Arabic critical edition of Ibn al-Haytham's *Qawl fī al-ḍawʾ* is contained in the *Majmūʿ al-Rasāʾil* as his epistolary collection, Ch. 2. The annotated French translation of this treatise is contained in Rashed, *Optique et mathématiques*, Ch. 5. See also: El-Bizri, 'Ibn al-Haytham et le problème de la couleur', pp. 201–26.
35   Ibn al-Haytham, *Optics*, I.3 [113–16], [114–66]; I.4 [129–31].
36   Aristotle, *De Anima*, ed. by Ross, 418a 31–32, 418b 9–11.

to be a phenomenon of reflection from a humid and dense air or cloud that has the properties of an arc segment of a concave spherical mirror. He remained close to the Aristotelian thesis at the conceptual level, but in epistemic and methodological terms, he mathematized this natural phenomenon as a model in catoptrics, and studied it via the geometrical reflective properties of the concave surface of a spherical mirror.[37]

The theories of Ibn Sīnā and Ibn al-Haytham remained dominant in attempting to explain the phenomenon of the rainbow, even though in Al-Andalus the philosopher Ibn Rushd was an exponent of Ibn al-Haytham's thesis, given that he repeats it in his own treatises, the *Jawāmiʿ kitāb al-āthār al-ʿulwīyya* and *Talkhīṣ al-āthār al-ʿulwīyya* in the context of his short and middle commentaries on Aristotle's *Meteorology*.[38] As these examples show, the materials available to medieval European scholars for study of the rainbow were patchy by comparison, and help to put Grosseteste's remarks on the phenomenon into a longer intellectual perspective.

Un-weaving the Rainbow: The Principle of Subalternation

On the Rainbow is divided into two distinct halves. The second concerns the applied optics of the rainbow while the first is an extended *resolutio* concerning the geometrical fundamentals of optics (or *perspectiva*). Grosseteste starts with a clear statement that the account of the rainbow given by Aristotle does not explain the causes of the phenomenon. The causal explanation can only be done in terms of optics:

> Propter hoc Aristoteles in libro metheorum non manifestavit propter quid, quod est perspectivi, sed ipsum quia, quod est physici, in brevi sermone coartavit. Ideoque, in presenti, ipsum propter quid, quod attinet ad perspectivum, pro modulo nostro et temporis opportunitate suscipimus explicandum.
>
>> (For this reason, Aristotle, in his book on meteorological phenomena, did not reveal the reason why [of the rainbow] which is the preserve of the student of optics, but condensed into a brief discourse the 'that' [of the rainbow] which belongs to the student of physics. And so, for the present we have, as far as our ability and availability of time has allowed, undertaken to explain the reason why, which pertains to optics.)[39]

Grosseteste moves then to further elaborate the science of optics, placing it in a broader disciplinary context:

> In primis igitur dicimus quod perspectiva est scientia que erigitur super figuras visuales; et hec subalternat sibi scientiam que erigitur super figuras quas continent

---

37 This is contained in Book V of Ibn al-Haytham's *Optics*, in Istanbul, Fatih, MS 3215, fol. 267; Istanbul, Aya Sophia, MS 2448, fol. 465. See also Rashed, 'Le modèle de la sphère transparente et l'explication de l'arc-en-ciel', esp. pp. 117–20.
38 See for instance Sayili, 'Al-Qarāfī and his Explanation of the Rainbow', p. 16.
39 Grosseteste, *De iride*, § 1.

linee et superficies radiose, sive proiecta sint illa radiosa ex sole, sive ex stellis, sive ex alio corpore radiante.

> (Therefore, we start by declaring that optics is a science that is founded on [geometrical] figures pertaining to sight, and this subalternates to itself the science that is founded on figures enclosed by radiant lines and surfaces, whether these radiants are projected from the sun, or from stars, or from some other radiant body.)[40]

Whereas the study of the rainbow is subalternated to optics, that is, the projection of rays of light issuing from a luminous body and proceeding to another point, optics is, in this scheme, in turn subalternated to geometry.[41] Grosseteste was aware of the controversy and ambiguities surrounding Aristotle's view that sight occurs by intromission of external light rays into the eye rather than their emission from it, which rests on the thesis that an unbroken causal chain connects the visible object to the eye through an actualization of the potentiality of transparency of the diaphanous medium between them (such as air) when it is illuminated.[42] At the same time Grosseteste also accepted that sight occurs by way of the emission of light rays from the eyes. In this way he accepted both the broadly Aristotelian and broadly Platonic teachings on vision. It is an important tenet of geometrical optics that the direction of the light ray is not important. In modern terms the Law of Reciprocity, a fundamental principle in optics, states that the same phenomena must occur when all the directions of rays of light are reversed in a system. In this way both extramission and intromission as models of sight can be used successfully within the same geometric framework.

In understanding the geometrical basis of optics, Grosseteste chooses to describe the behaviour of rays emerging from the eye, proceeding to the visual object. He then notes that optics can be sub-divided into three areas:

> Aut enim transitus radii ad rem visam est rectus per medium diaphani unius generis interpositum inter videntem et rem visam; aut transitus eius est secundum rectum ad corpus habens naturam huiusmodi spiritualis per quam ipsum est speculum, et ab ipso reflectitur ad rem visam; aut transitus radii est per plura diaphona diversorum generum in quorum contiguitate frangitur radius visualis et facit angulum, pervenitque radius ad rem visam non secundum transitum rectum sed secundum viam plurium linearum rectarum angulariter coniunctarum.

---

40 Grosseteste, *De iride*, § 1.
41 It is interesting to note the parallels between medieval optics and modern 'ray tracing' which forms a significant part of computer assisted optical component design. There is no contradiction between the geometrical optics of the ancient and medieval scholars and the understanding of light as a wave motion. Light rays in the geometrical interpretation are simply normals to the wavefronts in the wave interpretation. While the controversy over the wave or particle nature of light ran for a long time, in a quantum description the two views are compatible.
42 A thorough discussion of Aristotle's thesis in connection with *On the Soul* and the *On Animals*, along with the ambiguities in its conceptual reception in the European medieval milieu and through Latin translations that evoke the so-called 'visual species', is articulated in Lindberg's *Theories of Vision*, pp. 6–8, 113–16.

(For either the passage of the ray towards the thing seen is straight through the medium of something diaphanous of a single kind interposed between the one who sees and the thing seen; or its passage is along a straight line towards a body having a nature of a certain spiritual kind that makes it function as a mirror, and from this [the ray] is reflected towards the thing seen; or the passage of the ray is through several diaphanous [things] of different kinds, at whose point of contact the visual ray is refracted and makes an angle, and the ray arrives at the thing seen not by a straight progression, but by a route consisting of several straight lines connected at angles.)[43]

The first part of optics concerns sight, the second describes reflections from mirrors, but the third, which Grosseteste claims, 'apud nos intacta et incognita usque ad tempus hoc permansit' (has remained untouched and unknown among us until the present time) is concerned with refraction 'dioptrics'.[44] To take this at face value is to acknowledge the implication that while the studies of reflection (catoptrics) by Euclid and/or Ptolemy were known in northern Europe at this time, the principles of dioptrics were unknown, at least to Grosseteste and his confrères. In assessing knowledge of optics, the difficulties involved in ascertaining both Grosseteste's location and the nature of his readership come again to the fore.

Grosseteste and Dioptrics in *On the Rainbow*

Whatever were the sources available to him there is no doubt that Grosseteste had a thorough grasp of geometrical optics and refraction at interfaces. Before expounding the principles, he asserts that

> Hec namque pars perspective perfecte cognita ostendit nobis modum quo res longissime distantes faciamus apparere propinquissime positas, et quo res magnas propinquas faciamus apparere brevissimas, et quo res longe positas et parvas faciamus apparere quantum volumus magnas, ita ut possibile est nobis ex incredibili distantia litteras minutas legere, aut arenam, aut granum, aut gramina, aut quevis numerare.

> (if known perfectly, this part of optics shows us the way in which we may make things very far away appear as though placed very close by, and in which we may make large things placed close by appear very small, and in which we may make small things placed far away appear as large as we please, so that it becomes possible for us to read very small letters at an incredible distance, or count [grains of] sand, or seeds, or [leaves of] grass, or whatever you might want.)[45]

---

43 Grosseteste, *De iride*, § 3.
44 Grosseteste, *De iride*, § 4.
45 Grosseteste, *De iride*, § 4.

Here, Grosseteste appears to be concerned first with demagnification and then magnification, both of which can be achieved with a single lens (Fig. 2.1). Although there is no evidence that Grosseteste was familiar with lenses, or if he were, used them in this way, the coherence of the description with the properties of lenses is striking.[46] Roger Bacon would later repeat and expand on Grosseteste's position when investigating convex and concave surfaces in light through reading Ibn al-Haytham.[47]

Grosseteste's own explanation of these phenomena works on the basis that when a visual ray passes through several transparent (diaphanous) bodies, it is refracted at the point of contact between them. He cites the famous experiment recorded in (pseudo-)Euclid's *Catoptrica*, where an object, placed at the bottom of a vessel, is not visible along the line of sight until water is poured into the vessel, the object then becoming visible (Fig. 2.2).[48] It is not clear whether Grosseteste himself did this experiment, but he could easily have arranged to have done so, had he so desired. Interestingly, an early fourteenth-century copy of the treatise, now held in Merton College, University of Oxford, identifies the 'object' placed in the vessel as a coin (*denarius*), which suggests the possibility of experiment for at least one medieval scholar subsequently.[49]

Grosseteste's explanation of this phenomenon invokes the tension between the axiom that two diaphanous bodies of different kinds must necessarily be discontinuous with one another, and the axiom that the ray cannot be fully discontinuous with itself. The only way in which this can be resolved is for the 'medium autem inter plenam continuitatem et completam discontinuitatem non potest esse nisi punctus unus coniungens duas partes non directe sed angulariter' (the middle between full continuity and full discontinuity cannot be anything but a single point joining two parts not in a straight line, but at an angle).[50] The ray, in other words, is refracted.

Grosseteste then explains where the image is located after such refraction:

Res autem que videtur per medium plurium perspicuorum non apparet esse ubi ipsa est secundum veritatem, sed apparet esse in concursu radii egredientis ab oculo in continuum et directum protracti et linee ducte a re visa cadentis in superficiem secundi perspicui propinquiorem oculo ad angulos equales undique.

> (However, a thing seen through medium of several transparent [bodies] does not appear where it truly is, but it appears to be located at the convergence of the ray coming out of the eye and prolonged continuously and straight, and the line drawn from the thing seen, falling at right angles on all sides on a second transparent [body] on the surface closest to the eye.)[51]

---

46 White and others, 'Magnifying Grains of Sand'.
47 Roger Bacon, *Opus Maius*, ed. by Bridges, Bk v. Pt 3. ii.4.
48 *The Medieval Latin Translations of Euclid's Catoptrica*, ed. and trans. by Takahashi, Definition 6; see also Eecke, *Euclide L'Optique et la catoptrique*.
49 University of Oxford, Merton College, Manuscript 306, fol. 118ra. The portion of the *De iride* found in this manuscript is in the form of a fragment from the second half of the fourteenth century.
50 Grosseteste, *De iride*, § 5.
51 Grosseteste, *De iride*, § 7.

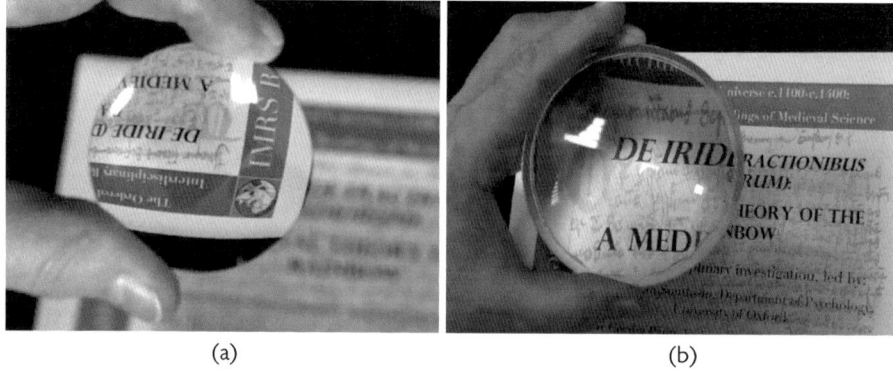

Figure 2.1. (a) Demagnification with lens further from the object than its focal length; (b) Magnification when lens is closer to the object than its focal length. Note that the demagnified image is inverted. Photos: Brian Tanner.

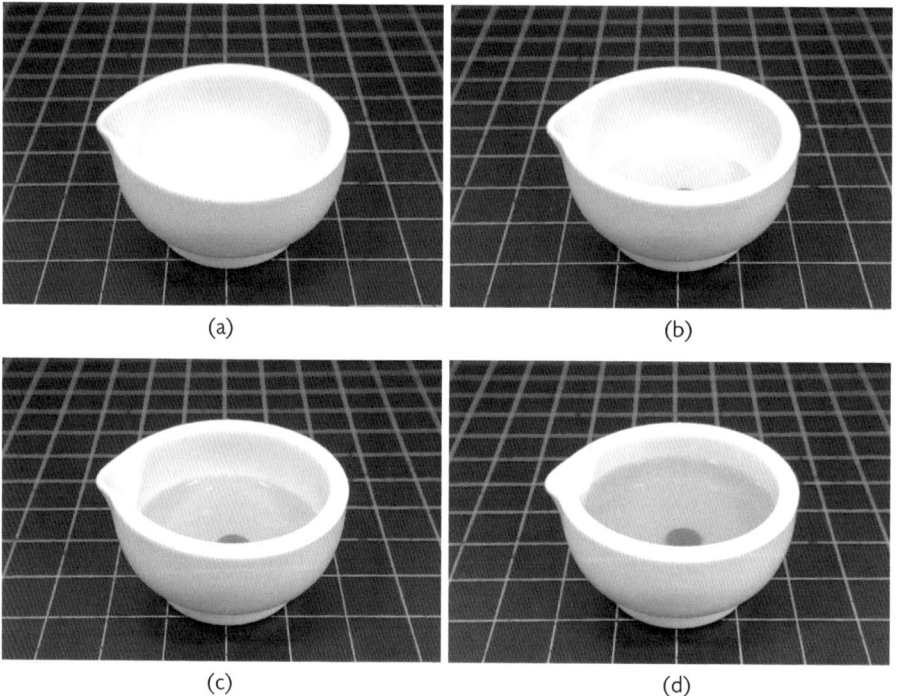

Figure 2.2. The experiment of the appearing coin. As water is poured in, (a) to (d), the image of the object rises due to refraction at the water-air interface, enabling it to become visible. Photos: Brian Tanner.

A geometric construction (Fig. 2.3) of exactly what Grosseteste is saying in the above quotation shows perfectly the explanation of the appearing coin experiment. The virtual image position, determined by the projection along the line of sight, is raised above the bottom of the vessel.

This understanding of the nature and position of the image seen through various transparent bodies enables Grosseteste to explain the working of imaging devices. In the final section of the *resolutio*, he introduces a fundamental principle of image formation, namely that:

> secundum quantitatem anguli sub quo videtur aliquid et situm et ordinem radiorum apparet quantitas et situs et ordo rei vise, et quod magna distantia non facit rem invisibilem.
>
> > (the size and position and ordering of the thing seen appears according to the size of the angle at which something is seen, and the position and ordering of rays, and that narrowness of the angle at which a thing is seen, is what makes that thing invisible, and not great distance.)[52]

Grosseteste understands that it is the angle at which rays leave the eye which determines the apparent size and position of the object being viewed.[53] A specific object brought nearer to the eye shows an increased angle (Fig. 2.4(a)) while smaller objects close to the eye can be apparently the same size as larger objects at a greater distance, (Fig. 2.4(b)). Accordingly, from Grosseteste's argument, the virtual image in Fig. 2.3 appears larger, as it is nearer, even though there is no actual magnification. Such insight, combined with the understanding of the position of the image, enables the role of a lens in magnifying to be determined (Fig. 2.5(a)) and demagnifying objects (Fig. 2.5(b)).

While Grosseteste does not specifically discuss lenses, he does explain that by awareness of these principles, it is possible to make imaging devices by shaping transparent media. Convex lenses were familiar items in antiquity, theoretical and experimental knowledge were refined by earlier medieval scholars from the Islamic world, such as Ibn al-Haytham, and reading stones for magnifying details of manuscripts being copied were certainly available in the thirteenth-century West, and possibly earlier.[54]

> et patens est eisdem modus figurandi diaphana ita ut illa diaphana recipiant radios egredientes ab oculo secundum quantitatem anguli quem voluerint in oculo facti,

---

52 Grosseteste, *De iride*, § 8.
53 With the visual ray reversed to become the ray of light from the object, modern scientists agree with Grosseteste's analysis that it is the angle at which the ray enters the eye that is key to the interpretation of distance and size.
54 On glass in antiquity see Beretta, ed., *When Glass Matters*, and a useful summary is provided in Ilardi, *Renaissance Vision*, pp. 33–50; on Ibn al-Haytham see El-Bizri, 'Grosseteste's Meteorological Optics', p. 31; the invention of an almost (but not quite) hemispherical glass giving magnification when placed on a surface is credited to Abbās Ibn Firnās (810–87): Vernet, 'Abbās ibn Firnās', vol. 1, p. 5; on the use of reading stones in the West see Ilardi, *Renaissance Vision*, pp. 8 and 40 (including discussion of the Visby Lenses: Schmidt, Wilms and Lingelbach, 'The Visby Lenses').

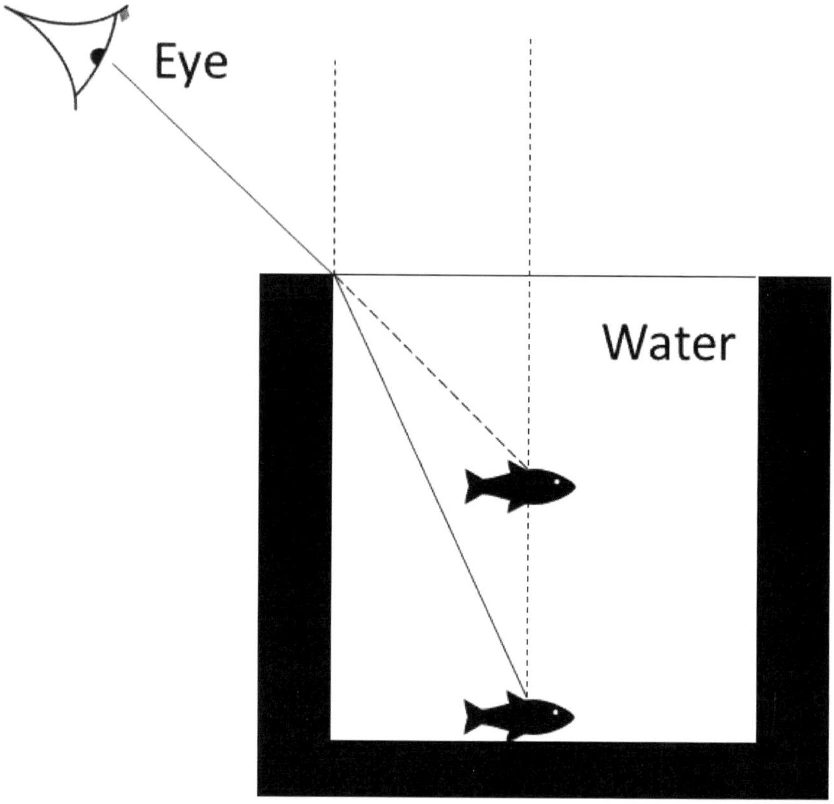

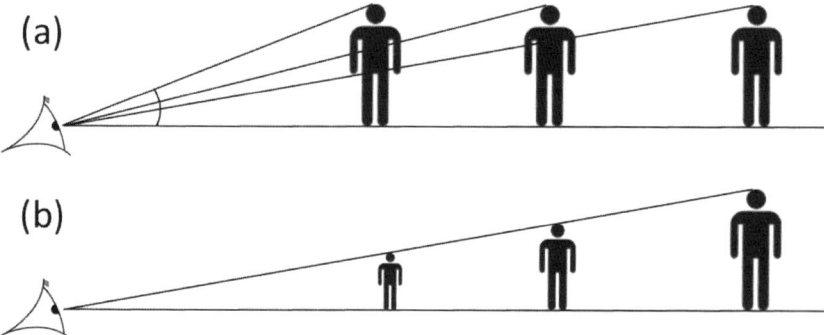

Figure 2.3 Graphical illustration of Grosseteste's description of image formation and the position of the virtual image of the object. Figure: Brian Tanner.

Figure 2.4 (a) Identically sized objects appear larger as the distance to the observer is reduced due to the increasing angle subtended at the observer; (b) Different size objects at different distances appear of similar size due an identical angle subtended at the observer. Figure: Brian Tanner.

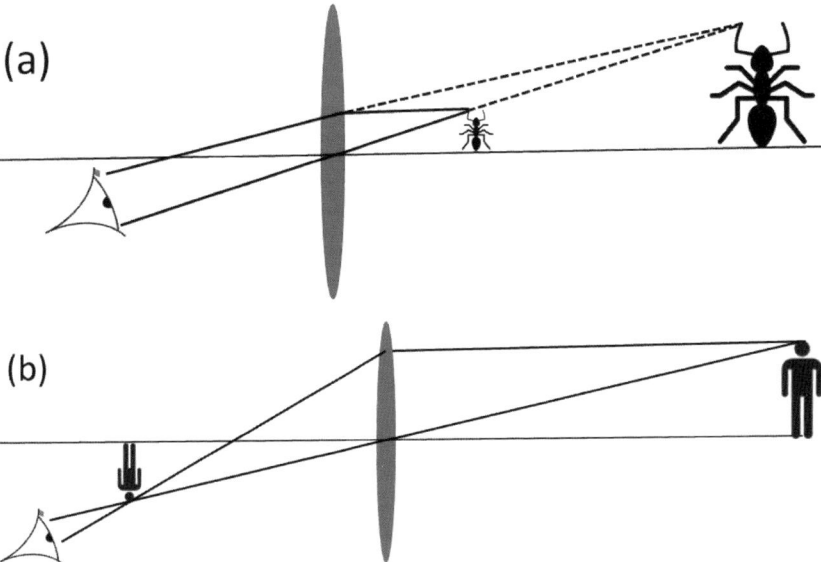

Figure 2.5 (a) Magnification of an object by formation of a virtual image on the far side (from the eye) of a lens; (b) Demagnification by formation of a real, inverted image on the same side of the lens as the eye. Figure: Brian Tanner.

> et refringant radios receptos quousque voluerint super res visibiles, sive fuerint res ille magne sive parve, sive longe sive prope posite; et ita appareant eis omnes res visibiles in situ quo voluerint, et in quantitate qua voluerint, et res maximas quam voluerint faciant apparere brevissimas, et e contrario brevissimas et longe distantes faciant apparere magnas et optime visu perceptibiles.
>
>> (It is also evident to these same [the perfect] how to shape diaphanous [objects] so that these diaphanous [objects] will receive the rays emitted from the eye according to an angle, made in the eye, of whatever size they want. And they will refract the rays received, to whatever extent they like, onto visible things, regardless of whether these visible things are large or small, and placed near or far away; and in this way all visible may appear to them [the perfect] in the place they would like and the size they would like, and they could make the largest things appears to be very small, and inversely they could make the smallest and most remote objects appear large and perfectly perceptible to sight.)[55]
>
> This is an intriguing statement since the effects described might appear at first glance to be similar to those associated since the seventeenth century with telescopes.[56]

---

55 Grosseteste, *De iride*, § 9.
56 White and others, 'Magnifying Grains of Sand'.

While it is manifest that Grosseteste had no knowledge of any such device the possibility that he was familiar with glass shaped for optical purposes should also be acknowledged. It is, for example, possible for a myopic person to achieve telescopic vision by arranging for the focal point of a single lens to be at the focal point of the lens and curved cornea combination in the eye. The principle of operation of the telescope is not to make the image larger, but moved to a position nearer to the observer. This is done by increasing the angle at which rays of light enter the eye, a principle with which Grosseteste's text agrees. Whether this interpretation conforms with anything Grosseteste thought or did is difficult to assert with any confidence. The coherence, however, remains.

Grosseteste and the Transmission of the Laws of Refraction

Grosseteste's account of the rainbow represents, in all probability, the last of his scientific *opuscula*, though he would continue to ponder the material world in later theological writings, and returned, at the end of his life to translate from Greek Aristotle's *On the Heavens*. To that extent the treatise *On the Rainbow* remains of its time of composition, and representative of the textual resources available to its author. As David Lindberg rightly pointed out, Grosseteste's 'famous "quantitative" law of refraction is remarkably primitive by comparison with Greek and Islamic theories of refraction, soon to be reproduced in the West'.[57] That Grosseteste was not familiar with the works of Ibn al-Haytham is evident from his treatment of the angle of refraction. Grosseteste took his cue from the postulate of Euclid that the angle of reflection from a mirror is equal to the angle of incidence and himself postulated that the angle (from the normal to the interface) of refraction is half the angle of incidence. Although Crombie rather dismissively suggested that a simple experiment would have shown Grosseteste the error in this assumption, within the limits of such a simple experiment using the air-glass interface, at small angles of incidence, the approximation does, in fact, hold reasonably well.[58] Grosseteste based his argument primarily on the a general notion attributed to Aristotle of the fundamental simplicity of physical phenomena, enunciated in the statement that: 'omnis operatio nature est modo finitissimo ordinatissimo brevissimo et optimo quo ei possibile est' (every operation of nature is in the most complete, most ordered, shortest, and best way possible for it).[59]

The subsequent development of the law of refraction in European circles was closely bound to the transmission of translations of the optical works of Ptolemy and Ibn al-Haytham. One particular example, that of the Silesian scholar, Witelo

---

57 Lindberg, *Theories of Vision*, p. 102.
58 Crombie, *Robert Grosseteste and the Origins of Experimental Science*, p. 124; Tanner and others, 'Unity and Symmetry in the *De luce* of Robert Grosseteste'. It does not hold so well for water, which has a lower refractive index.
59 Grosseteste, *De iride*, § 7.

(born c. 1230), is presented here as a case-study for the nature of textual transmission and scientific experiment: travelling optics. The transmission is multi-layered, and starts, long before Grosseteste proposed his law of refraction, with Ibn al-Haytham's re-discovery of the pioneering optical experiments of Ptolemy (c. 100–60). Ptolemy undertook experiments, and in this case systematically changed one parameter, the angle of incidence, and recorded quantitatively the associated behaviour of another parameter, (the angle of refraction). This is an extraordinary example of careful construction and use of apparatus specifically designed to test a hypothesis, namely that there existed a simple law relating the angles of incidence and refraction similar to that found in the case of reflection.[60] In the fifth book of *Optics*, Ptolemy presented tabulated numerical data relating these angles for the cases of air–water, air–glass and water–glass interfaces.

Ptolemy's *Optics* survives in a single twelfth-century Latin manuscript in poor condition, the Greek original and any Arabic translation having been lost.[61] This Latin translation, from an Arabic version, was made by Eugenius of Sicily in the second half of the twelfth century, eclipsed a century later by the circulation of the Latin translation of Ibn al-Haytham's *De Aspectibus* which made considerable use of Ptolemy, from a now-lost translation into Arabic. Book 7 of *De Aspectibus* includes a detailed description of how to construct an equivalent apparatus and make the measurements described by Ptolemy.[62] Although the reader is taken through the exact procedure for making the measurements, Ibn al-Haytham only describes them qualitatively. It is not until the end of the thirteenth century that quantitative, tabulated data are to be found, the work of Witelo. In his *Perspectiva* (or *Optica*), completed shortly after 1270 in Viterbo, Witelo gives a very similar set of instructions to Ibn al-Haytham as to how to construct apparatus for the measurement of angles of refraction.

As this indicates, the *Perspectiva* is based heavily on Ibn al-Haytham's work and the apparatus and experimental instructions are essentially the same, with a few incremental improvements. However, it is the numerical data that are telling. Witelo presents the results from three systematic sets of measurement of the incidence and refracted angles, exactly as described by Ptolemy. Most importantly, his results are identical, except for just two data points where there are small differences of value.[63] The question arises as to whether Witelo did in fact perform these experiments. It was, after all, hardly an unusual practice amongst medieval authors to quote sources without naming them.

In examining the data from air to water, the plausible assumption that both Witelo and Ptolemy used clear water should be adopted. The modern value of the refractive index $n = 1.33299 \pm 0.00001$ should therefore be the same as for the ancient

---

60 Smith, *Ptolemy's Theory*, pp. iii–v, vii–xi, 1–61, 63–261, 263–69 and 279–300; Smith, 'Ptolemy's Search for a Law of Refraction'.
61 Smith, *Ptolemy's Theory*, pp. 7–8; Berschin, *Greek Letters*, pp. 234–35.
62 Smith, *Alhacen on Refraction*, pp. 213–331, 333–97, 399, 401–51, 453, 455–91, 493, 495–535, 537–50.
63 For air to water, at 10° incidence, Witelo has 7.75° for the angle of refraction whereas Ptolemy gave 8°. For water to glass, at 30° incidence Witelo has 27.5° for the angle of refraction whereas Ptolemy gave 27°.

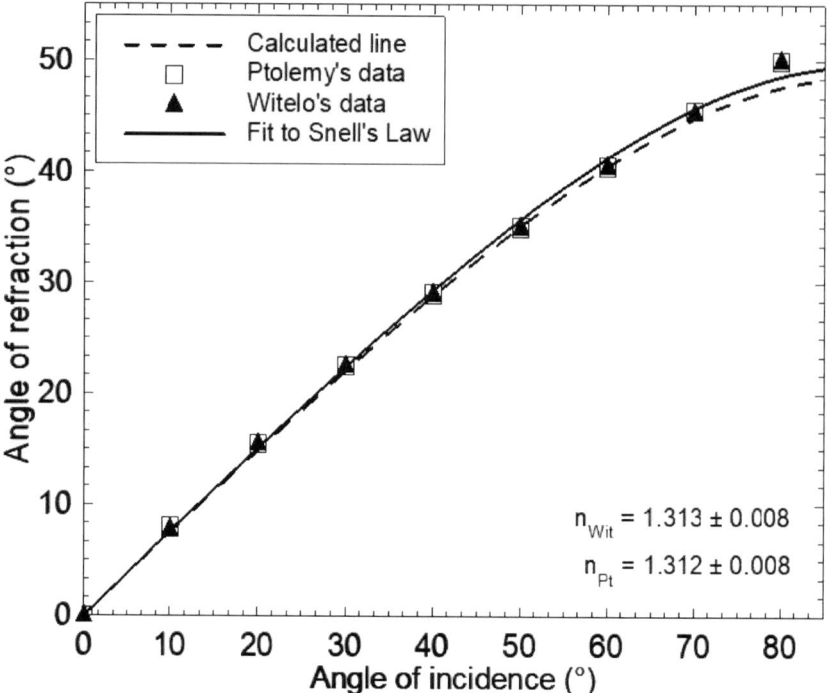

Figure 2.6. Data points of Ptolemy and Witelo with best fit to Snell's Law (solid line) for either data set and calculated (dotted) line for the air-water interface using the modern value for the refractive index. Figure: Brian Tanner.

and medieval experiments. Fig. 2.6 shows the data points of both Ptolemy and Witelo for the angles of incidence and refraction plotted against the dashed line calculated using Snell's Law.[64] Except for the highest data point (incidence angle = 80°) there is a satisfactory agreement within the assumption that the measurement precision is ¼°.[65] A numerical fit of the data to Snell's law gives $n_{Pt} = 1.312 \pm 0.008$ for Ptolemy's values and $n_{Wit} = 1.313 \pm 0.008$ for Witelo's numbers, identical within the experimental precision.

The data for the air–glass refraction are more interesting (Fig. 2.7). Noting that the data of Ptolemy and Witelo are identical, a fit to Snell's Law gives a refractive index of 1.504 ± 0.009. Here, the quality of the glass available to both authors is worth

---

64 The angle of incidence i (measured from the normal to the interface) is related to the angle of refraction r through Snell's Law given by $(\sin i / \sin r) = n$ where n is a constant for any material and called the refractive index.
65 Smith, *Ptolemy's Theory*, has argued that this point was placed artificially high by Ptolemy in order to derive a law of constant second differences. There is no mention by Ptolemy of such a law, and it may be a subsequent imposition of pattern on the data. However, the measurement at this high angle, which is close to grazing incidence, is difficult to perform accurately.

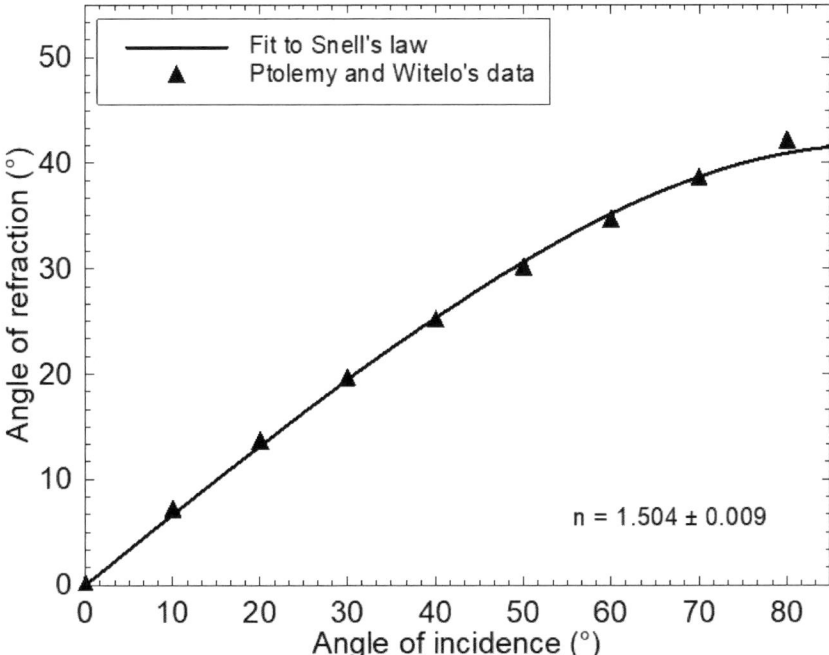

Figure 2.7. Refraction from air to glass data with best fit to Snell's Law. Figure: Brian Tanner.

considering. In the first millennium AD, glass was cast in primary workshops across the Middle East in batches of up to twenty tons before being broken up and shipped across the Roman Empire for re-melting and reworking.[66] It was a soda-lime-silica glass predominantly of silica and sodium carbonate from natron deposits in Lower Egypt, particularly from Wadi Natrun, about 100 km north-west of Cairo.[67] A typical composition for a colourless Roman glass from Cyprus was 71% SiO_2, 16% Na_2O, 6.8% CaO, 2.4% Al_2O_3, 1.5% MnO, 0.4% K_2O, 0.3% Fe_2O_3.[68] The potassium content was low. Taking this composition produces a refractive index of 1.495 ± 0.007.[69] Similarly, the composition of an Egyptian colourless glass from the second century was 66% SiO_2, 18.4% Na_2O, 8.6% CaO, 2.2% Al_2O_3, 0.8% MgO, 0.4% K_2O, 0.6% Fe_2O_3 giving a refractive index of 1.491 ± 0.007.[70] Within measurement precision, both these values

---

66 Freestone, Ponting, and Hughes, 'The Origins of Byzantine Glass'.
67 Freestone, 'Glass Production in Late Antiquity and the Early Islamic Period'.
68 Ceglia, Cosyns, Nys, Terryn, Thienpont, and Meulebroeck, 'Late Antique Glass Distribution'.
69 The refractive index n is related to the component weight fractions (WF) by $n - 1 = 0.460$ (WF SiO_2) $+ 1.158$ (WF TiO_2) $+ 0.581$ (WF Al_2O_3) $+ 1.09$ (WF Fe_2O_3) $+ 0.897$ (WF FeO) $+ 0.765$(WF MgO) $+ 0.903$ (WF MnO) $+ 0.795$ (WF CaO) $+ 0.505$ (WF Na_2O) $+ 0.495$(WF K_2O) $± 0.007$. Church and Johnson, 'Calculation of the Refractive Index of Sililcate Glasses from Chemical Composition'.
70 Rosenow and Rehren, 'Herding Cats'.

are in good agreement with that deduced from Ptolemy's data. It seems quite likely, then, that Ptolemy actually made these measurements. Even if he did make small upward adjustments of selected data points in order to obtain a Law of Equal Second Differences, as suggested by Smith, the overall trend could not have been predicted from his postulate of equal angle reflection in mirrors.[71] More telling, and unequivocal, is the analysis of the water-glass data. Particularly in view of the way his apparatus is described, it is extremely unlikely that Ptolemy would have used a different piece of glass for the water-glass measurements and the air-glass measurements. When refraction occurs at an interface other than with air, Snell's Law becomes modified to $\sin i / \sin r = n_2 / n_1$, where n_2 is the refractive index of the second medium and n_1 is the refractive index of the first. Thus, a value for the ratio n_2/n_1 can now be predicted, although this is not something that Ptolemy would have had any way of knowing.

The water-glass data are shown in Fig. 2.8. When the data of Ptolemy is taken against the modified Snell's Law, a ratio for the refractive indices of glass and water is obtained of 1.129 ± 0.005, in very satisfactory agreement with the values of 1.15 ± 0.01, obtained by combining the individual results of the air-water and air-glass measurements of Ptolemy, and the value of 1.12, obtained by using a refractive index value derived from a typical glass composition from antiquity and the modern value of the refractive index of water.

The evidence for Witelo performing independent, original measurements and happening to record almost identical values to those of Ptolemy is less secure. First, as Crombie pointed out there are data in the tables in *Perspectiva* that are simply impossible to obtain.[72] Fig. 2.9 shows the values quoted by Witelo for light passing from water to air, that is from the medium of higher refractive index to one of lower refractive index. A plot of Snell's Law for the refractive index of water of 1.313 (derived from the fit in Fig. 2.6) shows that at a critical angle of 49.6° the angle of refraction equals 90° and the ray travels parallel to the surface.[73] Beyond this, the ray is internally reflected, a phenomenon well known to swimmers who attempt to look out of the water when submerged. There is no way that Witelo could have obtained the values above 49.6° since internal reflection is mirror-like and the angle of internal reflection equals the angle of incidence. Even below the critical angle, the data are invented. There is significant deviation from the quoted values and those predicted by Snell's Law (Fig. 2.9). This arises because Witelo appears to have derived these numbers by simply adding to the angle of incidence, the difference in angles of incidence and refraction derived from the measurements made for the light passing from air to water. The value added does not correspond to the new angle of incidence and there is a substantial difference in result. It should be noted in this context that Ptolemy did not do the water-air experiment and it was one of the improvements claimed by Witelo. Similar evidence of fabrication is apparent in the Witelo's glass to air and glass to water data.

---

71 Smith, *Ptolemy's Theory*, p. 45.
72 Crombie, *Robert Grosseteste*, p. 225.
73 The value of the so-called critical angle is 48.6° if we use the modern value of 1.333 for the refractive index of water.

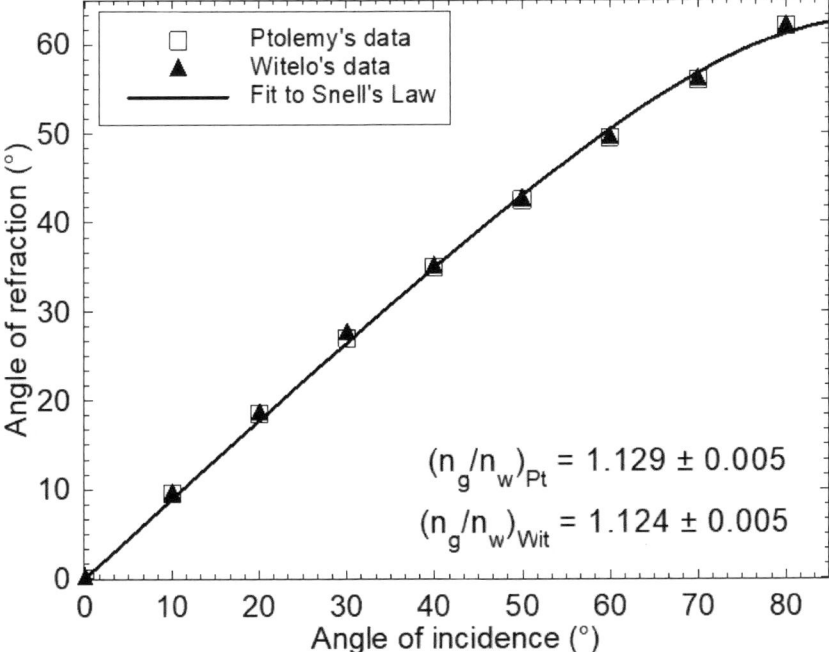

Figure 2.8. Data for water to glass refraction from Ptolemy and Witelo. Solid line is the best fit to the data using the modified Snell's Law. Figure: Brian Tanner.

On initial inspection, it would appear unlikely that Witelo would have used the same type of glass as that used by Ptolemy, resulting in scepticism about the originality of the data from the air-glass refraction contained in *Perspectiva*. In the ninth century the composition of glass changed as the natron deposits in Egypt became exhausted. Plant ash became the source of sodium, and this also added a significant quantity of potassium and magnesium to the composition. The refractive index of glass made using this process differs from that of the Antique glass. Plant ash was the common glass-making flux throughout the Islamic period and such a glass is most likely to have been used by Ibn al-Haytham in his detailed description of how to repeat of Ptolemy's experiments.[74] As Ibn al-Haytham did not tabulate his results, there can be no test of this assumption. While there is evidence that, in northern Italy, there was substantial recycling of Classical, natron based, glass between the tenth and twelfth centuries, by the end of the thirteenth century the transition to soda plant ash technology seems to have become complete.[75] The composition of a clear glass from the windows of Siena Cathedral, originating in the late thirteenth century, is 63% SiO_2, 12.6% Na_2O, 10% CaO, 2.2% Al_2O_3, 2.8% MgO, 0.7% MnO, 7.2% K_2O, 0.6% FeO giving a refractive index of 1.517.

---

74 Rehren and Freestone, 'Ancient Glass' and Smith, *Alhacen on Refraction*.
75 Silvestri and Marcante, 'The Glass of Nogara (Verona)'.

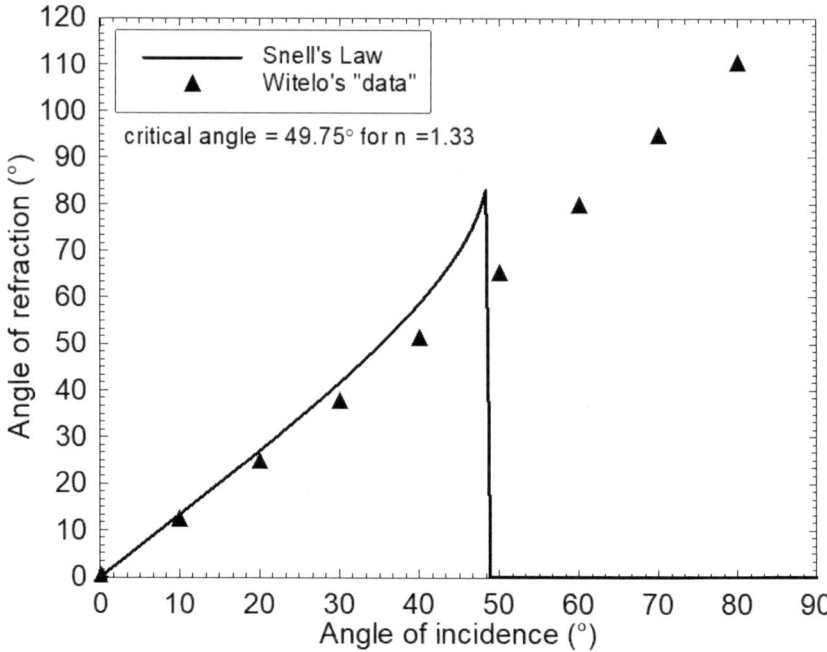

Figure 2.9. Plot of Witelo's supposed measurements of the angles of incidence and refraction for the water–air interface, where light is now passing from the denser to the less dense medium. Figure: Brian Tanner.

By the end of the thirteenth century, glass-making in northern Europe had moved to use of beech ash, as described in great detail by Theophilus earlier in the twelfth century.[76] These glasses were high in potassium and calcium and very low in sodium. The composition of clear glass from the windows of the church of St Michael and St Gudule in Brussels, probably dating from 1273, is reported to be 47.0% SiO_2, 0.95% Na_2O, 24.9% CaO, 2.3% Al_2O_3, 3.8% MgO, 15.7% K_2O, 0.9% MnO.[77] The refractive index for such a composition, typical of beech ash glasses is 1.547.[78] If it is assumed that Witelo's glass came from northern or central Europe, all the experimental data points lie above the predicted curve to the extent that the possibility of experiment can be ruled out with some confidence. Even with the soda plant ash glass such as in Siena Cathedral, the data points are systematically low, and it appears unlikely that a thirteenth-century glass was actually used by Witelo. It is therefore just possible that Witelo did make measurements for the air–water

---

76 Theophilus, *De diversis artibus*, ed. and trans. by Dodwell.
77 Schalm, de Raedt, Caen, and Janssens, 'A Methodology for the Identification of Glass Panes'.
78 Stern and Gerber, 'Potassium-Calcium Glass'; Wedepohl and Simon, 'The Chemical Composition of Medieval Wood Ash Glass'.

interface, but all other data appear to have either been copied from Ptolemy or invented. The evidence for his use of 'experiment', as opposed to 'experience', looks on this basis to be limited.

Concluding Reflections

Grosseteste would go on in the rest of his treatise *On the Rainbow* to use refraction in his explanation of how the rainbow is formed. Refraction at the point of contact between air and the cloud, and within the increasingly dense material in the cloud, eventually produces the rainbow arc. Then he moves to colours, recalling his treatise *On Colour*, and applying its terms of reference to the colours within and between rainbows. Some seventy years after Grosseteste's theory Theodoric of Freiburg and Kamāl al-Dīn al-Fārisī would independently produce far more sophisticated explanations of refraction and reflection within raindrops in their reformulation of Ibn al-Haytham's rainbow thesis.[79] Grosseteste stood, then, at the beginning of an intensive interest in optics from medieval European scholars, and did so without the full range of sources that would be available by the mid-thirteenth century. The rainbow stirred not only his scientific, but his literary imagination. The rainbow-clad throne at the centre of the *Château d'amour* offers a Christological image blending the arc in the sky, representing the Old Testament covenant between God and humanity after the flood, and the coloured rainbow of the New Testament Book of the Apocalypse.[80] It is the whole Christ, the *Christus integer* as Grosseteste would explain in his Hexaemeron who unites Creation with Creator, fulfils all covenants, and offers the chance of redemption and salvation.[81] As a metaphor of the bridge between the temporal and eternal, and as a natural phenomenon to be investigated through optics, Grosseteste's treatment of the rainbow represents an important moment in the

---

79   See Rashed, 'Le modèle de la sphère transparente'; El-Bizri, 'Grosseteste's Meteorological Optics'; El-Bizri, 'Ibn al-Haytham et le problème de la couleur' and Topdemir, 'Kamal Al-Din Al-Farisi's Explanation of the Rainbow'.
80   The Vulgate presents the rainbow shown by God to Noah as an arc in the sky with an intimate connection to clouds: Genesis 9.13–14: 'Arcum meum ponam in nubibus, et erit signum foederis inter me et inter terram. Cumque obduxero nubibus caelum, apparebit arcus meus in nubibus' (I will set my bow in the clouds, and it shall be the sign of a covenant between me, and between the earth. And when I shall cover the sky with clouds, my bow shall appear in the clouds). The vision of heaven in the New Testament Apocalypse of St John (Revelations) refers, by contrast, to the rainbow in the Greek-derived word Iris, the emphasis here placed on colour: Apocalypse 4.3: 'Et qui sedebat similis erat aspectui lapidis jaspidis, et sardinis: et iris erat in circuitu sedis similis visioni smaragdinae' (And he that sat, was to the sight like the jasper and the sardine stone; and there was a rainbow round about the throne, in sight like unto an emerald). Chapter 10.1 includes another account, this time of an angel wearing a rainbow, also identified as 'iris': 'Et vidi alium angelum fortem descendentem de caelo amictum nube, et iris in capite ejus, et facies ejus erat ut sol, et pedes ejus tamquam columnae ignis' (And I saw another mighty angel come down from heaven, clothed with a cloud, and a rainbow was on his head, and his face was as the sun, and his feet as pillars of fire).
81   Robert Grosseteste, *Hexaemeron*, ed. by Dales and Gieben, 1.i.1. English translation available at Robert Grosseteste, *On the Six Days of Creation*, trans. by Martin.

development of thinking on the subject, weaving many-travelled ideas together, from ancient Greece, the early Church, early medieval and contemporaneous expressions of Islam, along with his own reflections on the subject.

Works Cited

Manuscripts and Archival Sources

Istanbul, Aya Sophia, MS 2448
Istanbul, Fatih, MS 3215
London, British Library, MS Royal 11 B IX
Oxford, University of Oxford, Merton College, Manuscript 306

Primary Sources

Aristotle, *De Anima*, ed. by David Ross (Oxford: Clarendon Press, 1961)
—, *Meteorologica*, ed. and trans. by Henry D. P. Lee, Loeb Classical Library (Cambridge, MA: Harvard University Press, 1952)
Baur, Ludwig, *Die Philosophischen Werke des Robert Grosseteste, Bischofs von Lincoln* (Münster: Aschendorff, 1912)
Bede, *De natura rerum*, ed. by Charles W. Jones, Corpus Christianorum series Latina, 123A (Turnhout: Brepols, 1975)
—, *On the Nature of Things and on Times*, trans. by Calvin Kendall, and Faith Wallis (Liverpool: Liverpool University Press, 2010)
Epistles of the Brethren of Purity (Rasāʾil Ikhwān al-Ṣafāʾ). On the Natural Sciences. An Arabic Critical Edition and Translation of EPISTLES 15–21, ed. and trans. by Carmela Baffioni, with a Foreword by Nader El-Bizri (Oxford: Oxford University Press, 2013)
Ibn al-Haytham, *Alhacen on Refraction: A Critical Edition, with English Translation and Commentary, of Book 7 of Alhacen's 'De Aspectibus,' the Medieval Latin Version of Ibn al-Haytham's 'Kitāb al-Manāẓir'*, ed. by A. Mark Smith (Philadelphia: American Philosophical Society, 2010)
—, *Kitāb al-manāẓir*, ed. by Abdelhamid I. Sabra (Kuwait: National Council for Culture, Arts and Letters, 1983, 2002)
—, *Majmū ʿal-Rasāʾil* (Hyderabad: Osmania Oriental Publications Bureau, 1938–1939)
—, *The Optics of Ibn al-Haytham, Books I–III, On Direct Vision*, trans. by Abdelhamid I. Sabra (London: Warburg Institute, 1989)
Ibn Sīnā, *Al-Maʿādin wa al-āthār al-ʿulwīyya, Kitāb al-Shifāʾ* 5, ed. by Ibrahim Madkour and others (Cairo: al-Hayʾa al-miṣriyya al-ʿāmma liʾl-ṭibāʿa waʾl-nashr, 1965)
Isidore of Seville, *De natura rerum*, ed. and trans. by Jacques Fontaine (Paris: Institut d'études augustiniennes, 2002)
—, *On the Nature of Things*, trans. by Calvin Kendall and Faith Wallis (Liverpool: Liverpool University Press, 2016)
Kitāb al-āthār al-ʿulwīyya (Aristotle's *Meteorology*, trans. into Arabic by Yūḥannā ibn al-Biṭrīq), introduced and ed. by Casimir Petraitis (Beirut: Dār al-Mashriq, 1967)

Lindberg, David C., 'Robert Grosseteste: On the Rainbow', in *A Source Book in Medieval Science*, ed. by Edward Grant (Cambridge MA: Harvard University Press, 1974), pp. 385–88

Mackie, Evelyn A., 'Robert Grosseteste's Anglo-Norman Treatise on the Loss and Restoration of Creation, Commonly Known as Le Château d'Amour: An English Prose Translation', in *Robert Grosseteste and the Beginnings of a British Theological Tradition*, ed. by Maura O'Carroll (Rome: Instituto Storico dei Cappuccini, 2003), pp. 151–79

Matthew of Westminster, *Flores historiarum*, vol. 1: *The Creation to A.D. 1066*, ed. by Henry Richards Luard, Rolls Series (London: Eyre and Spottiswode, 1890)

The Medieval Latin Translations of Euclid's Catoptrica, ed. and trans. by Ken'ichi Takahashi (Higashi-ku, Fukuoka-shi: Kyushu University Press, 1992)

Murray, Jessie, *Le Château d'Amour de Robert Grosseteste, évêque de Lincoln* (Paris: Librairie Champion, 1918)

Nicholas Trivet, *Annales sex regum Angliae, 1135–1307*, ed. by Thomas Hog (London: English Historical Society, 1845)

Ptolemy, *Ptolemy's Theory of Visual Perception: An English Translation of the "Optics" with Introduction and Commentary*, trans. by A. Mark Smith (Philadelphia: American Philosophical Society, 1996)

Rashed, Roshdi, *Optique et mathématiques: Recherches sur l'histoire de la pensée scientifique en arabe* (Aldershot: Variorum, 1992)

Robert Grosseteste, *Colour and the Refraction of Rays: Robert Grosseteste's De colore 'On Colour' and De iride 'On the Rainbow'*, ed. by Giles E. M. Gasper and others (Oxford: Oxford University Press, forthcoming)

—, *Commentarius in Posteriorum Analyticorum libros*, ed. by Piero Rossi (Florence: Olschki, 1981)

—, *De colore*, ed. by Greti Dinkova-Bruun, in Greti Dinkova-Bruun, Giles E. M. Gasper, Michael Huxtable, Tom C. B. McLeish, Cecilia Panti, and Hannah Smithson, *Dimensions of Colour: Robert Grosseteste's De Colore; Edition, Translation and Interdisciplinary Analysis* (Toronto: Pontifical Institute of Mediaeval Studies, 2013)

—, *Hexaemeron*, ed. by Richard C. Dales and Servus Gieben, Auctores Britannici Medii Aevi, VI (Oxford: Oxford University Press, 1982)

—, *The Letters of Robert Grosseteste Bishop of Lincoln*, trans. by Frank A. C. Mantello and Joseph Goering (Toronto: University of Toronto Press, 2010)

—, *Moti, virtù e motori celesti nella cosmologia di Roberto Grossatesta. Studio ed edizione dei trattati 'De sphera', 'De cometis', 'De motu supercelestium'*, ed. by Cecilia Panti (Florence: SISMEL, 2001)

—, *On the Six Days of Creation*, trans. by C. F. J. Martin, Auctores Britannici Medii Aevi, 6.2 (Oxford: Oxford University Press, 1996)

—, 'Robert Grosseteste's *De luce*: A Critical Edition', ed. by Cecilia Panti, in *Robert Grosseteste and His Intellectual Milieu*, ed. by John Flood, James R. Ginther and Joseph W. Goering (Toronto: Pontifical Institute of Mediaeval Studies, 2013), pp. 193–238

—, *Roberto Grossatesta, La Luce*, ed. by Cecilia Panti (Pisa: Edizioni Plus, 2011)

Roberti Grosseteste episcopi quondam Lincolniensis epistolae, ed. by Henry Richards Luard, Rolls Series (London: Longman, Green, Longman and Roberts, 1861)

Roger Bacon, *Opus Maius*, ed. by J. H. Bridges (London: Williams & Norgate, 1900)
Theophilus, *De diversis artibus*, ed. and trans. by Charles R. Dodwell (Oxford: Oxford University Press, 1967)
Wharton, Henry, *Anglia Sacra*, vol. 2 (London: Richard Chiswell, 1691)

Secondary Studies

Beretta, D. Marco, *When Glass Matters: Studies in the History of Science and Art from Graeco-Roman Antiquity to Early Modern Era* (Florence: Leo S. Olschki, 2004)
Berschin, Walter, *Greek Letters and the Latin Middle Ages*, trans. by Jerold C. Frakes (Washington DC: Catholic University of America Press, 1988)
Boyer, Carl B., *The Rainbow: from Myth to Mathematics* (New York: Thomas Yoseloff, 1959)
—, 'Robert Grosseteste on the Rainbow', *Osiris*, 11 (1954), 247–58
Burnett, Charles, 'The Institutional Context of Arabic-Latin Translations of the Middle Ages: A Reassessment of the "School of Toledo"', in *Vocabulary of Teaching and Research between Middle Ages and Renaissance*, ed. by Olga Weijers (Turnhout: Brepols, 1995), pp. 214–35
Callus, Daniel A. 'Robert Grosseteste as Scholar', in *Robert Grosseteste. Scholar and Bishop*, ed. by Daniel A. Callus (Oxford: Oxford University Press, 1955), pp. 1–69
Ceglia, Andrea, Peter Cosyns, Karin Nys, Herman Terryn, Hugo Thienpont, and Wendy Meulebroeck, 'Late Antique Glass Distribution and Consumption in Cyprus: A Chemical Study', *Journal of Archaeological Science*, 61 (2015), 213–22
Church, B. N. and W. M. Johnson, 'Calculation of the Refractive Index of Silicate Glasses from Chemical Composition', *Geological Society of America Bulletin*, 91 (1980), 619–25
Clark, James G., 'Trevet, Nicholas (b. 1257x65, d. in or after 1334)', in *Oxford Dictionary of National Biography* (Oxford: Oxford University Press, 2004) <http://www.oxforddnb.com/view/article/27744>
Crombie, Arthur C., *Robert Grosseteste and the Origins of Experimental Science 1100–1700* (Oxford: Oxford University Press, 1953)
Dales, Richard C., 'Robert Grosseteste's Scientific Works', *Isis*, 52 (1961), 381–402
Eastwood, Bruce S., 'Grosseteste's "Quantitative" Law of Refraction: A Chapter in the History of Non-Experimental Science', *Journal of the History of Ideas*, 28 (1967), 403–14
—, 'Medieval Empiricism: The Case of Robert Grosseteste's Optics', *Speculum*, 43 (1968), 306–21
—, 'Robert Grosseteste's Theory of the Rainbow: A Chapter in the History of Non-Experimental Science', *Archives Internationales d'Histoire des Sciences*, 19 (1966), 313–32
Eecke, P. Ver, *Euclide L'Optique et la catoptrique* (Paris: Blanchard, 1959)
El-Bizri, Nader, 'Grosseteste's Meteorological Optics: Explications of the Phenomenon of the Rainbow After Ibn al-Haytham', in *Robert Grosseteste and the Pursuit of Religious and Scientific Learning in the Middle-Ages*, ed. by Jack P. Cunningham and Mark Hocknull (Basel: Springer, 2016), pp. 21–39
—, 'Ibn al-Haytham et le problème de la couleur', *Oriens-Occidens: Sciences, mathématiques et philosophie de l'antiquité a l'Age Classique*, 7 (2009), 201–26
—, 'A Philosophical Perspective on Alhazen's Optics', *Arabic Sciences and Philosophy*, 15 (2005), 189–218

Freestone, Ian C., Matthew Ponting, Michael J. Hughes, 'Glass Production in Late Antiquity and the Early Islamic Period: A Geochemical Perspective', in *Geomaterials in Cultural Heritage*, ed. by M. Maggetti and B. Messiga (London: Geological Society, 2002), pp. 201–16

—, 'The Origins of Byzantine Glass from Maroni Petrera, Cyprus', *Archaeometry*, 44 (2001), 257–72

Gasper, Giles E.M., Cecilia Panti, Hannah E. Smithson, Tom C. B. McLeish, Sigbjørn Olsen Sønnesyn, David Thomson, and others, *Knowing and Speaking: Robert Grosseteste's De artibus liberalibus* 'On the Liberal Arts' *and De generatione sonorum* 'On the Generation of Sounds' (Oxford: Oxford University Press, 2019)

Ginther, James, *Master of the Sacred Page: A Study of the Theology of Robert Grosseteste ca. 1229/30–1235* (Aldershot: Ashgate, 2004)

Glick, Thomas F., 'Science in Medieval Spain: The Jewish Contribution in the Context of Convivencia', in *Convivencia: Jews, Muslims, and Christians in Medieval Spain*, ed. by Vivian B. Mann, Thomas F. Glick, and Jerrilynn Denise Dodds (New York: Braziller, 1992), pp. 83–111

Goering, Joseph, 'Where and When did Grosseteste Study Theology?', in *Robert Grosseteste: New Perspectives on his Thought and Scholarship*, ed. by James McEvoy (Turnhout: Brepols, 1995), pp. 17–51

Gransden, Antonia, *Historical Writing in England c. 550 – c. 1307* (London: Routledge, 1974)

Ilardi, Vincent, *Renaissance Vision, From Spectacles to Telescopes* (Philadelphia: American Society of Philosophy, 2007)

Laird, W. Roy, 'Robert Grosseteste on the Subalternate Sciences', *Traditio*, 43 (1987), 147–69

Lee, Raymond L. and Alistair B. Fraser, *The Rainbow Bridge: Rainbows in Art, Myth, and Science* (University Park; Pennsylvania State University Press: 2001)

Lindberg, David C., 'Roger Bacon's Theory of the Rainbow: Progress or Regress?', *Isis*, 57 (1966), 235–48

—, *Theories of Vision from Al-Kindi to Kepler* (Chicago: Chicago University Press, 1976)

Mackie, Evelyn A., 'Scribal Intervention and the Question of Audience: Editing Le Château d'amour', in *Editing Robert Grosseteste* ed. by Evelyn A. Mackie and Joseph Goering (Toronto: University of Toronto Press, 2003), pp. 61–77

Mantello, Frank A. C., 'The Editions of Nicholas Trevet's Annales sex regum Angliae', *Revue d'histoire des textes*, 10 (1982 for 1980), 257–75

McEvoy, James, 'The Chronology of Robert Grosseteste's Writings on Nature and Natural Philosophy', *Speculum*, 58 (1983), 614–55

—, *Robert Grosseteste* (Oxford: Oxford University Press, 2004)

Ordered Universe Project: Interdisciplinary Readings of Medieval Science: Robert Grosseteste (c. 1170–1253) <https://ordered-universe.com/>

Panti, Cecilia, 'The Evolution of the Idea of Corporeity in Robert Grosseteste's Writings', in *Robert Grosseteste: His Thought and Its Impact*, ed. by Jack P. Cunningham (Toronto: Pontifical Institute of Mediaeval Studies, 2012), pp. 111–39

—, 'Robert Grosseteste and Adam of Exeter's Physics of Light, Remarks on the Transmission, Authenticity, and Chronology of Grosseteste's Scientific Opuscula', in *Robert Grosseteste and His Intellectual Milieu*, ed. by John Flood, James. R. Ginther, and Joseph W. Goering (Turnhout: Brepols, 2013), pp. 165–90

—, 'The Theological Use of Science in Robert Grosseteste and Adam Marsh According to Roger Bacon: The Case Study of the Rainbow', in *Robert Grosseteste and the Pursuit of Religious and Scientific Learning in the Middle-Ages*, ed. by Jack P. Cunningham and Mark Hocknull (Basel: Springer, 2016), pp. 143–63

Rashed, Roshdi, *Geometry and Dioptrics in Classical Islam* (London: al-Furqān Islamic Heritage Foundation, 2005)

—, *Les mathématiques infinitésimales du IXe au XIe siècle, Vol. IV: Ibn al-Haytham, méthodes géométriques, transformations ponctuelles et philosophie des mathématiques* (London: al-Furqān Islamic Heritage Foundation, 2002)

—, 'Le modèle de la sphère transparente et l'explication de l'arc-en-ciel: Ibn al-Haytham et al-Fārisī', *Revue d'histoire des sciences*, 23 (1970), 109–40

—, 'A Pioneer in Anaclastics: Ibn Sahl on Burning Mirrors and Lenses', *Isis*, 81 (1990), 464–91

Rehren, Thilo, and Ian C. Freestone, 'Ancient Glass: From Kaleidoscope to Crystal Ball', *Journal of Archaeological Science*, 56 (2015), 233–41

Rosenow, Daniela, and Thilo. Rehren, 'Herding Cats – Roman to Late Antique Glass Groups from Bubastis, Northern Egypt', *Journal of Archaeological Science*, 49 (2014), 170–84

Sayili, Aydia M., 'Al-Qarāfī and his Explanation of the Rainbow', *Isis*, 32 (1940), 14–26

—, 'The Aristotelian Explanation of the Rainbow', *Isis*, 30 (1939), 65–83

Schalm, Olivier, Ine de Raedt, Joost Caen, and Koen Janssens, 'A Methodology for the Identification of Glass Panes of Different Origin in a Single Stained-Glass Window: Application on two 13th Century Windows', *Journal of Cultural Heritage*, 11 (2010), 487–92

Schmidt, Olaf, Karl-Heinz Wilms, and Bernd Lingelbach, 'The Visby Lenses', *Optometry and Vision Science*, 76 (1999), 624–30

Schulman, Nicole M., 'Husband, Father, Bishop? Grosseteste in Paris', *Speculum*, 72 (1997), 330–46

Silvestri, Alberta, and Alessandra Marcante, 'The Glass of Nogara (Verona): A "Window" on Production Technology of mid-Medieval Times in Northern Italy', *Journal of Archaeological Science*, 38 (2011), 2509–22

Smith, A. Mark, *From Sight to Light: The Passage from Ancient to Modern Optics* (Chicago: Chicago University Press, 2015)

—, 'Ptolemy's Search for a Law of Refraction: A Case-Study in the Classical Methodology of 'Saving the Appearances' and its Limitations', *Archive for History of Exact Sciences*, 26 (1982), 221–40

Smithson, Hannah E., 'All the Colours of the Rainbow: Robert Grosseteste's Three-Dimensional Color Space', in *Robert Grosseteste and the Pursuit of Religious and Scientific Learning in the Middle-Ages*, ed. by Jack P. Cunningham and Mark Hocknull (Basel: Springer, 2016), pp. 59–83

Smithson, Hannah E. and others, 'A Color Coordinate System from a 13th Century Account of Rainbows', *Journal of the Optical Society of America A*, 29 (2014), A341–49

—, 'A Three-Dimensional Colour Space from the 13th Century', *Journal of the Optical Society of America A*, 29 (2012), A346–52

Southern, Richard W., *Robert Grosseteste The Growth of an English Mind in Medieval Europe*, 2nd edn (Oxford: Oxford University Press, 1992)

Stern, William B. and Yvonne Gerber, 'Potassium-Calcium Glass: New Data and Experiments', *Archaeometry*, 46 (2004), 137–56

Stevenson, Francis, *Robert Grosseteste* (London: Macmillan, 1899)

Tanner, Brian K., and others, 'Unity and Symmetry in the *De luce* of Robert Grosseteste', in *Robert Grosseteste and the Pursuit of Religious and Scientific Learning in the Middle-Ages*, ed. by Jack P. Cunningham and Mark Hocknull (Basel: Springer, 2016), pp. 3–20

Topdemir, Hüseyin Gazi, 'Kamal Al-Din Al-Farisi's Explanation of the Rainbow', *Humanity and Social Sciences Journal*, 2.1 (2007), 75–85

Turbayne, Colin. M., 'Grosseteste and an Ancient Optical Principle', *Isis*, 50 (1959), 467–72

Vernet, Juan, 'Abbās ibn Firnās', in *Dictionary of Scientific Biography*, ed. by Charles C. Gillispie (New York: Scribner, 1970–1980), I, 5

Wedepohl, Karl H. and Klaus Simon, 'The Chemical Composition of Medieval Wood Ash Glass from Central Europe', *Chemie der Erde*, 70 (2010), 89–97

White, Rebekah, and others, 'Magnifying Grains of Sand, Seeds and Blades of Grass: Optical Effects in Robert Grosseteste's *De iride* – On the Rainbow (*c.* 1228–1230)', *Isis*, forthcoming

Würschmidt, Joseph, 'Die Theorie des Regenbogens und des Halo bei Ibn al-Haitam und bei Dietrich von Freiberg', *Meteorologische Zeitschrift*, 31 (1914), 484–87

MICHELE CAMPOPIANO

Language and Wisdom

Mathematics and Astronomy in Bacon´s Edition of the Secretum secretorum

The translation movement of the twelfth and thirteenth centuries brought a wide range of scientific texts to the attention of Western scholars.[1] Some of these were Greek works, translated directly from Greek or from Arabic translations. A number of original Arabic texts were also translated. This translation movement heavily influenced the evolution of medieval science: Western scholars, such as the Franciscan Roger Bacon, benefitted greatly from the knowledge of these texts. Bacon also benefitted from a cultural environment in which the presence and influence of texts translated from Greek or Arabic had been steadily developing and where the interest in scientific themes was strong. Oxford, where he received part of his education, had seen the magisterium of Robert Grosseteste. Grosseteste, who would become archbishop of Lincoln in 1235, had been the first lector to the Oxford Franciscans between 1229 and 1231.[2] Grosseteste knew Greek and translated Greek texts (including the *Nichomachean Ethics* and parts of the *De caelo*), and he had played a pivotal role in developing an interest in sciences in England, especially optics and cosmology.[3] Cecilia Panti has recently shown how that the teaching of Grosseteste and of his successors in Oxford — such as the Franciscan Adam Marsh — was focused on theology and the exegesis of the Holy Writings, and natural philosophy was probably used mostly to illustrate ethical and religious themes; however, the influence of Grosseteste's scientific interests among the English Franciscans remains clear, not only in Bacon's works but also those of Adam of Exeter and the exegetic works of Thomas Docking.[4]

Bacon had been educated in Oxford (the dates are uncertain, but probably 1228–1236 or 1234–1242) and had been professor in Paris. He was acquainted with

---

1 Burnett, 'Humanism and Orientalism'; d'Alverny, 'Translations and Translators', and Haskins, *The Renaissance of the Twelfth Century*.
2 Southern, *Robert Grosseteste*, pp. 74–75.
3 See Gasper and others, this volume.
4 Panti, 'Scienza e teologia'.

Michele Campopiano • is Senior Lecturer (Associate Professor) at the University of York and von Humboldt Fellow (Experienced Researcher) at the Institute for History at the Technical University of Darmstadt. michele.campopiano@york.ac.uk

Arabic scientific works, such as Ibn al-Haytham's Kitāb al-Manāẓir (Book of Optics: *De Aspectibus* in Latin), which exerted a major influence on the development of his thought. He was particularly concerned with the role of languages in the transmission of knowledge. He also expressed his worries about the possibility that errors in translations and manuscript transmission of texts may have corrupted their meaning.[5] This is evident, for example, in the great care he took in preparing his edition of the *Secretum secretorum*, the Latin translation by Philip of Tripoli of the Pseudo-Aristotelic Arabic text *Sirr al-Asrār* (which translates as *The Secret of Secrets*).[6] The *Sirr al-Asrār* is a compilation of advice for Alexander the Great concerning politics, medicine, astrology and hermetic knowledge. During the Middle Ages, the works of Aristotle were translated into Arabic and Latin, and sometimes from Arabic into other languages. Other Arabic texts were attributed to Aristotle and translated into Latin from Arabic as authentic works of the Stagirite. A very well-known example is the metaphysical work known as *Liber de causis*.[7] This chapter will show how Bacon's study and editing of this treatise were crucial for his understanding of mathematics and astronomy. The *Secretum secretorum* was also crucial for the reception of ideas concerning hermetic sciences originating in the Islamic and pre-Islamic Middle East in Bacon's work. We must therefore understand how the *Sirr al-Asrār* was composed, and how different kinds of pre-Islamic traditions were included in the text.

The Origins of the *Secretum secretorum/ Sirr al-Asrār*.

Contrary to medieval assumptions, the *Sirr al-Asrār* is an Arabic text: there is no Greek original from which it has been translated. The text is known in both a short and a long version. A prologue to the long redaction ascribes the work to ibn al-Batrīq, the famous translator of Greek texts active between the eighth and ninth centuries, and explains that the *Secret* is a translation of a work found in the Temple of the Sun built by the philosopher Asclepius.[8] It is on the long redaction that we shall focus, since this is the version which would be edited and commented on by Bacon. There is much uncertainty as to the date of the *Sirr al-Asrār*. The oldest manuscripts of the long version date to the twelfth century. One is a manuscript once kept in Mosul (Madrasat Ğāmiʿal Bāshā, 55/134, its present location is unknown) and another is probably Oxford, Bodleian Library, MS Laud 210 of the (date and origins of this are unclear).[9] The sources used are more important to an understanding of when this work may have been written. The *Sirr al-Asrār* uses the *Rasā'il Ikhwan aṣ-ṣafā'*, the encyclopaedic texts attributed to the Iraqi secret society called Brethren of Purity

---

5 Lemay, 'Roger Bacon's Attitude toward the Latin Translations'.
6 Williams, 'Roger Bacon and the Secret of Secrets'; Williams, 'Roger Bacon and his edition of the Pseudo-Aristotelian *Secretum secretorum*'.
7 Vegas Gonzalez, 'Liber de Causis', Cristina D'Ancona, *Recherches*.
8 Bacon, *Secretum secretorum*, ed. by Steele, pp. 176–78.
9 Forster, *Das Geheimnis der Geheimnisse*, p. 16. See also: Grignaschi, 'Remarques sur la formation et l'interprétation du "Sirr al-'asrâr"'.

(in Arabic *Ikhwan aṣ-ṣafāʾ*), which were heavily influenced by Neoplatonism. The *Rasāʾil Iḫwan aṣ-ṣafāʿ* were written between the 930s — a period in which the works of al-Fārābī (who died in 950) were being disseminated — and the period between 954–60, when the *Rasāʾil Iḫwan aṣ-ṣafāʿ* were used by the author of the *Ġāyat al-ḥakīm*, a text on magic which would later circulate in a Latin translation known as *Picatrix*.[10] The *Sirr al-Asrār* is quoted for the first time by ibn Ǧulǧul in his *Ṭabāqāt al-ṭibbāʾ*, completed in 987.[11] The Italian Arabist Grignaschi argued that the *Sirr al-Asrār* was derived from the *Kitāb as-siyāsa*, a reworking of the *Rasāʾil Arisṭāṭālīsa ʾilā-l-Iskandar*, the fictional exchange of letters between Alexander and Aristotle.[12] Arabic texts were often attributed to important figures of antiquity: the *Kitāb al-nawāmīs* (known in the Latin translation as *Liber vacce* or *Liber aneguemis*) is ascribed to Plato, while the *Kitāb sirr al-Khalīqa* (Book of the Secret of Creation) was attributed to Apollonius of Tyana. The book describes, among other things, the relationship between the seven heavenly bodies and their associated metals.[13] It was translated as the *De secretis nature* by Hugo of Sanctalla at the beginning of the twelfth century.[14]

Aristotle was probably the figure that attracted the most identifications. He became a central figure in Arabic culture in the transmission of the hermetic sciences, continuing a tradition started in late Antiquity. In particular, he was connected with the work on magic stones and the creation of talismans, a craft that is discussed in detail in the *Sirr al-Asrār* and other works attributed to Aristotle.[15] Also attributed to Aristotle is the so-called *Lapidary of Aristotle*, which came to be known as a work of the Stagirite because it mentioned several stones included in the Alexander Romance. This text is connected to the *Kitāb al-ahjār*, a Latin translation of which is attributed to Gerard of Cremona.[16] It is not clear whether the *Lapidary of Aristotle* originates from the *Kitāb al-ahjār* or vice versa.[17] What is clear is that they both draw from a tradition of Persian works on magic stones, an example of which can be found in the *Riwāyāt* of the *Dādestan ī dēnīg*.[18] The *Lapidary of Aristotle* is probably one of the sources of the long version of the *Sirr al-Asrār*: the fact that they share a tradition is shown in particular by the three magic stones presented in Book x of the long version of the *Sirr al-Asrār*.[19] The concept of magic stones is not the only notion which may have been absorbed into the *Sirr al-Asrār* because of the influence of Persian heritage. The model of government and justice that the *Sirr al-Asrār* describes also has its origins in

---

10   Forster, *Das Geheimnis der Geheimnisse*, p. 18.
11   Forster, *Das Geheimnis der Geheimnisse*, pp. 31–32.
12   Gutas, *Greek Wisdom Literature*, pp. 436–51.
13   Weisser, *Buch über das Geheimnis der Schöpfung*; Weisser, *Das Buch über das Geheimnis*.
14   Pingree, 'The Diffusion of Arabic Magical Texts', p. 73.
15   Burnett, 'Arabic, Greek and Latin Works', pp. 84–96.
16   Pingree, 'The Diffusion of Arabic Magical Texts', p. 67.
17   Ulmann, *Die Natur – und Geheimwissenschaften im Islam*, p. 108.
18   Macuch, 'Pahlavi Literature', pp. 142–45, Ulmann, *Die Natur – und Geheimwissenschaften im Islam*, pp. 102–07.
19   *Das Steinbuch des Aristoteles*, ed. by Ruska, pp. 6–7; Ulmann, *Die Natur – und Geheimwissenschaften im Islam*, pp. 110–11.

the pre-Islamic period, in particular in Sassanian political culture.[20] This is exemplified by the Circle of Justice. This concept refers to a harmony between royal authority, the army, and subjects, which is guaranteed by justice. This harmony mirrors the order of the universe. In the *Sirr al-Asrār* Aristotle is quoted saying: *bi l-'adl qāmat as-samuwāt 'alā al arḍ* (it is through justice that the heavens stand over the earth).[21] The philosopher states that he invented for Alexander a diagram, divided according to the divisions of the heavenly spheres, which will inform him of *'ammā fī l-'ālim bi asrihi* (what is in the world) and which comprehends all classes of *ṭabaqa* (people) and the form of justice required for each of them.[22] Aristotle explains that all the forms of government (*tadābīr*, 'dispositions, managements') depend *'alā al-'ālim* (upon the world). He decided therefore to design a figure that illustrates the relationship between government and the world.[23]

In the *Sirr al-Asrār*, the Circle of Justice is defined as follows:

> 'Al- 'ālamu bustānun. Siyağuhū d-dawla; ad-dawlatu sulṭānuntaḥadjabuhu s-sunna, a s-sunnatu siyāsatun yasūsuhā l-malik; al-malik rā'in, ya'ḍuduhū l-ğayš; al-ğayš A'wānun, yakfaluhumu l-māl; al-mālu rizqun, tağma'uhū r-ra'īya; ar-ra'īyatu 'abīdun, yata'baduhum- yakfuluhumu l-'adl; al-'adlu ma'lūfun wa huwa ṣalāḥ- l-'ālami'
>
> (The world is a garden, the state its fence; the state is a power, tradition its principle; tradition is a policy, the king administers it; the king is a shepherd, the army supports him; the army are helpers, the treasury guarantees it; the treasury is a support, the subject contributes to it; the subjects are slaves, justice conditions them; justice is harmony, and it is the goodness of the world.)[24]

The *Sirr al-Asrār* is, together with the *'Ahd Ardashir* (*The Testament of Ardashir*), the most ancient text in which the Circle of Justice is described. The *Testament* was originally written in the sixth century.[25] The concept of Circle of Justice is clearly based on a connection between political and cosmology.

The cosmological aspects connected to the *Sirr al-Asrār* remind us that Pahlavi culture left important traces in Arabic astronomy and astrology. Many of the greatest astrologists who wrote in Arabic in the eighth and ninth centuries were actually Persian, and Middle-Persian language left a deep influence on Arabic vocabulary in astronomy and astrology.[26] Some Greek texts in these fields, such as the *Paranatéllonta* by Teucer of Babylon (probably first century AD), were translated in Arabic from a Middle Persian version.[27] Much remains yet to be understood, and more research should be dedicated to the study of the Persian roots of the *Sirr al-Asrār*.

---

20 Campopiano, 'A Philosopher between East and West'; Forster, *Das Geheimnis der Geheimnisse*, p. 56.
21 'Abd ar-Rahman Badawī, *Al-Uṣūl*, p. 125; Bacon, *Secretum secretorum*, ed. by Steele, p. 224.
22 'Abd ar-Rahman Badawī, *Al-Uṣūl*, p. 126; Bacon, *Secretum secretorum*, ed. by Steele, p. 226.
23 'Abd ar-Rahman Badawī, *Al-Uṣūl al*, p. 126; Bacon, *Secretum secretorum*, ed. by Steele, p. 226.
24 'Abd ar-Rahman Badawī, *Al-Uṣūl*, p. 128.
25 Askari, *The Medieval Reception of the Shāhnāma*, p. 155; Al-Azmeh, *Muslim Kingship*, p. 90 and Wiesehofer, *Ancient Persia*, p. 211.
26 Ulmann, *Die Natur– und Geheimwissenschaften im Islam*, pp. 296–97; Pingree, 'Astronomy and Astrology'.
27 Ulmann, *Die Natur – und Geheimwissenschaften im Islam*, pp. 278–79.

The cosmology, politics, and hermetic sciences are deeply connected in the *Sirr al-Asrār*. In this text Aristotle is quoted as describing methods for achieving political harmony through the hermetic sciences, and in particular by the use of talismans, which are described in particular in Book x.[28] The production of talismans depends on the position of celestial bodies. Aristotle says that the wisest men of the world entrusted him with this knowledge, and he transmits it to Alexander because the emperor finds it a worthy science. Aristotle explains that those *bimā ṭhahara lihim 'alā mā khafā 'anihim* (who can find out the hidden through that which is apparent to them), having reached the hidden truths of this deep and mysterious science, observe extreme caution and miserliness in communicating it to others, although it is of such universal benefit. They do so through fear that they might come to share this knowledge with those *man laysa lihi idrāku mā adrakūhu wa lā 'ilm mā 'alima* (who did not have comprehension of what they had comprehended and no knowledge of what they had known).[29] The knowledge remains an elite knowledge. In what does it consist? What we can perceive falls into two classes, matter and form. The cause of the permanence of forms is the reflection of their types from the planets. This reflection in turn undergoes continual changes, according to the motions of its planet.[30] On these motions depend the creation of talismans: all physical forms *aṣ-ṣūr al-arḍiyya* (terrestrial forms) are governed by their relative heavenly bodies *aṣ-ṣūr al-falakiyya* (celestial forms). This law of nature has given rise to the science of talismans *a'māl al-ṭilasmāt* (the actions of the talismans).[31] Among the talismans, there is one that will make subjects obedient to their lord and instill fear in one's enemies.[32] The creation of the talismans is due to the connection between celestial bodies and earthly forms. This line of thought would exercise a strong influence on Bacon, as we shall see in the rest of this chapter. Before moving to Bacon, however, we have to look at how the *Sirr al-Asrār* made its way to the West.

Sirr Al-Asrar in the West

The *Sirr al-Asrār* circulated even more widely in Europe than in the Middle East.[33] The prestige of the two key figures of this text was already enormous: Alexander the Great was at the centre not only of the very popular romance of the Pseudo-Callisthenes, translated into Latin by Leo of Naples and disseminated in the three re-elaborations known as *Historia de preliis*, but also one of the masterpieces of medieval poetry, the *Alexandreis* by Walter of Châtillon, which was translated into several vernaculars including Old Norse.[34] As for Aristotle, already by the twelfth century the English

---

28 Williams, *The Secret of Secrets*, pp. 10–11.
29 Bacon, *Secretum secretorum*, ed. by Steele, p. 254, see 'Abd ar-Rahman Badawī, *Al-Usūl*, p. 157.
30 Bacon, *Secretum secretorum*, ed. by Steele, p. 255.
31 Bacon, *Secretum secretorum*, ed. by Steele, p. 254; see Badawī, *Al-Usūl*, p. 156.
32 Bacon, *Secretum secretorum*, ed. by Steele, p. 257.
33 Williams, *The Secret of Secrets*, pp. 183–208; Grignaschi, 'La diffusion du *Secretum secretorum*'.
34 Tilliette, 'La traduction et l'adaptation'; Campopiano, 'Parcours de la légende d'Alexandre en Italie'; Lafferty, *Walter of Châtillon's* Alexandeis, pp. 13–29.

philosopher John of Salisbury in his *Metalogicon* reported that the common noun 'philosopher' had come to be reserved for him.[35] The translations of large parts of the Aristotelian corpus between the twelfth and thirteenth centuries — as well as the circulation of the works of first Avicenna and then Averroes, who were heavily influenced by the thought of the Greek philosopher and wrote commentaries on some of Aristotle's works — only strengthened the fame of the Stagirite.

A partial Latin translation of the short version of the *Sirr al-Asrār* was made around 1120 by Iohannes Hispalensis. It is not clear who this Iohannes Hispalensis was; it is, however, clear that the text was translated in an Iberian context in the twelfth century. We can be relatively specific with the date of the work, since it is dedicated to *Domine T. gratia Dei Hispaniarum regine*: this queen is probably Theresa/Tharasia of Portugal, who reigned in the years 1112–1128.[36] The long version was translated around 1232 in the Holy Land by Philip of Tripoli, with the title *Secretum secretorum*. Philip was born between 1195 and 1200; originally from Umbria, he was protected by his uncle Ranerius (Patriarch of Antioch between 1219 and 1225), and later saw service in the Holy Land, as a papal legate and chaplain. He died in 1269. His translation is dedicated to Guido bishop of Tripoli, who was probably bishop between 1228/1229 to 1232 or 1236/7.[37] Philip's translation is used by Michael Scot, who died in 1235.[38] Philip's interest in this surprising work is expressed by the enthusiastic words in the prologue of his translation. Philip describes this work as a 'philosophie pretiosissima margarita' (most precious pearl of philosophy).[39] He also says that the book was not found among the Latins and rarely found among the Arabs. The book originated from a request from Alexander the Great for knowledge of the secrets of arts such as the 'motum, operationem et potestatem astrorum in astronomia et artem alconomicam et artem cognoscendi naturas et operandi incantations et celimantiam et geomantiam' (movement, operation, and power of the celestial bodies in astronomy, the art of alchemy and the art of knowing the natures, of operating incantations, celimancy, geomancy).[40] This sentence shows how interest in the *Sirr al-Asrār* was largely motivated by interest in occult sciences. It also echoes the link between knowledge and power we have already seen in the Arabic text, a link suggested by the principle that knowledge operates actions in men and nature, and that this helps in the exercise of political power. The relationship between Aristotle and Alexander becomes a symbol of this connection.

This is clearly stated in the prologue, translated from the original Arabic text, which shows how Alexander became the ruler of the world by following Aristotle's

---

35 Iohannes Saresberiensis, *Metalogicon*, ed. by Hall and Keats-Rohan, II, 16.
36 Forster, *Das Geheimnis der Geheimnisse*, pp. 114–15.
37 Forster, *Das Geheimnis der Geheimnisse*, pp 121.
38 Forster, *Das Geheimnis der Geheimnisse*, p. 122.
39 For the translation by Philip of Tripoli, we still need to refer to the edition by Reinhold Möller in connection with the Middle High German translation by Hiltgard von Hürnheim, *Mittelhochdeutsche Prosaübersetzung*, p. 1.
40 Hiltgard von Hürnheim, *Mittelhochdeutsche Prosaübersetzung*, p. 2; Bacon, *Secretum secretorum*, ed. by Steele, p. 26.

advice: 'et ideo subiugavit sibi civitates et triumphans adquisivit cuncta regna et totius mundi solus tenuit monarchiam' (and in this way he subjugated the cities to himself, and triumphing he acquired all the kingdoms, and he gained the monarchy of the entire world).[41] The Latin version of the *Secret of Secrets* found a ready audience in the West.

The association between power and knowledge that these two figures symbolized would have favoured the dissemination of the *Secret of Secrets* in a place that was, in the years of the translation of the Philip, both a centre of learning and one of the major political powers of Europe: the court of the emperor Frederick II. The study of philosophy was a common occupation at the court of the Hohenstaufen and contact with Arabic and Jewish philosophy was intense. Also at his court was Michael Scot, who knew Arabic and was well versed in hermetic science and Aristotelian philosophy. Scott even achieved, somehow, legendary status as a mage, so much so that he was remembered in this way by Dante. Frederick's court represented one natural focus for the dissemination of the *Secretum secretorum*. Another was the papal court. The *Secretum secretorum* was used by a certain Dominus Castri Goet in a medical work, a version of which was dedicated to Innocent IV Pope (1243–1254). One of the earliest manuscripts of the *Secretum secretorum* is the personal notebook of the papal agent Albert Behaim, copied during his stay at the council of Lyon (1245). This was during a period in which the papacy fostered the formation of universities (and the intellectualization, as the historian Steven Williams has called it, of the mendicant order).[42] Although the court of the emperor in Sicily and the Apostolic See promoted the early dissemination of this work, the text also found its fortune among the mendicants.

Bacon's Edition

It is well known that Roger Bacon paid attention to the flow of translations from Greek and Arabic to Latin.[43] He was also a critical voice in several aspects of high medieval translation. In his masterpiece, the *Opus maius*, he says that the translator should have knowledge of the science he wants to translate, and of the languages from which and into which he wants to translate; many translators lacked of all these skills.[44] Knowledge of the languages in which these texts were originally written remains crucial, since some properties of a text cannot be adequately maintained when it is translated into a new language.[45] Bacon pays particular attention to Arabic

---

41 Hiltgard von Hürnheim, *Mittelhochdeutsche Prosaübersetzung*, p. 14; Bacon, *Secretum secretorum*, ed. by Steele, p. 37.
42 Campopiano, 'A Philosopher'; Williams, *The Secret of Secrets*, pp. 112–16; Zamuner, 'La tradizione romanza del "Secretum secretorum" pseudo-aristotelico'.
43 Pereira, *Arcana sapienza*, pp. 138–44; Hackett, 'Roger Bacon on Astronomy-Astrology'; Hackett, 'Roger Bacon on *Scientia experimentalis*'; and Newman, 'An Overview of Roger Bacon's Alchemy'.
44 Bacon, *Opus Maius*, III, 67–68.
45 Roger Bacon, *The 'Opus Maius'*, ed. by Bridges, III, 66–67.

science and philosophy: he writes that the philosophy of Aristotle had been brought out of darkness by the work of Averroes and Avicenna.[46]

An interest in languages and science is evident in the work of Grosseteste, Lector of the Franciscans. He shows particular interest in the *Secretum secretorum*, which he considers an authentic work of Aristotle. As Thomas Maloney has written, Bacon was a pioneer in Aristotelian studies: he also knew Greek (and some elements of Hebrew and Arabic) and made an intensive study of Greek works, often integrating Arabic commentary.[47] To him, the new scientific and philosophical trends did not conflict with the Christian faith. He explains in the *Opus Maius* that philosophy has its origins in the divine revelation of wisdom to the prophets and patriarchs: the *Secretum secretorum* is one of the texts he quotes to support this statement.[48]

The *Secretum secretorum* had a major impact on Bacon's vision of science.[49] According to Easton, the study of the *Secretum* directed Bacon's research — crucially — towards the establishment of a universal science, and to a practically oriented *scientia experimentalis*, a judgment that was essentially shared by Maloney.[50] More cautious was Williams's position on this issue, who sees the importance of the *Secretum* in Bacon's intellectual biography, but does not attribute to it the path-breaking role in the development of his philosophical and scientific thought.[51] The discussion of the role of the *Secretum* is closely related to the date of Bacon's editing of this work, which Easton argued had taken place in the 1240s, but which scholars now tend to date sometime after 1280.[52]

Bacon prepared an edition of the text on the basis of different manuscripts, and he completed it with an introduction and a commentary. The commentary shows particular interest in several aspects of the representation of Aristotle that derive from the Arabic tradition of Alexander and Aristotle. Bacon comments on the epithet of Alexander 'Qui Alexander dicitur duo cornua habuisse [...] cornua dua significant duo regna, scilicet, Grecie et Asie que optinuit' (And this Alexander is said to have had two horns [...] the two horns mean the two kingdoms, i.e. Greece and Asia, which he gained).[53] This description originates from the identification of Alexander with a qur'anic figure, the *Dhū Al-Qarnain* (the one with the two horns). Bacon's text has in common with the *Sirr al-Asrār* the idea that knowledge has a direct connection to political power. Echoing a famous sentence by John of Salisbury, he underlines the importance of the education of the king: 'Nota hic mirabilem sapienciam quam reges antique adimpleverunt quia fuerunt instructi in philosophia, set nunc ut Henricus

---

46 Roger Bacon, *The 'Opus Maius'*, ed. by Bridges, II.XII, p. 55.
47 Roger Bacon, *Compendium studii theologiae*, ed. by Maloney, p. 4; see also Power, *Roger Bacon*, p. 47; Hackett, 'Roger Bacon and Aristotelianism'.
48 Roger Bacon, *The 'Opus Maius'*, ed. by Bridges, II.VIII, pp. 44–46.
49 Hackett, 'Roger Bacon and the Moralization of Science'; Easton, *Roger Bacon*; Williams, 'Roger Bacon and the Secret of Secrets'.
50 Easton, *Roger Bacon*, pp. 70–86; Roger Bacon, *Compendium studii theologiae*, ed. by Maloney, pp. 4–5.
51 Williams, 'Roger Bacon and the Secret of Secrets' pp. 369–72.
52 Power, *Roger Bacon*, pp. 78–79.
53 Bacon, *Secretum secretorum*, ed. by Steele, p. 36.

filius Willelmi regus qui dicebatur "bastardus" solebat dicere patri et fratribus "rex illiteratus est asinus coronatus"' (Notice here the wonderful wisdom that the ancient kings acquired because they were learned in philosophy, but now how Henry son of king William who was called 'the Bastard' was used to saying to his father and brothers, 'an illiterate king is a crowned ass').[54]

Bacon shows a clear interest in the use of the hermetic sciences for political goals and tries to adapt these to a Christian vision of the cosmos. Chiara Crisciani has shown that, of Bacon's glosses of the work (which in her view account for relatively few), those that address cosmological and metaphysical aspects tend to prevent possible readings of this text in support of emanationist necessarianism.[55]

Astronomy and Cosmology in the *Secretum Secretorum*

Crucial to Bacon's 'Christianization' of the *Secretum* is the discussion of judicial astronomy. This is a key topic in the Franciscan's work, since it is present not just in the *Secretum secretorum* but also in what is the most important of Bacon's works: the *Opus Maius*.[56] The discussion of judicial astronomy connects with the issue of free will. In his *Hexaemeron* Grosseteste, who had a major influence on Bacon, followed Augustine in condemning the application of judicial astronomy to human affairs:

> si stelle vel natura vel voluntate ad malum cogunt, vel malum persuadeant, male sunt. Et si natura male sunt, conditor eorum Deus malus esse convincitur, quod blasfemum est dicere. Si vero voluntate male sunt, est in celestibus peccatum et error. Item si valeret constellacio, sicut fingunt astronomi, periret, ut dictum est, libertas arbitrii
>
> > ([I]f stars by nature or will constrain to evil, or persuade to evil, they are evil. And if they are bad by nature, their creator God is demonstrated to be evil, which is a blasphemy. If they are bad by will, there is sin and error among the celestial bodies. So, if the constellation has influence, as the astronomers pretend, the free will would die, as it has been said.)[57]

To avoid this interpretation, Bacon distinguishes the true from the false mathematicians.[58] Astrologers were often called *mathematici*, and attacked for their necessitarian views.[59] The false mathematicians believed that all events depend on necessity

---

54 Bacon, *Secretum secretorum*, ed. by Steele, p. 58; see also: Iohannes Saresberiensis, *Policraticus*, ed. by Keats-Roahn, IV, 6.
55 Crisciani, 'Ruggero Bacone e l' 'Aristotele''.
56 Hackett, 'Roger Bacon on Astronomy-Astrology'.
57 Robert Grosseteste, *Hexaemeron*, ed. by Dales and Gieben, v. ix, 1–2, pp. 165–66; see also: Dales, 'Grosseteste's Views on Astrology'.
58 The distinction between true and false astrology is already in Robert Kilwardby: see Hackett, 'Roger Bacon on Astronomy-Astrology', p. 183; Robert Kilwardby, *De ortu scientiarum*, ed. by Judy, pp. 41–48.
59 Molland, 'Roger Bacon's Knowledge of Mathematics', p. 152; see also Molland, 'Roger Bacon's *De Laudibus Mathematicae*: A Preliminary Study'.

and destiny.⁶⁰ In both the *Opus Maius* and the *Opus tertium*, Bacon attempted to distinguish *mathesis* — with an aspiration and short middle syllable, representing true mathematics — from *matesis*, without aspiration and long middle syllable, representing the false mathematics:⁶¹

> Theologi igitur multa invenerunt a sanctis effusa contra mathematicos, et aliqui eorum propter ignorantiam mathematicae verae et mathematicae falsae nesciunt distinguere veram a falsa, et ideo tanquam auctoritate sanctorum culpant veram cum falsa. Vocabulum enim verae mathematicae scribi per t aspiratum, et ab hoc nomine mathesis media correpta, quod scientiam designat, derivari a multis refertur auctoribus, et certum est ex Graeco; quia matheo verbum idem est quod disco, et mathetes est discipulus, et mathesis disciplina. Unde mathematica est disciplinalis scientia et doctrinalis, sicut Cassiodorus dixit superius. Sed vocabulum falsae mathematicae sine aspiratione scribi asseritur ab eisdem auctoribus et a mathesi media producta, quod divinationem notat, descendere, vel quod certius est, a mantos vel a mantia, quae sunt idem quod divinatio, sicut Hieronymus dicit in originali ix Isaiae. Quicquid vero sit de hac scriptura et derivatione, tamen falsa mathematica est ars magica.⁶²

> (Theologians then have found many statements scattered abroad by the sacred writers against mathematicians, and some of these men, owing to their ignorance of the true mathematics and of the false, do not know how to distinguish the

---

60 Bacon, *Secretum secretorum*, ed. by Steele, p. 3; see also David C. Lindberg, 'On the Applicability of Mathematics to Nature'.
61 See also Bacon's *Communia mathematica*, which Easton dates to 1257, Molland to after 1267: 'Una est pars philosophia et alia inter magicas et erroneas stulticias computatur. Nam illa que erronea est nec est pars philosophie nec alicuius sapiencie (p. 2). imponit enim necessitatem libero arbitrio, ut homo natus in tali constellacione sit castus necessario, alius sit luxuriousu in alia constellacione natus, et sic de aliis moribus et fortunis et de omnibus actibus docet, quod a virtute stellarum necessitas sit confirmata, ut nichil valeat racio nec consilium nec gracia dei, nec temptacio dyaboli noceat homini in hac vita. Item de singulis futris et presentibus occultis et preteritis et naturalibus et aliis certitudinem repromittit quorum utrumque nephas et non solum contrarium fidei set contra philosophicam veritatem. Et hec mathematica dicitur a "mantos" bel "mantia" quod est divinacio, ut in tractatut meo Grece gramatice explicavi. Sed mathematica vera que considerat quantitatem et eius species neutrum ponit nisi in hiis que in celo renovantur, ut sunt eclipses et motus stellarum et hiusmodi, potest omnino certificare, et potest ac debet veraciter iudicare. Sed de certificacione inferiorum reru omnium et de iudicio infallibili circa omnes (aliter omnia), et maxime crica humana opera non presumit. Veumptamen certificat celestia que sunt cause horum inferiorum naturalium et occasiones humanorum per que potest homo peritus in rebus mundi et temperatus in iudiciis quamplurima veraciter et laudabiliter iudicare. Sed de hiis in Metaphysica certificatum est quantum as eam pertinet Et cito inferius exponetur hec difficultas, sciliet quando fiet mencio de partibus mathematice, atque in tractatu Astronomie iudiciarie complebitur de hoc sermo. Et hec mathematica scribitur cum aspiracione, et dicitur a "mathesis" quod est disciplina, quia habet mediam productam, ut in mean gramatica Greca exposui, licet totum vulgus Latinorum credat quod media sit correpta secundum quod fiunt versus falsi: scire facit mathesis sed divinare mathesis et consimiles'; *Communia mathematica fratris Rogeri partes prima et secunda*, ed. by Steele I.2, pp. 2–3; Molland 'Roger Bacon's Knowledge of Mathematics', p. 153; Easton *Roger Bacon*, p. 88.
62 Roger Bacon, *The 'Opus Maius'*, ed. by Bridges, pp. 239–40.

true from the false, and therefore condemn the true with the false as though by authority of the sacred writers. For the word for the true mathematics is said by many authors to be written with an aspirated t, and to be derived from this word mathesis with short middle syllable, meaning knowledge, and it is certainly derived from the Greek; because the verb *matheo* is the same as *disco* [learn], and *mathetes* is the same as *discipulus* [learner], and *mathesis* is the same as *disciplina* [instruction]. Whence *mathematica* [mathematics] is instructional and theoretical knowledge, as Cassiodorus said above. But the word for false mathematics is said by the same authorities to be written without an aspirate and to be derived from *matesi* with long middle syllable, meaning divination, or with greater certainty, from *mantos* or from *mantia*, which are the same as *divinatio* [divination], as Jerome states in the original, Isaiah, Chapter IX. But whatever it may be as regards its writing and derivation, false mathematics is the art of magic.)[63]

The sacred doctors of the past did not make use of the sciences of philosophy, in part because many works had not been translated yet, in particular concerning mathematics.[64] Bacon discusses in several places the importance of mathematics to religion.[65] Christians seek to understand the heavens, because they know their souls will dwell there.[66] Astronomical problems need to be understood in order to study the Bible.[67] Since the scriptures are set in a geographical context, we need to know mathematical geography and locate the places mentioned in the Bible on earth.[68] Bacon also reflects on the influence of the planets' position and nature on the events on earth and on earthly bodies. True mathematicians study the influence of the stars on earthly bodies.[69] He explains the effect of the planets on the earth in his theory of *species*, the first effects of natural things, for example the light of the sun on the earth.[70] In his *De multiplicatione specierum*, he explains that:

> transumitur hoc nomen ad designandum primum effectum ciuiuslibet agentis naturaliter. Et, ut in exemplo pateat hec species, dicimus lumen solis in aere esse speciem lucis solaris que est in corpore suo.
>
> (this name is meant to designate the first effect of any naturally acting thing. And to explain this meaning of species with an example, we say that the lumen of the sun in the air is the species of the solar lux in the body of the sun.)[71]

---

63 Roger Bacon, *Opus Maius*, trans. by Belle Burke, p. 261.
64 Hackett, 'Roger Bacon on Astronomy-Astrology', p. 177.
65 Woodward and Howe, 'Roger Bacon on Geography and Cartography', pp. 202–04.
66 Roger Bacon, *The 'Opus Maius'*, ed. by Bridges, I, 180.
67 Roger Bacon, *The 'Opus Maius'*, ed by Bridges, I, 183.
68 Roger Bacon, *The 'Opus Maius'*, ed, by Bridges, I, 183.
69 Gautier Dalché, 'Vers une Perfecta Locorum Doctrina:'; see also: Trifogli, 'Roger Bacon and Aristotle's Doctrine of Place'.
70 Gautier Dalché, 'Vers une Perfecta Locorum Doctrina'; Gautier Dalché, *La Géographie*, pp. 95–113; Lindberg, 'Roger Bacon on Light'.
71 Bacon, *De multiplication specierum*, in *Roger Bacon's Philosophy of Nature*, ed. by Lindberg, I.1, pp. 2–3.

The study of the transmission of the effect of species is strictly related to the study of mathematics, and in particular of geometry, in order to understand the line of physical actions on the world.[72] For example, Bacon explains:

> sicut prius dictum est, ad omne punctum terrae incidit conus unius pyramidis virtuosae a toto coelo. Et coni isti sunt {diversi} in natura, et pyramides similiter, quia diversas habent bases propter diversitates horizontum, quoniam quilibet punctus terrae est centrum proprii horizontis […] Et ideo oportet omnium rerum diversitatem magnam ex hac causa oriri, etiam quantumcunque [propinquae] sunt, ut gemelli in eodem utero; et sic de omnibus, prout videmus quod a duobus punctis terrae proximis oriuntur herbae diversae secundum speciem. Et hic sumit astronomus fundamenta sui iudicii, et merito, quia diversitas plena rerum per coelum sic invenitur. Quapropter potest astronomus peritus non solum in naturalibus sed in humanis rebus multa considerare de praesenti et futuro et praeterito, et ideo saltem super regna et civitates potest iudicare per coelestia et secunda coelestium quae per virtutes speciales coelorum renovantur, ut sunt cometae et huiusmodi, quia facilius iudicium est super communitate quam super singulari persona.[73]

> (as has already been stated, on every point of the earth there occurs the vertex of a pyramid with full of force from the whole heavens. These virtues are different in nature, and the pyramids likewise, because they have different bases owing to the differences in their horizons, since every point of the earth is the center of its own horizon. […] Therefore, of necessity a great difference in all things arises from this cause, no matter how close they are, like twins in the same womb; and such is the case in all things, just as we see that, from two points very close to each other on the earth, plants spring that differ according to species. On this principle the astronomer rests the foundations of his judgment, and rightly so, because the complete diversity in things is thus discovered to be due to the heavens. Wherefore the skillful astronomer is able to consider, not only in matters of nature but also in human affairs, many things regarding the present and the future and the past, and therefore at least as regards kingdoms and states is he able to judge by means of the heavenly bodies and of the secondary members of the same, which are renewed by special forces of the heavens, such as comets and the like, because it is an easier judgment in regard to a community than in regard to an individual.)[74]

The necessity of separating the true from the false mathematicians while supporting the necessity of studying the influence of the heavenly bodies on earthly forms dominates Bacon's edition of the *Secretum secretorum*, and his argument draws considerable support from this text. True mathematicians can understand aspects

---

72 Molland, 'Roger Bacon's Knowledge of Mathematics', p. 171.
73 Roger Bacon, *The 'Opus maius'*, ed. by Bridges, IV, 250–51.
74 Roger Bacon, *Opus Maius*, trans. by Belle Burke, pp. 272–73.

of the influence of heavenly bodies on natural events and human characteristics.[75] A real mathematician, says Bacon in his introduction to the *Secretum*:

> Set veri mathematici considerant situs et loca planetarum, et quas fortitudines habent in signis diversis, et quos respectus habent adinvicem et ad stellas fixas, et sic veraciter possunt iudicare de alteracionibus corporum inferiorum in terra et aqua et aere, secundum possibilitatem, ut dictum est, et quod una pars contradiccionis eveniet, set non de necessitate.
>
>> (considers the sites and the positions of the planets, and which strengths they have under the different signs, and which positions they have in respect to each other and to the fixed stars, and in this way they can judge truly of the alterations of the inferior bodies on earth and water and air, according to this possibility, as has been said; and that one part of contradiction happens, but not as necessity.)[76]

This knowledge allows them to act in accordance to celestial influences:

> Set mathematici veri [...] servant omnino veritatem philosophie et fidei, et pro debitis constellacionibus et stellarum fixarum possunt, Dei disposicione, multa fieri per naturam et per Artem iuvantem naturam, et utuntur dictis et factis certis, sed non carminibus magicis nec vetularum set secundum graciam datam philosophis.
>
>> (But true mathematicians [...] thoroughly maintain the truth of philosophy and faith, and they can recognize through the due constellations the times in which many things can happen by the virtues of planets and fixed stars, through the divine order, and through nature and arts supporting nature, and they use specific spells and actions — not, however, the magic evocations or the spells of old women, but according to the grace granted to the philosophers.)[77]

Bacon gave cautious indications that he thought occult arts might profitably be employed to improve people's morals as well as to defend Christendom from its enemies. In the *Secretum secretorum* we read that Alexander wrote to Aristotle saying that he found many people who wished to dominate other populations: for this reason, he proposed to kill all of them. Aristotle replied: 'Si non potes illius terre mutare aerem et aquam, insuper et disposicionem civitatum, imple tuum propositum. Si potes dominari super eos cum bonitate, exaudies eos cum benignitate' (If you cannot change the air and water of that land, and moreover alter the disposition of the cities, fulfil your intention. If you can rule them with goodness, listen to them with benignity).[78] Bacon commented:

> Hic tangit maximum secretum. Vult enim quod Alexander deberet mutare malas qualitates terre et aeris illarum regionum in bonas, ut hominum complexio mala

---

75 Hackett, 'Roger Bacon on Astronomy-Astrology', pp. 185–86.
76 Bacon, *Secretum secretorum*, ed. by Steele, p. 4.
77 Bacon, *Secretum secretorum*, ed. by Steele, p. 8.
78 Bacon, *Secretum secretorum*, ed. by Steele, p. 38; see also Power, *Roger Bacon*, pp. 248–51.

> mutaretur in bonam, et ut sic mali mores mutarentur in bonos. Per qualitates enim regionis cujuslibet invenitur complexio, et per complexionem excitatur homo, ad mores, licet non cogatur, set ut gratis velit ea ad que qualitates regionis et complexio inclinant, ut in principio expositum est.
>
>> (He [Aristotle] touches here upon the greatest secret. He desires that Alexander should change the qualities of the land and air in those regions for the better, so that the bad complexion of the men is turned into good, and so that the bad customs are turned into good ones. For the complexion is found through the qualities of any region, and the human being is excited by the complexion to good customs, but not constrained, but so that he may want gratefully those things to which the qualities of the region and complexion tend, as it is explained at the beginning.)[79]

Bacon adds: 'Set qualiter deberent qualitates regionis immutari docet alibi in hoc libro' (but he teaches elsewhere in this book how the qualities of the regions should be changed).[80] Here Bacon probably refers to what, in his edition, is the third part of the *Secretum secretorum*, which describes which heavenly powers rule the worldly forms and how matter assumes form in relation to the movements of planets. In this way it is possible to produce talismans.[81] According to Bacon, this knowledge and these practices are not in contradiction with the doctrine of free will:

> nam secundum Ptolomeum in libro de Disposicione sphere qui alio nomine vocatur Introductorius in Almagesti, duplex est pars philosophie de futuris cognoscendis secundum possibilitatem, ut dictum est. Una est Astronomia naturalis, quia de rebus naturalibus in hoc mundo inferiori iudicat, ut dictum est, scilicet medio modo inter necessarium et impossibile, sicut Ptolomeus docet in Centilogio et in Quadripartito, ita quod nullam necessitatem imponunt astronomi libero arbitrio, neque rebus contigentibus, ut prius expositum est.
>
>> (For according to Ptolemy in the book on the Disposition of the sphere, which is also called by the name Introduction to the Almagest, the part of philosophy concerning the knowledge of future things according to possibility is twofold, as has been said. One part is natural astronomy, because it judges of the natural things in this inferior world, as has been said; that is, in a way that mediates between the necessary and the possible, as Ptolemy teaches in the *Centiloquium* and in the *Quadripartitum*, so that the astronomers impose no necessity on free will, nor to contingent things, as it has been previously explained.)[82]

Skies and stars influence the complexions of earthly bodies, but this does not mean that they override free will:

---

79 Bacon, *Secretum secretorum*, ed. by Steele, pp. 38–39.
80 Bacon, *Secretum secretorum*, ed. by Steele, p. 39.
81 Bacon, *Secretum secretorum*, ed. by Steele, pp. 114–17.
82 Bacon, *Secretum secretorum*, ed. by Steele, p. 9.

De ista doctrina Aristotilis sciendum est hic, quod sicut dictum est in principio libri, quod virtutes celorum et stellarum non cogunt liberum arbitrium set mutant complexiones corporum, ad quarum mutacionem excitatur mens, ut sine coaccione velit gratis illud ad quod excitatur; sic est hic de virtutibus lapidum et vegetabilium

> (Of this doctrine of Aristotle's, it must be known here that, as it is said in the beginning of the book, the virtues of the skies and stars do not constrain free will but change the complexions of the bodies, by the mutation of which the mind is excited, so that without coercion it may freely want that to which it is excited; so it is here regarding the virtues of the stones and vegetables.)[83]

Bacon explains also in his edition of the *Secretum* that some of the misunderstandings concerning false magic and true philosophy were due to linguistic misunderstandings:

> cum igitur ge Grece sit Terra Latine, et mancia divinacio, hoc est, iudicium de futuris, una erit geomancia magica de qua dictum fuit superius, et alia castigato nomine est pars philosophie, que, scilicet, considerat signa in animalibus et ceteris rebus terrestribus super futura, nec habet aliquid falsitatis si bene intelligatur.
>
> (Since *ge* in Greek is earth in Latin, and *mancia* is divination, that is to say, judgment about future things, the one will be magic geomancy, about which it was written above, and the other, having corrected the name, is part of philosophy, which, of course, considers signs in animals and other earthly things concerning the future, and does not imply falsehood if it is correctly understood.)[84]

Bacon also refers to the fact that the doctrines on the *qualitates regionis* (qualities of regions) have been erased or not copied, *quia translator nomina aliqua ponit quibus utuntur magici* (because the translator puts some names which the wizards use).[85] Bacon, however, considered the prophets and learned Jewish men to be the origin of this wisdom: 'Consederandum est quod Aristotiles et ceteri magni philosophi legerunt Vetus Testamentum et edocti sunt a prophetis et ceteris sapientibus Hebreis' (It should be considered that Aristotle and the other great philosophers read the Old Testament and are taught by the prophets and by other learned Jewish men).[86]

The same applies to the geometry and arithmetic:

> Sciendum ergo quod infidels philosophi non invenerunt hanc scienciam nec alias partes Mathematice, ut Geometriam et Arithmeticam, nec alias sciencias, set Deus dedit eas suis sanctis et iustis Hebreis, a quibus omnes philosophi infidels habuerunt omnium scienciarum principia.

---

83 Bacon, *Secretum secretorum*, ed. by Steele, p. 121.
84 Bacon, *Secretum secretorum*, ed. by Steele, p. 12.
85 Bacon, *Secretum secretorum*, ed. by Steele, p. 39.
86 Bacon, *Secretum secretorum*, ed. by Steele, p. 56; Molland, 'Roger Bacon's Knowledge of Mathematics', p. 154.

(It must be known that the unfaithful philosophers did not discover this science [astronomy] nor other parts of mathematics, like geometry and arithmetic, nor other sciences, but God granted them to his pious and just Jews, from whom all faithless philosophers took the principles of all the sciences.)[87]

Bacon's interpretation of the *Secretum* is able to maintain the idea of the influence of heavenly bodies on earthly forms while reconciling them to the doctrine of free will, but it is also able to attribute the ultimate source of this knowledge to divine revelation. Bacon similarly maintains the relationship between cosmology and political order expressed in the text. He clearly connects the structure of the universe and human justice in his commentary:

> Dividitur ergo in duas divisiones circulares, etc. scilicet, celestes et elementares que sunt, circulares et sperice partes mundi que a Deo create sunt ordinate iusticia naturali. Et alia pars iusticie est in rebus contentis in eis et precipue inter homines et prosequitur de utraque parte justicie, scilicet, naturali et legali sive equo nichil est preciosius, videlicet, circulum firmamenti cum aliis circulis celestibus et angelicis spiritibus qui sunt in celis quando fuerunt ordinaciones sive regimina tam in inferioribus quam in superioribus ad conservanciam huius mundi, visum est mihi debere incipere tali modo in mundo, et hec est utilitas huius libri et hec est ejus figura.

> (It [the universe] is divided in two circular divisions, this is to say, the heavenly and the elemental, circular and spherical parts of the world, which are created by God and ordered according to natural justice. And the other part of justice is in the things contained in them and particularly among men, and derives from both parts of justice, his is to say, natural and legal justice, than which nothing is more precious; this is to say, the circle of the firmament with the other heavenly circles and the angelic spirits who are in the skies when there were the ordnances or regimens as much in the inferior things as in the superior things for the conservation of this world, it seemed to me to have to start in this way in the world, and this is the usefulness of this book and this is its figure.)[88]

This passage is faithful to the notion of Circle of Justice as expressed in the Arabic text. Natural justice (the creation and ordination of the world) and civil justice are connected. The notion of the Circle of Justice is introduced by the title *De creacione primordialis materie* (On the creation of primordial matter). The paragraph in Bacon does not appear to have been modified, but Bacon's commentary makes the relationship between the order of the universe and human justice much clearer than in the original.[89] The following chapter is introduced by the title: 'Capitulum sextum de creatis, in quibus attenditur iusticia naturalis et quot sunt celi' (Sixth chapter

---

87 Bacon, *Secretum secretorum*, ed. by Steele, p. 62.
88 Bacon, *Secretum secretorum*, ed. by Steele, p. 125.
89 Hiltgart von Hürnheim, *Mittelhochdeutsche Prosaübersetzung*, pp. 122–23.

on created things, in which natural justice is discussed and of what number are the heavens), which discusses metaphysics and the creation of the *simplex spitirualis substancia*, called *intelligencia*, and the *substancia minor gradu suo*, called anima, and then the *hyle* (matter), and describes then the heavenly spheres.[90]

Conclusion

In his work, Bacon develops his interest in the development of scientific thinking and the necessity of connecting it with Christian theological reflection, which had characterized the work of English thinkers such as Grosseteste, particularly on themes such as free will. Bacon's edition of the *Secretum* emphasizes the importance of the cosmological vision of the Arabic text, and the possibility of using this knowledge to produce practical effects. True mathematicians can create these effects by understanding the connection between heavenly bodies and earthly forms. This interpretation of his edition of the *Secretum* corresponds to Lemay's suggestion that Bacon's criticism of translating trends at the end of the twelfth and the beginning of the thirteenth centuries points largely to an interpretation of Aristotle that must have been conceived within the Persian/Arabic tradition.[91] Arabic and Persian texts frequently saw Aristotle as a master of hermetic sciences. On the other hand, Bacon's efforts are also clearly directed towards the necessity of Christianizing these sciences. The vision of the cosmos that emerges from his comments is one that is surprisingly faithful to the cosmological and astrological doctrines elaborated within Persian and Arabic traditions, but at the same time acceptable to Christian readers, since it does not conflict with the notion of free will and the possibility of claiming an ultimately divine origin for this lore. Bacon's views in his *Opus Maius* show a clear connection to his interpretation of the *Secretum* and the vision of the cosmos that emerges from this text. The *Secretum secretorum* is therefore a key text from which Bacon can elaborate both his position on the translation movement of the twelfth and thirteenth centuries, and his interpretation of Aristotle.

Works Cited

Primary Sources

'Abd ar-Rahman Badawī, *Al-Uṣūl al-yunaniya li-nazariyat al-siyasiya fī al-Islam* (Cairo: n.p., 1954)

Das Steinbuch des Aristoteles, ed. by Julius Ruska (Heidelberg: Winter, 1914)

Hiltgart von Hürnheim, *Mittelhochdeutsche Prosaübersetzung des "Secretum secretorum"*, ed. by Reinhold Möller (Berlin: Akademie Verlag, 1963)

---

90 Bacon, *Secretum secretorum*, ed. by Steele, pp. 127–28; compare with pp. 122–24.
91 Lemay, 'Roger Bacon's Attitude toward the Latin Translations', pp. 37–47.

Iohannes Saresberiensis, *Metalogicon*, ed. by John Barrie Hall and Katharine S. B. Keats-Rohan (Turnhout: Brepols, 1991)

—, *Policraticus I–IV*, ed. by K. S. B. Keats-Rohan (Turnhout: Brepols, 1993)

Robert Grosseteste, *Hexaemeron*, ed. by Richard C. Dales and Servus Gieben (Oxford: Oxford University Press, 1983)

Robert Kilwardby, *De ortu scientiarum*, ed. by Albert G. Judy (London: The British Academy, 1976)

Roger Bacon, *Communia mathematica fratris Rogeri partes prima et secunda*, ed. by Robert Steele (Oxford: Clarendon Press, 1940)

—, *Compendium studii theologiae*, ed. by Thomas S. Maloney (Leiden: Brill, 1988)

—, *De multiplicatione specierum*, in *Bacon's Philosophy of Nature. A Critical Edition, with English Translation, Introduction, and Notes of* De multiplicatione specierum *and* De speculis comburentibus, ed. by David Lindberg (Oxford: Clarendon Press, 1983)

—, *Opus Maius*, vol. 1, trans. by R. Belle Burke (New York: Russell & Russell, 1962)

—, *The 'Opus Maius'*, ed. by John Henry Bridges (Oxford: Clarendon Press, 1897)

—, *Secretum secretorum cum glossis et notulis, tractatus brevis et utilis ad declarandum quedam Obscure dicta, accedunt versio Anglicana ex Arabico edita per A.S. Fulton, versio Vetusta Anglo-Normanica nunc primum edita*, ed. by Robert Steele (Oxford: Clarendon Press, 1920)

Secondary Studies

d'Alverny, Marie-Thérèse, 'Translations and Translators', in *Renaissance and Renewal in the Twelfth Century*, ed. by Robert L. Benson, Giles Constable, and Carol D. Lanham (Oxford: Clarendon Press, 1982), pp. 421–62

Askari, Nasrin, *The Medieval Reception of the Shāhnāma as a Mirror for Princes* (Leiden: Brill, 2016)

Al-Azmeh, Aziz, *Muslim Kingship: Power and the Sacred in Muslim, Christian and Pagan Politics* (London: I. B. Tauris, 2001)

Burnett, Charles S. F., 'Arabic, Greek and Latin Works on Astrological Magic Attributed to Aristotle', in *Pseudo-Aristotle in the Middle Ages. The Theology and other Texts*, ed. by Jill Kraye, William E. Ryan, and Charles B. Schmitt (London: Warburg Institute, 1986), pp. 84–96

—, 'Humanism and Orientalism in the Translations from Arabic into Latin in the Middle Ages', in *Wissen über Grenzen. Arabisches Wissen und lateinisches Mittelalter*, ed. by Andreas Speer and Lydia Wegener (Berlin: Walter de Gruyter, 2006), pp. 22–31

Campopiano, Michele, 'Parcours de la légende d'Alexandre en Italie. Réflexions sur la réception italienne de l'Historia de Preliis, recensio J2 (XIIe-XVe siècles)', in *L'historiographie médiévale d'Alexandre le Grand*, ed. by Catherine Gaullier-Bougassas (Turnhout: Brepols, 2011), pp. 65–83

—, 'A Philosopher Between East and West: Aristotle and the "Secret of Secrets"', *Lampas*, 46.3 (2013), 282–89

Crisciani, Chiara, 'Ruggero Bacone e l' "Aristotele" del "Secretum secretorum"', in *Christian Readings of Aristotle from the Middle Ages to the Renaissance*, ed. by Luca Bianchi (Turnhout: Brepols, 2011), pp. 37–64

Dales, Richard C., 'Grosseteste's Views on Astrology', *Medieval Studies*, 29 (1967), 357–63

D'Ancona, Cristina, *Recherches sur le* Liber de Causis (Paris: VRIN, 1995)
Easton, Stewart C., *Roger Bacon and his Search for a Universal Science: A Reconsideration of the Life and Work of Roger Bacon in the Light of his own Stated Purposes* (New York: Russel and Russel, 1952)
Forster, Regula, *Das Geheimnis der Geheimnisse: Die arabischen und deutschen Fassungen des pseudo-aristotelischen Sirr al-asrar / Secretum Secretorum* (Wiesbaden: Reichert, 2006)
Gautier Dalché, Patrick, *La Géographie de Ptolémée en Occident (IVe-XVIe siècle)*, Terrarum Orbis, 9 (Turnhout: Brepols: 2009)
—, 'Vers une Perfecta Locorum Doctrina: Lieu et espace géographique selon Roger Bacon', in *Représentations et conceptions de l'espace dans la culture médiévale: Colloque Fribourgeois 2009*, ed. by Tiziana Suárez Nani and Martin Rohde (Berlin: De Gruyter, 2011), pp. 9–44
Grignaschi, Mario, 'La diffusion du *Secretum secretorum* (Sirr al- 'asrar) dans l'Europe Occidentale', *Archive d'histoire doctrinae et littéraire du Moyen Age*, 47 (1980), 7–70
—, 'Remarques sur la formation et l'interprétation du "Sirr al-'asrâr"', in *Pseudo-Aristotle, The Secret of Secrets. Sources and Influences*, ed. by William F. Ryan and Charles B. Schmitt, The Warburg Institute Surveys and Texts, 9 (London: Warburg Institute, 1982), pp. 3–33
Gutas, Dimitri, *Greek Wisdom Literature in Arabic Translation. A Study of the Graeco-Arabic Gnomologia* (New Haven: American Oriental Society, 1975)
Hackett, Jeremiah, 'Roger Bacon and Aristotelianism. Introduction', *Vivarium*, 35.2 (1997), 129–35
—, 'Roger Bacon and the Moralization of Science: from *Perspectiva* through *Scientia Experimentalis* to *Moralis Philosophia*', in *I francescani e le scienze: atti del XXXIX Convegno internazionale: Assisi, 6–8 ottobre 2011* (Spoleto: CISAM, 2012), pp. 371–92
—, 'Roger Bacon on Astronomy-Astrology the Sources of the *Scientia experimentalis*', in *Roger Bacon and the Sciences: Commemorative Essays*, ed. by Jeremiah K Hackett (Leiden: Brill, 1997), pp. 176–98
—, 'Roger Bacon on *Scientia experimentalis*', in *Roger Bacon and the Sciences: Commemorative Essays*, ed. by Jeremiah K Hackett (Leiden: Brill, 1997), pp. 277–315
Haskins, Charles Homer, *The Renaissance of the Twelfth Century* (Cambridge, MA: Harvard University Press, 1927)
Lemay, Richard, 'Roger Bacon's Attitude toward the Latin Translations and Translators of the Twelfth and Thirteenth Centuries', in *Roger Bacon and the Sciences: Commemorative Essays*, ed. by Jeremiah K Hackett (Leiden: Brill, 1997), pp. 25–47
Lafferty, Maura K., *Walter of Châtillon's* Alexandreis. *Epic and the Problem of Historical Understanding* (Turnhout: Brepols, 1998)
Lindberg, David C., 'On the Applicability of Mathematics to Nature: Roger Bacon and His Predecessors', *The British Journal for the History of Science*, 15 (1982), 3–25
—, 'Roger Bacon on Light, Vision, and the Universal Emanation of Force', in *Roger Bacon and the Sciences: Commemorative Essays*, ed. by Jeremiah K Hackett (Leiden: Brill, 1997), pp. 243–75
Macuch, Maria, 'Pahlavi Literature', in *The Literature of Pre-Islamic Iran. A History of Persian Literature, I*, ed. by Ronald E. Emmerick and Maria Macuch (London: I. B. Tauris, 2009), pp. 116–96
Molland, George, 'Roger Bacon's *De Laudibus Mathematicae*: A Preliminary Study', in *Texts and Contexts in Ancient and Medieval Science Studies on the Occasion of John E. Murdoch's*

Seventieth Birthday, ed. by Edith Sylla and Michael McVaugh (Leiden: Brill, 1997), pp. 68–83

—, 'Roger Bacon's Knowledge of Mathematics', in *Roger Bacon and the Sciences: Commemorative Essays*, ed. by Jeremiah K Hackett (Leiden: Brill, 1997), pp. 151–74

Newman, William R. 'An Overview of Roger Bacon's Alchemy', in *Roger Bacon and the Sciences: Commemorative Essays*, ed. by Jeremiah K. Hackett (Leiden: Brill, 1997), pp. 318–36

Panti, Cecilia 'Scienza e teologia agli esordi della scuola dei Minori di Oxford: Roberto Grossatesta, Adamo Marsh e Adamo di Exeter', in *I Francescani e le scienze. Atti del XXXIX Convegno internazionale. Assisi, 6–8 ottobre 2011* (Spoleto: CISAM, 2012), pp. 311–51

Pereira, Michela, *Arcana sapienza. Storia dell'alchimia occidentale dalle origini a Jung*, 2nd edn (Carocci: Rome, 2019)

Pingree, David, 'Astronomy and Astrology in India and Iran', *Isis*, 54.2 (1963), 229–46

—, 'The Diffusion of Arabic Magical Texts in Western Europe', in: *La diffusione delle scienze islamiche nel medio evo europeo (Roma 2–4 ottobre 1984)*, ed. by Biancamaria Scarcia Amoretti (Roma: Accademia dei Lincei, 1987), pp. 57–102

Power, Amanda, *Roger Bacon and the Defence of Christendom* (Cambridge: Cambridge University Press, 2013)

Southern, Richard, *Robert Grosseteste: The Growth of an English Mind in Medieval Europe* (Oxford: Oxford University Press, 1986)

Tilliette, Jean-Yves, 'La traduction et l'adaptation aux sources de la création sur Alexandre dans les littératures européennes: l'Alexandreis de Gautier de Châtillon', in *La fascination pour Alexandre le Grand dans les littératures européennes (Xe – XVIe siècle). Réinventions d'un mythe*, ed. by Catherine Gaullier-Bougassas (Brepols: Turnhout, 2014), pp. 179–97

Trifogli, Cecilia, 'Roger Bacon and Aristotle's Doctrine of Place', *Vivarium*, 35.2 (1997), 155–76

Ulmann, Manfred, *Die Natur – und Geheimwissenschaften im Islam* (Leiden: Brill, 1972)

Vegas Gonzalez, Serafin, 'Liber de Causis', *Revista española de filosofía medieval*, 7 (2000), 115–25

Weisser, Ursula, *Buch über das Geheminis der Schöpfung und die Darstellung der Natur* (Aleppo: Institute for the History of Arabic Science, 1979)

—, *Das Buch über das Geheminis der Schöpfung von Pseudo-Apollonius von Tyana* (Berlin: De Gruyter, 1980)

Williams, Steven J, 'Roger Bacon and his Edition of the Pseudo-Aristotelian *Secretum secretorum*', *Speculum*, 69 (1994), 57–73

—, 'Roger Bacon and the Secret of Secrets', in *Roger Bacon and the Sciences: Commemorative Essays*, ed. by Jeremiah K Hackett (Leiden: Brill, 1997), pp. 365–94

—., *The Secret of Secrets: The Scholarly Career of a Pseudo-Aristotelian Text in the Latin Middle Ages* (Ann Arbor: University of Michigan Press, 2003)

Wiesehofer, Josef, *Ancient Persia from 550 BC to AD 650* (London: I. B. Tauris, 2001)

Woodward, David, Howe, Herbert, 'Roger Bacon on Geography and Cartography', in *Roger Bacon and the Sciences: Commemorative Essays*, ed. by Jeremiah K. Hackett (Leiden: Brill, 1997), pp. 199–22

Zamuner, Ilaria, 'La tradizione romanza del "Secretum secretorum" pseudo-aristotelico. Regesto delle versioni e dei manoscritti', *Studi Medieval*, 46.1 (2005), 31–116

STEN EBBESEN

Wisdom's Trips to Denmark

In the twelfth and thirteenth centuries, Paris was *the* place for Danes to acquire higher education. Some returned home, but what did they do with their acquired wisdom once they were back in Denmark? I shall discuss the cases of Archbishop Andrew Sunesen of Lund († 1228) and Bishop Gunner of Viborg († 1251), adding some remarks about evidence of high-level teaching in a Franciscan convent in the fourteenth century. Finally, I shall try to draw some general conclusions about academic learning in late-medieval Denmark.[1]

Wisdom does not literally travel, but people and books who carry it do. It was a lot of travelling that made it possible for twelfth- and thirteenth-century Paris to develop into the Latin world's undisputed centre for philosophical and theological studies with Aristotle's writings as the basic textbooks for most subjects in the faculty of arts and the Bible and selected bits of Augustine and other church fathers in the faculty of theology.

For the development of the arts faculty it was of primary importance that some Westerners travelled to the Greek world, Constantinople in particular, and brought back both a knowledge of Greek and manuscripts of Aristotle and Aristotelian commentaries, which they would subsequently translate into Latin so that they could be used in teaching. At the beginning of the twelfth century, the Latin Aristotelian library contained only a part of the *Organon* and some companion books, all of them about logic; by 1275 virtually the whole *Corpus Aristotelicum* and several Greek commentaries had become available, as had a number of important Arabic companions to Aristotle.[2]

---

[1] The content of this chapter is to a considerable extent a duplication of information that I have published in Danish in Ebbesen, *Dansk middelalderfilosofi ca. 1170–1536*. There is also some overlap with Ebbesen, 'The Danes, Science, Scholarship, and Books in the Middle Ages', which is in English.

[2] The story of the translation movement has often been told. For a survey in tabular form, see Dod, 'Aristoteles Latinus', pp. 74–79 and the slightly revised version of his table in *The Cambridge History of Medieval Philosophy*, vol. 2, ed. by Pasnau, pp. 793–97.

Sten Ebbesen • is Professor emeritus of the Saxo Institute at the University of Copenhagen, and Fellow of The Royal Danish Academy of Sciences and Letters as well as Academia Europaea. se@hum.ku.dk

Medieval Science in the North: Travelling Wisdom, 1000–1500, ed. by Christian Etheridge and Michele Campopiano, KSS 2 (Turnhout: Brepols, 2021), pp. 97-109

For all faculties, student travel was a *sine qua non*. And students were very mobile. By 1150 Paris city was teeming with student life, youngsters from all over Western Christendom gathering there to learn the 'arts', i.e. linguistics (called *grammatica*), logic, and various other philosophical disciplines. A fair portion of those who graduated as masters would then teach the arts for a few years. A minority would proceed to immerse themselves in theology. When returning to their native countries, those students would carry with them hitherto-unimagined wisdom from Paris to Bruges, Magdeburg, Kraków, Aarhus, or Lund. Some of it they would carry in books, but most of it would be in their heads. In the following I shall present some of the learned men who brought foreign wisdom to Denmark.

Absalon

The first Dane known to have studied in Paris and to have returned to Denmark is the Absalon who was in the city of cities for some years round 1150 and later became bishop of Roskilde (1158–1192) and archbishop of Lund (1178–1201).[3] We do not know with whom he studied, though it is a fair guess that at least he got a serious introduction to the theory of grammar and logic. It is sometimes taken for granted that he also studied theology. This is quite possible, but the claim rests on the anachronistic assumption that people who became high prelates had studied divinity. In the twelfth century, some had, but many had not. Absalon is likely to have spent some of his time on extra-curricular reading, as he is known to have owned copies of Valerius Maximus's *Facta et dicta mirabilia* and Justin's *Breviarium Historiarum Philippicarum Pompei Trogi*.[4]

---

3 We owe this information to the Life of St William of Æbelholt, which in passing mentions that the reason Absalon invited William to come to Denmark in 1161 was that they had struck up a friendship during Absalon's studies in Paris. Anonymous, *Sancti Willelmi abbatis vita et miracula*, 10.5: Reminiscitur tandem [sc. Absalon] familiaritatis et amicicie, quam cum Willelmo, uiro religioso, olim pepigerat, cum Parisius studendi gracia moraretur. Munk Olsen, 'Absalons studier i Paris', p. 57 estimates that Absalon (born about 1128) would have arrived in Paris at the age of seventeen or eighteen in about 1145 and left about a decade later, as some ten years of studies would appear to have been standard at the time. This is all very uncertain, however. The best supported date is that of 1145, because the normal enrolment age for advanced studies in the Middle Ages seems to have been early adolescence, fourteen-year-olds being the youngest. There is no way to know whether Absalon belonged to those who took an advanced decade-long education or to those who returned home after a few years.

4 In his will Absalon orders Saxo [Grammaticus] to return two books he had borrowed to the monastery of Sorø. See *Diplomatarium Danicum*, 1.4, ed. by Skyum-Nielsen, No. 32, p. 62: 'Saxo debet duos libros, quos archiepiscopus ei concesserat, ad monasterium de Sora referre'. The two MSS can hardly have been any but the actual *GKS 450, 2°* in the Royal Library, Copenhagen, which contains Justin, and a Valerius Maximus MS that perished in a fire in 1728. On Saxo's use of the two MSS, see Friis-Jensen 'Saxo's Books': I. lxviii–lxxiii.

Andrew Sunesen

A generation after Absalon's return to Denmark, his cousin Sune Ebbesen sent two of his sons to Paris.[5] One of them (Peter Sunesen, d. 1214) apparently did not excel in learning, but at least acquired sufficient Latin for him to fill the jobs of a bishop and a royal chancellor in Denmark. The other, Andrew Sunesen (d. 1228), absorbed not just the arts but also theology, and functioned as a *magister regens in theologia* (a teacher of theology) for a while. Sometime in the 1190s he became royal chancellor, probably in order that he should head the Danish team in the difficult negotiations with the pope and the French king Philip Augustus, who, after the wedding night had refused to have anything to do with his wife Ingeborg, a sister of the king of Denmark. Philip Augustus even tried to have the marriage annulled.

Back home in Denmark Andrew inherited the archbishopric of Lund from uncle Absalon (d. 1201). In his homebound luggage Andrew had two hexameter poems of his own production, one called *Hexa(e)meron* and another *On the Seven Sacraments of the Church*. Together they constituted a poetical *Summa Theologiae* covering the whole of contemporary theology. Only the *Hexaemeron* has been preserved.[6] It contains no less than 8040 verses. Investigations by Lars Boje Mortensen and myself have established that works of Stephen Langton (d. 1228), of *Magna Carta* fame, were the foundation of Andrew's theology. He must have been Langton's pupil in Paris in the 1180s.

Andrew's theological wisdom hardly had any great impact in Denmark. Very few copies of his works were produced, all of them, apparently, donated by himself or his family to cathedrals and monasteries, where they cannot have gained any noticeable readership. *Hexaemeron* is only comprehensible to someone steeped in Langton's theology, and in his time, Andrew was probably the only person in Denmark with that sort of training. A generation or two later, theological discourse had changed so much that it would be hard to find anyone with the requisite knowledge even in Paris. But at least the donation of copies is evidence of a wish to spread Andrew's wisdom.

Andrew also produced a Latin version of the *Law of Scania*, i.e. the law of the province and diocese that was under his immediate episcopal jurisdiction.[7] We have vernacular versions of the law that must be pretty close to what Andrew knew. Therefore, it is possible to compare the model and the translation or paraphrase, as it is often called.

First and foremost, Andrew's version stands out by being in Latin, which raises a question to which I have only a very tentative answer: who were the intended readers? One possible group might be foreign clerics who had settled in Denmark,

---

5 For the sources of the following section about the Sunesens, see Ebbesen and Mortensen 'Introduction' and Ebbesen and Mortensen 'A Partial Edition of Stephen Langton's Summa and Quaestiones'. The basic information about their lives, and about the fate of Andrew's *Hexaemeron*, is found in Ebbesen and Mortensen 'Introduction'.
6 *Andreae Sunonis Filii Hexaemeron*, ed. by Ebbesen and Mortensen.
7 For the 1933 edition, see *Skånske Lov*, ed. by Aakjær and Kroman. It is reprinted with an English translation in *The Latin version of the Law of Scania*, ed. by Tamm and Vogt.

as had a group of four French Augustinian canons whom Absalon had imported to Zealand in the 1160s,[8] and as did representatives of the new Dominican order who arrived in Lund about 1222, just about the time Andrew retired — if, that is, the first mendicants in Lund were not Danes who had entered the new order while abroad. Andrew may also have intended all of his literate clergy to read the Latin law and try to make people believe that these were the ancestral rules, whereas in fact Andrew had, under the impression of canon law, inserted quite a few rules without a basis in pre-existing vernacular law.

Whoever the intended audience, Andrew expected his readers to have well-developed Latin, for he did his best to raise the law to a stylistic level far above the down-to-earth vernacular text.[9] The law is in a moderately classical Latin, with a preference for the classical accusative with infinitive over the medieval *quod* clause as objects of verbs. Its vocabulary is basically classic, but it is bigger than necessary because of a constant resort to *variatio sermonis*. Even important legal notions are referred to in different words at different places. Some words are quite rare in prose but have a background in Roman law or ancient poetry. Finally, the law contains several pretty long periods and exploits several rhetorical figures, such as *hyperbaton, polyptoton, isocolon* and *antithesis*. Andrew waxes particularly eloquent in two additions of his own to the law. One is a short preamble to the section about manslaughter.

> Instigante humani generis inimico, quia proni semper fuerunt homines in nostris partibus ad homicidium perpetrandum, pacem angelicam deserentes et sedicionem dyabolicam amplexantes, diuersis temporibus diuersa sunt iura prodita super tanti reatus per multam pecuniariam castigacionem, quatenus et tantus excessus aliquatenus refrenari et amissionis dampnum quoquo modo posset satisfaccionis pecuniarie tristi solacio compensari.
>
> (Because, at the instigation of The Enemy of the Human Race, people in our part of the world have always been prone to leave angelic peace in favour of diabolic strife and commit homicide, at various times various laws have been issued about punishing this enormous offence with a money fine, in order that such enormous excess might be somewhat restrained and that the damage of loss be in some way compensated by the sad comfort of a pecuniary indemnification.)[10]

Another addition to the law is occasioned by a paragraph prescribing that if a wife accused of adultery fails to pass an ordeal by fire, she is to be thrown out of her husband's house and bed:

> Verum in parte constat huic humane legi, velud famule obsequenti, velud pedisseque sequenti domine sue vestigia, per diuine legis preminenciam derogari, que

---

8 The story about Absalon's import of the four French monks is told in Anonymous, *Sancti Willelmi abbatis vita et miracula* §§ 10–11.
9 For a more thorough analysis of the style of the law, see Ebbesen, 'Andrew Sunesen's Language'.
10 *Skånske Lov*, ed. by Aakjær and Kroman, § 43.

matrimonia iubet non fori, sed poli, non curie secularis, sed ecclesie spiritualis examini atque regimini subiacere, nec permittit eciam separacionem thori per igniti ferri iudicium celebrari.

> (It is, however, a fact that this human law, being in the position of an obedient maidservant or handmaid who follows in the footsteps of her mistress, is partly overruled by the preeminence of divine law, which bids that marriage be subject not to examination and rule by the Thing, but by Heaven, not by a secular court, but by the spiritual church, and which also forbids a separation of bed to be determined by ordeal by fire.)[11]

Again, whoever Andrew's intended audience was, he clearly wanted what he had learned about canon law to become accepted law in Scania.

A Letter to the Pope

Another case of imported wisdom being used for a practical purpose is a petition sent in 1226 from the Danish government to the pope, asking him to absolve King Valdemar II from an oath sworn under duress while in captivity. The central part of the letter is preserved in the pope's answer, which reproduces it verbatim, only changing 'I' into 'You'. The relevant part of Honorius III's answer runs:

> W. illustri regi Datie. Petitio tua nobis exhibita continebat quod, cum Henricus comes de Zverinc vasallus tuus te cepisset et teneret carceri mancipatum, a te iuramentum extorsit quod ei solveres quandam non modicam pecunie quantitatem ac interim filios tuos obsides ei dares. Quare nobis humiliter supplicasti ut, cum adimpletio promissionis, qua diu ante captionem tuam te sollempniter obligaras ad subsidium terre sancte magnifice impendendum, per hoc impediatur omnino, et dicte pecunie quantitas tuas facultates excedat, ac per hoc compelli non debeas ad solutionem eiusdem, cum nemo ad impossibile sit cogendus, idemque iuramentum a te fuerit per metum et vim manifestam extortum, ac per hoc ad observationem non tenearis eiusdem, cum ea que vi metusve causa fiunt non debeant esse rata, dictusque comes te capiendo vinculum fidelitatis abruperit, propter quod fidem servare non teneris eidem, cum fidem non servanti fides non debeat observari, te a predicto iuramento ad cautelam absolvere dignaremur.

> (To W., the illustrious king of Denmark.
>
> We have received a petition from you, the content of which was that when count Henry of Schwerin, your vassal, had captured you and kept you imprisoned, he extorted an oath from you that you would pay him a considerable sum of money, and in the meantime, give him your sons as hostages. Therefore, you humbly beseeched us that as

---

11   *Skånske Lov*, ed. by Aakjær and Kroman, § 127.

1. the implementation of a promise by which before your capture you had solemnly obliged yourself to provide a sizeable sum in support of the Holy Land is made completely impossible by this;
2. and (*b*) the amount of money involved exceeds your resources, and (*c*) you therefore should not be compelled to pay it, as (*a*) nobody should be forced to do what is impossible;
3. and (*b*) the said oath was extorted from you by threats and manifest violence, and (*c*) you therefore are not bound to observe it, as (*a*) what happens as a result of violence and threats and should not be considered valid;
4. and (*b*) the aforementioned count by capturing you has broken his bond of faithfulness, wherefore (*c*) you are not obliged to keep you faith towards him, as (*a*) faith should not be kept towards one who does not keep his faith,

we would, as a safeguard, deign to absolve you from the aforementioned oath.)[12]

The letter is a little masterpiece, in impeccable Latin with a juridically impeccable argumentation presented in three equally impeccable syllogisms (2–4), rhetorically disguised by following the order *minor (b), conclusion (c), major (a)* instead of *major, minor, conclusion*, which would reek too much of the classroom.[13]

As the matter was of the highest political importance, Valdemar and his councillors will have looked for the most learned man they could find. The royal chancellor at the time was bishop Nicholas of Schleswig, and if he wrote the petition, he was a very well-trained man, almost certainly someone who had studied abroad. I have a hunch, though, that a messenger had been sent to Sunesen, who at the time lived as a retiree in a small, but stylish, palace that he had built on Ifö, an easily defensible island in a lake in Scania. Very little is left of Sunesen's last home, but enough to show that it was a brick building of high quality. The letter, by the way, achieved its purpose. Honorius III granted the king's petition.

Bishop Gunner and the Young Masters

In 1251 one Gunner, bishop of Viborg in Jutland, died. A biography composed some fifteen years later relates that he had been a *collega et socius* in Paris of a cardinal legate who visited Denmark in 1222 and made Gunner a bishop because he knew he

---

12 *Diplomatarium Danicum*, 1.6, ed. by Skyum-Nielsen, No. 59, pp. 78–79: Honorius's letter is dated 26 June 1226. Some of the formulations of the Danish petition are also recognizable in a slightly earlier letter (dated 9 June 1226) from Honorius to count Henry of Schwerin; see *Diplomatarium Danicum*, 1.6, ed. by Skyum-Nielsen, No. 56, pp. 74–76.
13 I owe my awareness of the peculiarity of this letter to my late teacher and friend Jan Pinborg (1937–1982), who, to the best of my knowledge, was the first scholar to notice the syllogistic structure of the three arguments. For my conjectural connection of the letter to Andrew Sunesen he carries no responsibility.

was conversant with the liberal arts and was generally competent.[14] This obviously means that the two had studied or taught together, and what little is known about their careers indicate that this happened in the 1180s or 1190s. In another passage, the biographer relates the following:

> In arcium liberalium et sacrarum sciencia scripturarum tam perspicaci calluit intellectu, ut, cum clerici satis subtiles et acuti de Parisiensi studio redissent, ad unam sepius questionem eos cornuto concluderet sillogismo. Nam dicebat eis ioco suo, sicut solitus erat, gracioso: 'Vos in etate iuvenili, qui multa cum lectione sudastis in studio facultatum et egressi estis quasi noviter de camino, docete nos senes, qui vix ab olim tenuem istarum rerum noticiam retinemus. Respondete michi: Quid vobis videtur? Duo et tria sunt quinque, et non plura?' Ad talem et consimilem questionem ipsi, perpendentes perplexos nodos proposti sillogismi et quid in sophistica arte consequi potuisset, licet vulgo vocarentur magistri, in conspectu tamen eius parum potuerunt respondere. Sepius, cum (cum Ebbesen: tamen [quando] Gertz) familiaris esset cum clericis suis, pro ioco suo eis proposita questione ad disputandum eos provocauit; eis plurimum super hiis inter se altercantibus, postremo ipse subridendo dixit: 'Taliter et taliter deberetis respondisse; talis est istius questionis solucio'.

> (He [Gunner] had such a penetrating understanding of the liberal arts and theology that when some quite subtle and smart clerics had returned from the university of Paris, he caught them in a dilemma with a single question, saying in his usual elegantly jocular way. 'You youngsters, who have been sweating over much reading in the study of the disciplines and have now recently come out of the oven, as it were, please teach us old-timers, who barely remember anything about those matters from long ago. Please tell me: What do you think: Are two and three five and no more?' They wrecked their brains over the tangled knots of the syllogism with which they were presented and what could be done in the art of sophistics, and although they bore the title of master, they could not in front of him give a decent answer to this and similar questions. He also repeatedly in relaxed conversation with his clerics for fun put forward a question and called upon them to debate it. When, then, they had had a longish discussion of the matter, he would bring it to an end, saying with a smile 'You should have answered thus and thus, such is the solution of that question'.)[15]

The biographer's Latin is not bad, but he clearly had not had a top education like Sunesen, and it is not quite clear whether he himself fully understood the anecdote he was telling, but the text contains enough technical terminology to reveal that what Gunner actually did was take up the role of a master of arts, treating the youngsters

---

14 Anonymous, *Vita Gunneri*, Ch. 1, 266.4–7: 'quia ipse cardinalis et abbas memoratus college fuerunt aliquando Parisius et socii, eum per omnia novit, quod de septem liberalibus aliquam partem haberet et noticiam atque competentem ad hec prudenciam et discrecionem'.
15 Anonymous, *Vita Gunneri*, Ch. 4, 268.17–269.4.

like students in a class on *ars sophistica*, which was a special sort of logical exercise in the analysis of complex sentences called *sophismata*. One standard sophisma, known to the scholastics from Aristotle's *Sophistical Refutations* Ch. 4, was 'Two and three are five'. A discussion of it would have a format like this:

> Duo et tria sunt quinque. § Probatur sic: Duo et tria sunt aliquid, et non sunt plura quam quinque nec pauciora, igitur duo et tria sunt quinque. § Improbatur sic: Sequitur 'duo et tria sunt quinque, igitur tria sunt quinque'; consequens est falsum, ergo et antecedens. § Solutio: Prima est multiplex secundum compositionem et divisionem. In sensu compositionis fit copulatio inter terminos, et sic est vera, et est categoria. In sensu divisionis, sic est falsa, et est copulativa, et utraque pars falsa.
>
>> (Two and three are five.
>> 1. This [proposition] is proved as follows: Two and three are something, and neither more nor fewer than five; therefore, two and three are five.
>> 2. Is disproved as follows: 'Two and three are five' entails 'Three are five'; but the consequent is false, therefore the antecedent is false.
>> 3. Solution: The original proposition is ambiguous by composition and division. In the composed sense there is a joining of terms, and so the proposition is true and categorical. In the divided sense it is false, as it is then a conjunctive proposition both parts of which is false.)[16]

The solution amounts to saying that 'Two and three are five' can be taken in two senses, (a) $[2+3] = 5$, and (b) $[2 = 5]$ & $[3 = 5]$, in the first of which the proposition is true, while it is false in the second sense. The logical lesson to be drawn from the example is that the operator 'and' can in some cases have an ambiguous scope, either joining two terms as in case (a) or two propositions as in sense (b).

It is known from later thirteenth-century texts that often the teacher would provide the sophismatic proposition, the proof and the disproof, while a graduate student (a bachelor) was assigned the job of solving the riddle.

By failing to solve satisfactorily a standard *sophisma* like 'Two and three are five and no more' the would-be masters visiting Bishop Gunner had shown themselves to be less skilled in logic than a decent graduate student.

The second part of the anecdote, according to which Gunner used to lay a question before his clerics to debate, and after their debate would show them how they ought to have solved it, shows him in the role of a presiding master at a disputation. In its simplest variety, a scholastic disputation has format much like that of the sophisma above, involving (0) a question that can be answered with a Yes or a No, (1), arguments for the No answer, (2) arguments for the Yes answer, (3) a reasoned solution (*solutio* or *determinatio*) of the problem accompanied by (4) refutations of all arguments from steps (1) or (2) that disagree with the solution.

---

16 Walter Burleigh, *De Puritate Artis Logicae, Tractatus Brevior*, ed. by Boehner, p. 242 (my punctuation). This text is from the early fourteenth century, but the sophisma would have been treated in the same way a century earlier. For more detailed information about the formats of sophismatic disputations, see Ebbesen 'How to Build your own Sophism'.

This, the simplest form of the scholastic *quaestio*, was a staple of thirteenth-century courses on Aristotle's logic and other authoritative textbooks. The teacher would himself be responsible for at least steps 0, 3 and 4, often also for steps 1–2. But the anecdote about Gunner and his clerics points to a more developed variant of the *quaestio*, one known to have been used in thirteenth-century Paris in connection with the discussion of sophismata. After the bachelor's solution of the *sophisma*, one or more questions somehow related to the sophisma would be chosen for discussion. At first the disputation would proceed according to the regular format, with the bachelor taking the role of the teacher in steps 3–4. But then members of the audience would attack the bachelor's determination of the question. After each attack he would try to defend himself, and a rather lively debate between the opponent and the bachelor respondent could take place, at the end of which a new opponent could enter the fray. Finally, when everybody was tired of fighting, the presiding master would offer *his* determination of the question and refute such arguments from the debate as did not agree with his preferred solution. Gunner seems to have attempted to teach his clerics in Viborg at least the fundamentals of university disputation techniques.

Teaching in a Franciscan Convent

My final example of learning having reached Denmark is an anonymous text from the fourteenth century called *In Defence of Ockham*.[17] Actually, the text does not mention William of Ockham, who was a very controversial person, but the title, found in the one and only extant manuscript, is not off the point, for the work must reflect a course in Ockhamist logic-cum-ontology given by a *lector* in a Franciscan convent somewhere in Denmark.

That Ockham's *Defender* taught in a Franciscan convent is pretty obvious from the contents of his work. It is very difficult to imagine any other setting for a fourteenth-century work like this, defending Ockham's semantics and ontology against John Duns Scotus's rival theories. The Danish setting is given by a couple of examples which, by tradition, require the name of the place where one is lecturing and the name of some far-away place. Someone teaching in London might use London or England for 'here' and Rome for 'far away'; in fact, Ockham himself uses both London–Rome and England–Rome. His defender uses Denmark–Rome, and that ought to settle the question of where he was giving his course.[18]

Defensorium Ockham is a fairly long text, some ninety pages in the edition, and while it is not of the highest philosophical sophistication, it is still a relatively complicated

---

17 Edition in Andrews, 'The Defensorium Ockham: An Edition'.
18 Andrews, 'The Defensorium Ockham: An Edition', p. 190 thought it more probable that the work was written by a Dane in Rome, but, as I have shown in Ebbesen, 'A Note on Ockham's Defender', that is due to a misunderstanding of the role of 'Denmark' and 'Rome' in the two examples in which it occurs in the text.

text composed by a competent teacher, who must have attended either a university or some other high-level school. Moreover, its level of difficulty presupposes an audience with quite a bit of previous training, which, presumably, they would have acquired in their convent.

It is well known that mendicant houses could sometimes offer philosophical and theological training of a high standard. Due to the Lutheran reformation of the Danish church from 1536 onwards, the archives and libraries that might have thrown light on such teaching in Denmark have disappeared, but *Defensorium Ockham* offers us a glimpse of the advanced level that could sometimes be reached.

Ockham's *Defender* may have been a Dane with a higher education, or he may have been a friar of foreign origin sent to Denmark by the order to cover a need for a competent *lector*. Anyway, he took part in a philosophical debate that originated in England and was spread all over Europe by travelling friars and manuscripts.

Conclusion

There is ample evidence of the presence of Danish students and masters in Paris from the twelfth century onwards, and later also at other centres of learning, German universities being the Danes' favoured haunts toward the end of the Middle Ages.[19] Presumably, the majority of those who had gone abroad for studies returned home, but only rarely do we have information to confirm this assumption on an individual basis. A fair number of *magistri* occur in Danish documents, but usually without indication of where they had obtained the title, and, at least before the fifteenth century, it is rare that a master known from a Danish document can be matched with one known from foreign university records.

We have some learned works produced by Danish masters during their years abroad, but only in three instances do we know for certain that the author returned to Denmark. One is Andrew Sunesen. Another is Martin of Dacia, who after a Paris career in arts and theology became royal Danish chancellor about 1287 but returned to Paris about 1300 and died there in 1304. The last is Thuo of Viborg (*c.* 1405/10–1472), M.A. and D.Theol. from Erfurt, who in 1443 was elected archbishop of Lund. In all three cases we can say with great certainty that their departure from their respective universities also meant the end of their career as writers of learned books, though it is just possible that Martin resumed his learned activities when late in life he returned to Paris.[20] With the astronomer Peter Philomena de Dacia, we are on less certain ground. He seems to have been in Bologna in the early 1290s, in Paris in the late 90s,

---

19 For some information about Danish students abroad towards the end of the Middle Ages, see Pinborg 'Danish Students 1450–1535' and Mornet 'Le voyage d'études des jeunes nobles danois'.
20 Martin's academic works have been edited in *Opera*, ed. by Roos. The evidence for his career is presented in the introduction, pp. xxxi–xxxv. Thuo's academic works from his Erfurt period have been edited in *Opera*, ed. by Tabarroni and Ebbesen; what is known about his life prior to his election to the archbishopric is reported in the introduction, pp. vii–xiii.

and, probably, in Roskilde, Denmark, in 1303. None of his writings *must* have been composed in Denmark.[21]

I suppose that normally a return to the native soil meant an end to the production of learned books, although some returnees may have been able to continue to produce, if not cutting-edge research, at least up-to-date manuals for use in cathedral schools and in the houses of learned religious orders (Ockham's *Defender* being a possible representative of the latter class).

New opportunities for continuing an academic production in Denmark arose with the foundation of the University of Copenhagen in 1479.[22] The first professors — probably fewer than ten in all, we do not know the exact number — were recruited from the University of Cologne. Leading among them was a certain Petrus Alberti, who was probably a Dane. One of the others is known to have been a Dane, but the company also included a Dutchman, two Germans, and a Scot. Travelling wisdom, indeed!

The new university was certainly not a first-class one, but it is a reasonable assumption that it was no worse than the Swedish University of Uppsala, established in 1477, whose philosophy courses in the very first years are well documented.[23] The documents from Uppsala reveal an entirely respectable level of philosophical instruction, comparable to that of contemporary German universities (and way above the level in sixteenth and seventeenth-century post-reformation universities in Lutheran lands).

The ravage caused by the 1536 Lutheran reformation in Denmark, combined with the loss of almost the entire holdings of the University Library of Copenhagen in a 1728 fire, has left us with precious little source material to evaluate the transfer of knowledge from the great intellectual centres of the Middle Ages to the kingdom of Denmark. Yet, the scarce glimpses that we get of such transfer suggest that it was taking place all the time, every year seeing the return of a number of Danes with academic experience from foreign universities, while learned mendicants would be in continuous contact with their confrères abroad both through correspondence and through the orders' policy of moving personnel around.

While medieval Denmark possessed no intellectual centre of international importance, it could probably at any time boast a decent number of inhabitants with some high-level training in the theoretical disciplines, and some, though fewer, who were abreast of the latest developments in philosophy, theology and legal theory. Old-timers like Bishop Gunner might occasionally get the better of young cockerels who had recently returned from abroad and thought they were mightily learned. But old-timers who were still intellectually flexible would also have the opportunity to learn from returning young men about the latest advances in *scientia*.

---

21 For what little is known about Peter Philomena's life see the introduction to the edition in *Petri Philomenae de Dacia*, ed. by Saaby Pedersen, I, 35–39.
22 On the foundation of the University of Copenhagen and its first teachers, see *Universitas studii Haffnensis*, ed. by Pinborg and trans. by McGuire (in Danish and English) and Ebbesen, *Dansk middelalderfilosofi ca. 1170–1536*, pp. 235–41 (in Danish).
23 See *Studium Upsalense*, ed. by Piltz.

Works Cited

Primary Sources

Anonymous, *Sancti Willelmi abbatis vita et miracula*, in *Vitae Sanctorum Danorum*, ed. by M. Cl. Gertz (Copenhagen: Selskabet for Udgivelse af Kilder til Dansk Historie, 1908–1912)

Anonymous, *Vita Gunneri Episcopi Viburgensis*, in *Scriptores minores historiæ Danicæ medii aevi* 2, ed. by M. Cl. Gertz (Copenhagen: Selskabet for Udgivelse af Kilder til dansk Historie: Copenhagen, 1922), pp. 265–78

Andreae Sunonis Filii Hexaemeron, ed. by Sten Ebbesen and Lars Boje Mortensen, Corpus Philosophorum Danicorum Medii Aevi, XI.1–2 (Copenhagen: DSL, 1985–1988)

Diplomatarium Danicum I.4, ed. by N. Skyum-Nielsen (Copenhagen: Munksgaard, 1958)

Diplomatarium Danicum I.6, ed. by N. Skyum-Nielsen (Copenhagen: Munksgaard, 1979)

The Latin version of the Law of Scania, ed. by Ditlev Tamm and Helle Vogt, Routledge Medieval Translations: Medieval Nordic Laws (London: Routledge, 2017)

Martin of Dacia (Martinus de Dacia), *Opera*, ed. by H. Roos, Corpus Philosophorum Danicorum Medii Aevi, II (Copenhagen: DSL, 1961)

Petri Philomenae de Dacia et Petri de S. Audomaro opera quadrivialia, ed. by Fritz Saaby Pedersen, Corpus Philosophorum Danicorum Medii Aevi, 10, 2 vols (Copenhagen: DSL, 1983–1984)

Saxo Grammaticus, Gesta Danorum, ed. by Karsten Friis-Jensen, Oxford Medieval Texts (Oxford: Clarendon Press, 2015)

Scriptores minores historiæ Danicæ medii aevi II, ed. by M. Cl. Gertz, Selskabet for Udgivelse af Kilder til dansk Historie (Copenhagen: Gad, 1922)

Skånske Lov, Anders Sunesøns Parafrase, Skånske Kirkelov m.m., ed. by Svend Aakjær and Erik Kroman, Danmarks Gamle Landskabslove, 1 (Copenhagen: DSL, 1933)

Studium Upsalense. Specimens of the Oldest Lecture Notes Taken in the Mediaeval University of Uppsala, ed. by Anders Piltz, Skrifter rörande Uppsala universitet. C. Organisation och historia, 36 (Uppsala: Acta Universitatis Upsaliensis, 1977)

Thuo of Virborg (Thuo de Vibergia), *Opera*, ed. by A. Tabarroni and S. Ebbesen, Corpus Philosophorum Danicorum Medii Aevi, 13 (Copenhagen: DSL, 1983)

Universitas studii Haffnensis: Stiftelsesdokumenter og Statutter 1479, ed. by Jan Pinborg, trans. by Brian Patrick McGuire (Copenhagen: University of Copenhagen, 1979)

Vitae Sanctorum Danorum, ed. by M. Cl. Gertz, Selskabet for Udgivelse af Kilder til Dansk Historie (Copenhagen: Gad, 1908–1912)

Walter Burleigh, *De Puritate Artis Logicae, Tractatus Longior. With a Revised Edition of the Tractatus Brevior*, ed. by Philotheus Boehner, Franciscan Institute Publications Text Series, 9 (New York: Franciscan Institute Publications, 1955)

Secondary Studies

Anders Sunesen – Stormand, teolog, administrator, digter. Femten studier, ed. by Sten Ebbesen (Copenhagen: Gad, 1985)

Andrews, Robert, 'The Defensorium Ockham: An Edition', *Cahiers de l'Institut du Moyen-Âge Grec et Latin*, 71 (2000), 189–273

Dod, Bernard G. 1982, 'Aristoteles Latinus', in *The Cambridge History of Later Medieval Philosophy*, ed. by N. Kretzmann, A, Kenny, and J. Pinborg (Cambridge: Cambridge University Press, 1982), pp. 45–79

Ebbesen, Sten, 'Andrew Sunesen's Language', in *The Liber legis Scaniae: The Latin Text with Introduction, Translation and Commentaries*, ed. by Ditlev Tamm and Helle Vogt (London: Routledge, 2017), pp. 159–73

——, 'The Danes, Science, Scholarship, and Books in the Middle Ages' in *Living Words & Luminous Pictures: Essays*, ed. by Erik Petersen (Copenhagen: Det Kongelige Bibliotek, 1999), pp. 119–26

——, *Dansk middelalderfilosofi ca. 1170–1536*, Den Danske Filosofis Historie, 1 (Copenhagen: Gyldendal, 2002)

——, 'How to Build your own Sophisma (Late 13th- Early 14th-Century Style)', in *Sophismata. Histoire d'une pratique philosophique*, ed. by Alain de Libera, Laurent Cesalli, and Frédéric Goubier (Paris: Vrin, forthcoming)

——, 'A Note on Ockham's Defender', *Cahiers de l'Institut du Moyen-Âge Grec et Latin*, 71 (2000), 275–77

Ebbesen, Sten and Lars Boje Mortensen, 'Introduction' in *Andreas Sunonis filii Hexaemeron*, Corpus Philosophorum Danicorum Medii Aevi, 10.1 (Copenhagen: DSL, 1985)

——, 'A Partial Edition of Stephen Langton's Summa and Quaestiones with Parallels from Andrew Sunesen's Hexaemeron', *Cahiers de l'Institut du Moyen-Âge Grec et Latin*, 49 (1985), 25–224

Friis-Jensen, Karsten, 'Saxo's Books', in *Living Words & Luminous Pictures: Essays*, ed. by Erik Petersen (Copenhagen: Det Kongelige Bibliotek, 1999), pp. 96–110

Mornet, Elisabeth, 'Le voyage d'études des jeunes nobles danois du XIVᵉ siècle à la Réforme', *Journal des Savants*, 4.1 (1983), 287–318

Mortensen, Lars Boje, 'The Sources of Andrew Sunesen's *Hexaemeron*, *Cahiers de l'Institut du Moyen-Âge Grec et Latin*, 50 (1985), 113–216

Munk Olsen, Birger, 'Absalons studier i Paris', in *Absalon fædrelandets fader*, ed. by F. Birkebæk, T. Christensen, and I. Skovgaard-Petersen (Roskilde: Roskilde Museums Forlag, 1996), pp. 57–72

Pasnau, Robert, ed., *The Cambridge History of Medieval Philosophy*, vol. 2 (Cambridge: University of Cambridge Press, 2010)

Petersen, Erik, ed., *Living Words & Luminous Pictures: Essays*, ed. by Erik Petersen (Copenhagen: Det Kongelige Bibliotek, 1999)

Pinborg, Jan, 'Danish Students 1450–1535 and the University of Copenhagen', *Cahiers de l'Institut du Moyen-Âge Grec et Latin*, 37 (1981), 70–122

CHRISTIAN ETHERIDGE

Medieval Scientific Book Fragments Held in Swedish and Finnish Archives

The Tantalizing Remains of a Greater Scientific Corpus

Introduction

This chapter focuses on the important but overlooked fragments of medieval scientific manuscripts from Sweden and Finland. The borders of the Kingdom of Sweden in the Middle Ages were different from modern Sweden: the southern counties of modern Sweden were then a part of the Kingdom of Denmark, several of the western counties of Sweden belonged to the Kingdom of Norway, while the pagan and unconquered Sami occupied the northern part of modern-day Sweden and Finland. The archdiocese of Uppsala divided the Kingdom into seven dioceses, with the diocese of Turku consisting of modern-day southern Finland and parts of the Karelia Republic and Leningrad Oblast in modern-day Russia.

Within the borders of the Kingdom of Sweden there were seven cathedrals, forty-one religious houses and around 1500 churches.[1] These ecclesiastical foundations used, produced, and bought tens of thousands of manuscripts during the course of the Middle Ages. The religious requirements for the practice of the medieval Catholic faith of these foundations meant that the majority of these manuscripts were liturgical in nature. Law texts were important for both secular and ecclesiastical governance and so formed a significant number of manuscripts that were used: in contrast scientific manuscripts formed only a small but significant portion of the whole. Following the Reformation in Sweden and the subsequent adoption of Lutheran Christianity, these manuscripts were plundered from the Catholic cathedrals, religious houses and churches. Many were destroyed, while a large proportion were cut up to provide bindings for paper government documents in the sixteenth and seventeenth centuries. Many of these fragmented medieval

---

1 Brunius, *From Manuscripts to Wrappers*, pp. 13–20.

Christian Etheridge • is a Mads Øvlisen Novo Nordisk Postdoctoral Fellow at the National Museum of Denmark. He recently defended his doctoral thesis, 'The Transmission and Reception of Science in Medieval Scandinavia 1100–1525' at the University of Southern Denmark. Christian.etheridge@natmus.dk

manuscripts have managed to survive today still bound to these early modern government documents.²

There are around 23,000 surviving fragments that are held today in institutions in Sweden, mainly in Stockholm, with a further 6000 in Helsinki and 1700 that are held in the collection of the British Library in London. This last group of manuscript fragments was donated by the English philologist George Stephens (1813–1895) who had collected them during his time in Scandinavia.³ Together these over 30,000 fragments stem from the remnants of roughly 13,000 individual manuscripts.⁴ Of these fragmentary manuscripts, thirty-six contain scientific material. This material is an important testimony to the study and reception of medieval science in the Kingdom of Sweden, due to the high rate of manuscript loss during the early modern period.

This investigation of Swedish and Finnish scientific fragments falls into three parts. The first part looks at the history and previous study of the fragment collections. The second part catalogues the individual scientific fragments and places each one into a context of its content and reception. The final part analyses the place of science in the Kingdom of Sweden during the Middle Ages by comparing the fragments with other evidence such as that documented in contemporary library inventories, testaments, and complete manuscripts that somehow managed to escape the fate of the fragments.

The Destruction of Medieval Manuscripts in the Early Modern Period

The beginning of the fragment collection can be traced back to the formative 1527 parliament held by King Gustav I Vasa (r. 1523–1560) in Västerås Castle. This parliament signalled the start of the Reformation in the Kingdom of Sweden which then began with a large confiscation of church incomes and the exiling of the Catholic bishops. In 1529 a Lutheran handbook officially replaced Catholic works of liturgy and by 1540 all the mendicant houses in the Kingdom of Sweden had been abolished. In 1544, again at Västerås, the Catholic Mass, pilgrimages, and the cult of saints were all prohibited.⁵ By the death of Gustav Vasa, only three nunneries remained out of all the religious houses that had existed at the beginning of his reign, and these too would be closed by the end of the century during the reigns of his successors.

During the reign of Gustav Vasa, the mechanism of state bureaucracy was vastly increased. The number of bailiffs in the realm increased from thirty-six at

---

2   For the Swedish manuscript fragments see Brunius, 'The Recycling of Manuscripts in Sixteenth-Century Sweden'; see also Brunius, *From Manuscripts to Wrappers* and Abukhanfusa, *Mutilated Books*. For the Finnish manuscript fragments see Heikkilä, 'From Fragments Towards the Big Picture'.
3   Brunius, *From Manuscripts to Wrappers*, pp. 34–41.
4   For the latest research on medieval fragments in Scandinavia see Ommundsen and Heikkilä, eds, *Nordic Latin Manuscript Fragments*.
5   For the background to this see Brunius, *Vasatidens samhälle* and Brunius 'The Recycling of Manuscripts in Sixteenth-Century Sweden'.

the beginning of his reign to 174 at the end.⁶ Each of these bailiffs were responsible for the running and administration of royal castles, manors, royal estates, old monastic lands, and mining districts. The bailiffs were rotated every three to five years to make sure that they did not build up any local loyalties and so would theoretically only be loyal to the Crown. Once every year each bailiff travelled to Stockholm to have their accounts audited by the Chamber of the Royal Palace. The Chamber officials were also responsible for the military accounts and were directly supervised by the King. This Chamber System lasted from 1530 until 1630 and during this time produced thousands of government documents. These documents were written on paper and required good binding material to prevent damage. There was a ready supply of this material at hand, this being the tens of thousands of parchment manuscripts that had recently been confiscated from the Catholic ecclesiastical foundations in the Kingdom of Sweden. These manuscripts were then systematically cut up and used to bind the official documents of the Swedish government. Most were used as covers and survive as bifolia, whereas others were cut into strips to be used as binding for the spines of documents. Sometimes only a pair of leaves survive from a manuscript; in other cases, dozens of folios survive, but can be scattered over a number of documents and repositories. The preservation of the fragments is on the whole particularly good due to their important use as covers of important government records. However, these documents with their medieval passengers were stored away after use and largely forgotten about until the nineteenth century.⁷

The Rediscovery of the Fragments in the Nineteenth Century

In 1809, the forces of the Russian Empire finally defeated the Kingdom of Sweden in the Finnish War (1808–1809). The Treaty of Fredrikshamn saw that Finland was detached from the Swedish realm and instead incorporated into the Russian Empire as the Grand Duchy of Finland. The Russian Empire then ordered Sweden to send all historical documents concerning Finland to the city of Turku, then the official capital of the Grand Duchy. Most of the account books from the sixteenth and seventeenth centuries that dealt with Finland were sent to the new Grand Duchy from Stockholm along with the thousands of medieval fragments that were still attached to them.⁸ After Helsinki became the capital of the Grand Duchy in 1812, the account books were sent to the Senate archives which formed the nucleus of the later National Archives of Finland. During the 1840s and 1850s all the medieval fragments were removed from the account books and taken to Helsinki University

---

6 Brunius, *From Manuscripts to Wrappers*, p. 21, for a more detailed overview of the bureaucracy of the Vasa regime see Brunius, *Vasatidens samhälle*, pp. 13–23.
7 For the above see Brunius, *From Manuscripts to Wrappers*, pp. 24–33.
8 For a more detailed overview see Heikkilä 'From Fragments Towards the Big Picture'.

Library, now the National Library of Finland. Assembling the *Fragmenta membranea* collection was the work of the archivist Edward Grönblad (1814–1864) who was also responsible for removing the fragments from the account books.[9] Meanwhile in Sweden the first scholarly work on the fragment collection had begun in the 1860s by the head of the Royal Library in Stockholm, Gustav Klemming (1823–1893), who was interested in reconstructing some of the earliest Swedish printed books with help of the fragments.[10] It was during this time in the latter part of the nineteenth century that George Stephens removed part of the collection and donated it to the British Library.

Cataloguing the Fragments

The first systematic effort to catalogue the medieval fragments was made on the part of the Finnish collection by the musicologist Toivo Haapanen (1889–1950) in the 1910s. In the following decades he catalogued most of the liturgical materials while the church historian Aarno Maliniemi (1892–1972) focused on the calendar fragments.[11] Between 1912–1914 the noted Swedish philologist, Isak Collijn (1875–1949) made a preliminary inventory of the fragment collection held in Sweden.[12] Work systematically developed on cataloging the Swedish collection, first under Toni Schmid (1897–1972) between 1930–1968 and then under Oloph Odenius (1923–1987) from the 1950s to 1985.[13] The Swedish project was given the title *Catalogus Codicum Mutilorum* (the Catalogue of Mutilated Books, abbreviated to CCM).

In Finland in the 1970s Anja Inkeri Lehtinen began to catalogue the theological and legal materials — this work is still ongoing. In the 1980s Jyrki Knuutila began the cataloging of the Finnish liturgical material, this work continued with Ilkka Taitto who published his catalogue on Finnish antiphoners in 2001.[14] In Sweden from 1995–2003 the project *Medeltida Pergament Omslag* (Medieval Parchment Covers, abbreviated to MPO) was initiated and worked on by Jan Brunius, Gunilla Björkvall, and Anna Wolodarski.[15] The MPO project was a continuation of the earlier CCM project, but with a more ambitious aim: namely to finally catalogue all the Swedish fragment material with an accompanying online database with many of the fragments being digitized. The results of the project are now online as the *Riksarkivets databas över medeltida pergamentomslag* (National Archive of Sweden's database of Medieval Fragments).[16] In Finland the project *Literary Culture*

---

9 Lehtinen, 'From Fragments into Codices', p. 109.
10 Brunius, *From Manuscripts to Wrappers*, p. 41.
11 For early Finnish cataloguing see Heikkilä, *The Fragmenta Membranea Collection*
12 Brunius, *From Manuscripts to Wrappers*, p. 41.
13 A more detailed description of the project can be found in Schmid, 'Undersökningen av medeltida svenska bokfragment'.
14 Brunius, *From Manuscripts to Wrappers*, p. 47.
15 For more on the project see Abukhanfusa, *Mutilated Books*, pp. 14–15.
16 *Swedish Fragment Collection*.

in Medieval Finland led by Tuomas Heikkilä between 2006 and 2012 produced a preliminary catalogue of previously uncatalogued materials in the *Fragmenta membranea* collection. In 2008, following Heikkalä's initiative, the National Library of Finland began to construct a digital research database of the medieval fragments which is now the *Fragmenta membranea* collection online database, with all the fragments digitized.[17] The majority of the 30,000 fragments from the medieval Kingdom of Sweden therefore can now be searched for in these two online collections. The George Stephens collection held in the British Library has been given an overview by Anja Inkeri Lehtinen but as of 2021 has not yet been properly catalogued.[18]

Number of Manuscripts Represented in the Fragments

In total 12,586 manuscripts from both Sweden and Finland are represented in the fragment collection. Of these manuscripts 10,394 are liturgical, 926 law, 526 theological, 421 biblical, 160 varia, and 112 hagiographical.[19] Of these 160 varia manuscripts, thirty-six contain scientific material. Twenty-eight of these come from Sweden, five from Finland and three are divided between the collections. These scientific manuscripts therefore form less than one per cent of the total collection showing the specialist interest that this subject represented, usually for people with a university education. The fragments are catalogued in the Appendix. In the Swedish collection, the fragments are given the full catalogue title of Stockholm Riksarkivet (SRA) followed either by *Var* for *Varia* for a collection of manuscript fragments or *Fr* for *Fragmenta* for singletons. The fragment is then allocated its number in the collection. The Finnish collection gives the full catalogue title of Helsinki, National Library/ Helsinki, Kansalliskirjasto followed by F.m. for *Fragmenta membranea* and then the fragment number in the collection.

Content of the Scientific Works in the Fragment Collection

The most common scientific works among the collection are those of Aristotle (384–322 BCE) and commentaries on these same works. Pseudo-Aristotelian texts associated with the philosopher also form a part of this corpus. This predominance of Aristotelian material is mirrored in the content of other medieval European manuscript collections, due to his central role in the curriculum of scientific teaching in the faculty of the arts at medieval universities. Other genres of medieval science found in the collection are works on astronomy, computus, medicine, and encyclopaedic compilations.

---

17 *Finnish Fragment Collection*.
18 Lehtinen, 'From Fragments into Codices', pp. 114–15.
19 Statistics from Brunius, *From Manuscripts to Wrappers*, pp. 37–40.

The Wider Context of the Manuscript Fragments in Medieval Sweden: The Twelfth Century

The oldest fragments in the collection are religious texts that come from England and Germany and date from the eleventh century. This date fits in well with what is known of the earliest missionary work in Sweden that was carried out from England and Germany.[20] The earliest works of science in a Swedish context, however, date from the twelfth century, these being the computistical material in the possibly German calendar Kal 19 and the medical work, the *Practica* of Alexander of Tralles found in the manuscript F.m.V.Var.28. Computistical material also appears in one of the earliest surviving books produced in the Kingdom of Sweden, the *Vallentuna Missal* (Stockholm, Statens historiska museum, MS 21288) from Vallentuna Church in the diocese of Uppsala dating from 1198. This manuscript is written in Latin and contains computus verse, an Easter table, and a description of the unlucky days of the calendar known as *dies Aegyptiaci* (Egyptian days).[21]

The early surviving booklist from an ecclesiastic in the Kingdom of Sweden, that of Siward, or Sigurd, bishop of Uppsala from *c.* 1123–1134, provides us with a glimpse into the scientific works that he owned. Siward was expelled from Sweden after the political fallout from the Battle of Foteviken in 1134, and after travelling in England and Ireland for some time, he became abbot of the Benedictine abbey of Rastede near Oldenburg from 1147 until his death in 1157. It is unclear whether these books were acquired from his time as bishop in Uppsala or as abbot of Rastede. They could instead represent a culmination of his book acquisition formed throughout his whole ecclesiastical career. Siward left twenty-six volumes to his abbey in his testament, including several scientific texts. The first of these scientific texts in the testament is a herbal and lapidary in a single book titled *Herbarium, lapidarium in uno volumine*. The herbal may be either the *Herbarium* of Pseudo-Apuleius (fl. fourth century CE) or the *De viribus herbarum* of Macer (fl. eleventh century), as both were popular works in the twelfth century. The lapidary would most likely be the *Liber lapidum* of Marbod of Rennes (*c.* 1035–1123), the most popular lapidary of the period. The second of the texts listed in Siward's testament is six untitled medicinal works titled *Medicinales sex*. These six volumes could possibly refer to the medical compilation titled the *Articella* as this was in six parts: the *Isagoge* of Johannitus, the Hippocratic *Aphorisms* and *Prognostics*, the *De urinis* of Theophilus, the *On Pulses* of Philaretus and Galen's *Tegni*. The third text is an early bestiary titled *Physiologus*. The final text is listed as *Platonem*. The only work known of Plato before the twelfth century was the *Commentary on Timaeus* by Calcidius (fl. fourth century CE). During the twelfth century, this commentary was itself subjected to a lengthy tradition of commentary

---

20 For an overview see Brunius, *From Manuscripts to Wrappers*, pp. 49–53; for more details on the English fragments see Gullick, 'Preliminary Observations on Romanesque Manuscript Fragments'; for the German fragments see Niblaeus, 'German Influence on Religious Practice in Scandinavia'.
21 Helmfrid, *Vallentuna anno domini 1198*, contains an edition, facsimile, Swedish translation, and essays on this important early manuscript.

that often focused on the astronomical nature of Plato's work. The *Glosae super Platonem* by William of Conches could well be the treatise that Siward's testament refers to. Siward on his death owned treatises on medicine, bestiaries, and astronomy and is indicative of the depth of learning of a scholar in Scandinavia during the twelfth century.[22]

The University of Paris

In the thirteenth century, scholars from the Kingdom of Sweden, like their contemporaries elsewhere in Europe, stopped attending foreign cathedral schools and instead started to attend the early universities such as Paris and Bologna. It was through these new centres of learning that they were taught the scientific subjects that now formed part of the curriculum of the arts degree. Through the faculty of the arts new students would learn the fundamentals of arithmetic, geometry, astronomy, and natural philosophy. Medicine was taught in its own faculty to those students who had the ability to pursue this study and had already progressed through the arts curriculum. Some of the students from the Kingdom of Sweden showed great aptitude in these subjects and as magisters taught them to others. The most popular university for students from the Kingdom of Sweden to study in during the thirteenth and fourteenth centuries was the University of Paris. According to the University of Paris statutes of 1366 a student who wished to obtain the *licentiate*, or degree, had to have read and studied the major Aristotelian works of *Physica*, *De generatione et corruptione*, *De caelo et mundo*, and *Metaphysica*. This curriculum also included some of the *Parva Naturalia* treatises, these being *De sensu et sensato*, *De somnis et vigiliis*, *De memoria et reminiscentia* and *De longitude et breviate vite* (On Longevity and Shortness of Life).[23] The curriculum, if followed and understood, would give the prospective student a broad understanding of Aristotelian concepts of the natural world, physics, and astronomy. All these treatises, apart from the last named, appear in the fragment collection and show scholars from the Kingdom of Sweden having a familiarity with Aristotle's scientific works, most likely obtained from studying the arts curriculum at the University of Paris.

There are three scholars from the Kingdom of Sweden who we know passed through the curriculum and who ended up teaching scientific subjects at the University of Paris. The first of these was Canon Hemming from Uppsala Cathedral. He taught astronomy at Paris at the end of the thirteenth century. Hemming's important testament, which contains several scientific volumes will be discussed further below in the section on cathedral libraries. The next second scholar was Canon Suno Karoli from Linköping Cathedral who became magister at the University of Paris in August 1340.[24] Like all Scandinavians, Suno was a member of the English Nation and by November had become the Nation's procurator. In early 1341 he was responsible for the nation's licentiate

---

22 The booklist is edited in Kleberg, *Medeltida Uppsalabibliotek*, I, 21–42.
23 *Chartularium Universitatis Parisiensis*, ed. by Denifle and Chatelaine, p. 151.
24 For more on Suno see Schück, 'Svenska Pariserstudier under medeltiden', p. 152.

exam at St Geneviève and later that year became the English nation's treasurer. In 1342 he became procurator again. Suno taught Johannes de Sacrobosco's (*c.* 1195–1256) *De sphaera* (On the Sphere) at Paris from 1340–1344.[25] This was a classic primer on astronomy and was widely used in teaching the subject. The final scholar that we know of who taught scientific subjects at the University of Paris was Canon Jacobus Petri de Röd from Turku Cathedral. He first studied at the University of Leipzig and by 1417 he was at Paris where he obtained his bachelor and teaching license in 1418. Jacobus became magister in 1419. Like Suno, Jacobus served as Procurator of the English Nation 1425–1427 and was Treasurer 1426–1427.[26] In 1427 Jacobus is recorded in the *Diarium Bibliothecae Sorbonae* as taking the *De scientia astrorum* (On the Science of the Stars) by al-Battani (*c.* 858–929) from the Sorbonne College library. Jacobus probably used it to prepare his lectures on planetary theory held in the studia of the Carmelites in Paris.[27]

There are several manuscript fragments that could well orginate from the University of Paris. These fragments are Var 2 (Aristotle: *De memoria* see Fig. 5.1, *De motu animalium*, *De sensu* and *Metaphysica*; Pseudo-Aristotle, *Liber de causis*); Var 3 (Aristotle: *De anima* and *Metaphysica*); Var 21 (Aristotle: *De caelo*); Var 25 (Bartholomeus Anglicus: *De proprietatibus rerum*); Var 26 (Bartholomeus Anglicus: *De proprietatibus rerum*); Var 28 (Avicenna: *Canon*); Var 33 (Pseudo-Aristotle: *De coloribus, De lineis insecabilibus, De physionomia* and *Secretum secretorum* see Fig. 5.2); Var 41 (Arzachel: *Canones* see Fig. 5.3); Var 53 (Aristotle: *De anima*); F.m.V.Var 4 (Aristotle: *De sensu*); Fr 9399 (Averroes: *De substantia orbis* see Fig. 5.4) and Fr 9728 (Aristotle: *De memoria* and *De sensu*). These are all thirteenth- and fourteenth-century manuscripts that due to their subject matter and commentary, as well as some extensive marginal and interlineal glossing, seem most likely to stem from a university environment. During this time period Paris was the main university for students from the Kingdom of Sweden, therefore it is highly likely that the manuscripts came from booksellers associated with the university. Two other manuscript fragments are definitely from France. The first is the manuscript of *De animalibus* by Albertus Magnus, from 1367–1433, that is now extant in the fragment collections SRA Var 1 and F.m.V. Var 2. The second manuscript, now SRA Fr 7714 and dated from 1449, is from the environment of the University of Paris and contains Aristotle's *De caelo* and *De generatione*.[28]

Other Universities

The University of Bologna was very important for Scandinavian students in the Middle Ages, mainly those studying law.[29] Two manuscripts from the fragment collection

---

25 Heyman, 'Suno Karoli de Suecia', pp. 213–14.
26 Heininen, 'Finland: Medeltiden', p. 72.
27 Lehtinen, 'Suomalaisia teologeja Sorbonnen kollegion kirjastossa 1400-luvulla'.
28 This was written by a Henricus de Ansbeke from the Picardy Nation at the University of Paris, *Nationis Picardie per manus Henrici de Ansbeke*, see Brunius, *From Manuscripts to Wrappers*, p. 143.
29 For Scandinavian students studying in Bologna see Sällström, *Bologna och Norden intill Avignonpåvedömets tid*.

that come from Italy are the *Practica* of Alexander of Tralles from the eleventh to the twelfth century found in F.m V Var.28 and the *Summa conservationis et curationis* by Guillelmus de Saliceto found in F.m.V.Var.7, SRA Var 15 and London, British Library, Add. 34390. The *Practica* is a Salernitan text and may have originated from there but possibly also via Bologna, while the *Summa conservationis et curationis* is most likely to stem from Bologna.

From the late fourteenth century, Swedish students started to attend the newly created universities in the Holy Roman Empire. By the fifteenth century, these universities had eclipsed Paris as the place for scholars from the Swedish realm to study and teach. There is considerable surviving information from the fifteenth century on Swedish magisters at the universities of Leipzig and Vienna which has recently been collated by Olle Ferm in two edited volumes on Swedish scholars at these universities. Using these sources, it is possible to trace lectures and disputations given by Swedish magisters dating between 1426–1489.[30] Of these the most common teaching matter on scientific subjects were the works of Aristotle, these being *De anima*; *De caelo et mundo*; *De generatione et corruptione*; *Metaphysica*; *Parva Naturalia*; and *Physica*. All the works of Aristotle are represented in the fragments collection. Other subjects taught were the university primers in geometry, optics, and astronomy. These were respectively the *Elements* of Euclid, the *Perspectiva communis* of John Pecham and the *De sphaera* of Sacrobosco. Another subject taught by Swedish masters was advanced arithmetic. This is represented by the *Arithmetica speculativa* (Theoretical Arithmetic) of Johannes de Muris (c. 1290–1355) an abridgement of Boethius's *De arithmetica*; Nicholas Oresme's (c. 1325–1382), *Latitudines formarum* (On the Latitude of Forms) which included diagrams of the velocity of an accelerating object against time and the *Proportiones breves* (Short text on Proportions) by Thomas Bradwardine (c. 1300–1349) that worked with exponential growth. Finally, minerology was taught via the *De mineralibus et lapidibus* (On Minerals and Precious Stones) by Albertus Magnus on the nature of minerals and metals. Of manuscript fragments that have survived from this period, the *Physica novae translationis cum notis* in F.m.V.Var.3 has been dated from 1434–1466 and from either Austria or Germany.

Three volumes of the *De proprietatibus rerum* of Bartholomeus Anglicus in the fragment collection originate in England. These are SRA Var 26 from the fourteenth century; SRA Var 27 and F.m.V.Var.6 from 1301–1333 and SRA Var 6 and F.m.V.Var.5. from 1301–1333. There were only a few known scholars from the Kingdom of Sweden studying at English Universities. Some of them were seculars, while others were mendicants. The Dominicans' *studium generale* in Oxford allowed foreign students from 1261 onwards. Each Dominican province had the right to have two students at Oxford. Cambridge had a *studium particulare* from the mid-thirteenth century and from 1316 a *studium generale* that allowed two students from each province.

Tideman of Närke, one of the seculars, studied at Cambridge in the 1430s and 1440s. His autograph manuscript survives as Cambridge Peterhouse 188. It contains

---

30 For the University of Leipzig see Ferm 'Swedish Students and Teachers', pp. 25–54. For the University of Vienna see Ferm, 'Swedish Students in Vienna 1389–1491', pp. 55–79.

his own treatise the *Termini naturales*, largely based on William of Heytesbury's *Termini physicales*, as well as other treatises that he copied.[31] The Franciscan Peter Pauli, from the friary in Nyköping, was one of the mendicants active at Oxford in the second half of the fifteenth century. He wrote a manuscript, now preserved as Oxford, Corpus Christi College, MS 227, which included the commentary on *De anima* by John Duns Scotus.[32] The evidence for students from the Kingdom of Sweden at English universities is fragmentary, but surviving examples such as those above show that there were several scholars who wrote treatises and were part of the learned environments of Cambridge and Oxford. It could have been one of their forebears that brought the English volumes of the *De proprietatibus rerum* of Bartholomeus Anglicus to the Kingdom of Sweden.

Medieval Swedish Repositories

As shown above, the scientific manuscripts represented in the fragment collection mostly came from universities abroad and were taken to various libraries in the Kingdom of Sweden, usually situated in cathedrals and religious houses. Below I will give some examples of these repositories and how some of the scientific volumes in their collection ended up there. For reasons of space this list is not conclusive but gives a taste of the collections of medieval libraries in the Kingdom of Sweden.

Uppsala Cathedral Library

In 1299 the aforementioned Canon Hemming of Uppsala passed away in Paris. He donated the books he had used in his university career to Uppsala Cathedral, including many works on law but also on science.[33] The first of these scientific volumes listed in his testament was a *mappa mundi*; this was followed by the *Metaphysica* and *Physica* of Aristotle. The next entry was a combined volume that contained a *Theorica planetarum*, *Canones* and *Tabulae Tables*. The *Theorica planetarum* described the motions of the planets while astronomical *canones* were instructions for using astronomical tables such as the *Toledan Tables*, The *Toledan Tables* were originally created by Abū Isḥāq al-Zarqālī (1029–1087), an Andalusian astronomer known as Arzachel in the Latin West, translated by Gerard of Cremona in the twelfth century and then adapted further by scholars in the court of Alfonso X in the thirteenth century.[34] The testament also included Euclid's *Elements* with commentary from

---

31 For an edition and description of the treatise, see Andrews, 'Tideman of Närke's *Termini naturales*'.
32 Peter Pauli is titled *frater de Nicopia*, this could also refer to the friary in Nykøbing Falster in Denmark. For this debate and more on Peter's career see Ferm, 'Swedish Students in Cambridge and Oxford', pp. 16–19.
33 The testament is preserved in Stockholm Riksarkivet SDHK 1903, and edited in Kleberg, *Medeltida Uppsalabibliotek*, II, 29–39.
34 For further on this treatise see Pedersen, 'The Origins of the Theorica Planetarum'.

the mathematician and astronomer Johannes Campanus of Novara (c. 1220–1296) and the *De arithmetica* (On Arithmetic) of Boethius (c. 480–524). Further down the list is a title that reads *Liber quadrantis et ipse quadrans*. This would be a work on the astronomical instrument known as the quadrant, also including an actual quadrant. There were several treatises on quadrants known from the late thirteenth century and it is unsure which of these it could refer to.[35] The testament concluded with a book that incorporated the *Algorismus*, *Computus* and *De sphaera* of Johannes de Sacrobosco in a single volume. These works of Sacrobosco were used as primers for the sciences in medieval universities from the mid-thirteenth century onwards.[36] The volumes owned by Hemming point to him being very learned in the astronomical sciences and certainly qualified to teach them to students at the University of Paris. The collection of *Theorica planetarum*, astronomical *canones*, *Toledan Tables*, book on the use of a quadrant as well as the instrument itself, point to a strong possibility that Hemming conducted astronomical observations. Therefore, it would seem likely that Hemming was both a practical and theoretical astronomer. These books would then have entered the Uppsala Cathedral library and could have been used by university-educated scholars there.

Skara Cathedral Library

The library of Skara Cathedral had a large book collection in Scandinavian terms that numbered between 300–500 volumes at the beginning of the sixteenth century. Of these books, 138 were donated by Bero Magni de Ludosia (c. 1409–1464) in his testament. These included works on Aristotle, astronomy, geometry, and arithmetic. Bero was from Lödöse, close to modern day Gothenburg in the diocese of Skara and received his bachelor's degree at the University of Vienna in 1431, becoming a magister two years later in 1433. During his time in Vienna Bero also wrote his own commentary on *De anima*.[37] He was elected as bishop of Skara in 1461 but did not return from Vienna; therefore, the election had to be nullified by Pope Paul II. Between 1465, when Bero passed away, and 1475 the library of Skara Cathedral received part of his personal library numbering 138 volumes.[38] The library of Skara was partially destroyed by the Danish invasions of 1566 during the Northern Seven Years' War and finally again in 1612 during the Kalmar War, when what remained of the collection was then scattered.[39] Most of Bero's library consisted of works on theology. Those that are of

---

35   For the ongoing debate on the authorship of this text see Knorr, 'The Latin Sources of Quadrans Vetus'.
36   Kleberg, *Medeltida Uppsalabibliotek*, II, 29–39.
37   See Bero Magni de Ludosia, *Disputata super libros De anima*, ed. by Andrews. For the companion volume to the treatise see Andrews, *Bero Magni de Ludosia Questions on the Soul*. This treatise and Bero's other works aim to be published in the forthcoming series *Corpus Philosophorum Suecorum medii aevi*.
38   For more on this see Kihlman, 'Bero Magni de Ludosia', p. 90.
39   For further on the history of the collection see Kihlman, 'Bero Magni de Ludosia', p. 123.

a scientific nature consisted of twenty-four volumes. These include several works of Aristotle including two copies of *Metaphysica*, two copies of *Meteora* (Meteorology), *De caelo et mundo*, and *De generatione et corruptione*. The commentary on *Meteorologica* by Jean Buridan (c. 1295–1363), the *Questiones Buridani super libros Metheororum*, was also in the collection. Bero donated two copies of the *De proprietatibus rerum* of Bartholomeus Anglicus to Skara as well as two copies of the *Physiologus*, along with Euclid's *Elements*. Finally, the donation included anonymous astronomical tables, a calendar, a work on astronomy, a treatise on arithmetic, a universal philosophy, a treatise on chiromancy, and five books on medicine.[40] Like the library of Uppsala Cathedral, the library of Skara Cathedral received a substantial donation of scientific volumes. These would have proved useful for the learned bishops and canons who had been university educated.

Sigtuna St Mary's Dominican Friary Library

The friary at Sigtuna was the most important of all the religious houses founded by the Dominican Order in the Kingdom of Sweden. Founded in 1237, the friary was in the hands of the Dominican Order for around 300 years until 1529. In that year Gustav Vasa ordered that all Sigtuna's parishes were to be merged and that the Dominican church was to be used as the new Lutheran parish church. Of the once extensive library eighteen manuscripts survive, of which three contain scientific material.[41] The first manuscript, Uppsala UUB C 381, dates from the fourteenth century and contains a small treatise on computus in Old Swedish. The second manuscript, Uppsala UUB C 620, is a thirteenth-century Latin volume written in France that contains a commentary on the *Timaeus* of Plato, most likely the *Glosae super Platonem* of William of Conches (c. 1090–1154). This volume was bequeathed in 1296 by Carolus Erlandi, canon of Uppsala Cathedral, to his brother Israel Erlandi, prior of the Sigtuna friary. The third manuscript, Uppsala UUB C 579, is a fourteenth-century Latin manuscript that contains the *Summa de modo medendi* (Summary of the Way of Healing) of Gerard of Cremona or Montpellier (fl. thirteenth century). This volume was donated by a Frater Olaus Petri, an otherwise obscure Dominican friar.[42]

As well as the extant manuscripts, there is also the evidence from a surviving book list. On 3 September, 1446, the Dominican lector Olaus Pauli was issued a receipt from the Provincial Prior of Dacia for books that he was to deliver to Sigtuna Friary. The booklist, which survives today as Stockholm Riksarkivet SDHK 25006, contains many works on logic, canon law, theology, and a *Lincolniensis super phisicorum librum*;

---

40 The testament is edited in Kihlman, 'The Inventory of Bero's Library'..
41 For Sigtuna's book collection see Schmid, 'Sigtunabrödernas böcker och böner'.
42 UUB C 579 and UUB C 620 are described in *Mittelalterliche Handschriften der Universitätsbibliothek Uppsala: Bd. 6. C 551–935*, ed. by Andersson-Schmitt, Hallberg, and Hedlund, on pp. 58–59 and pp. 136–37 respectively. UUB C 381 is described in Andersson-Schmitt and Hedlund, *Mittelalterliche Handschriften der Universitätsbibliothek Uppsala: Bd. 4. C 301–400* on pp. 504–12.

Robert Grosseteste's *Notes on the Physics*.⁴³ The above-mentioned manuscripts and the booklist of Olaus Pauli show an interest in scientific works at Sigtuna Friary, which is typical for the learned Dominican Order in the Middle Ages.⁴⁴

Stockholm Franciscan Friary Library

The Stockholm Franciscan friary was founded 1268/1270 and was the centre for the Order's Custody of Stockholm. As an indication of the learned environment here during the fifteenth century many of the Stockholm Franciscan lectors were known to have studied abroad. Lars Jonsson, Peter Holmgersson, and Kanutus Johannis had all studied at the University of Greifswald, while the lectors Erik Olsson and Lars Nilsson had studied at the University of Erfurt. Finally, Bernhardinus Kempe had obtained his doctoral degree in Bologna. One of the best known of all learned figures from the medieval Kingdom of Sweden was a Franciscan, Kanutus Johannis, who was born to a wealthy burgher family in Stockholm and entered the Order in 1460. In 1467 he left Stockholm to study in Greifswald and Strasbourg and afterwards became lector at the Franciscan friary in Randers. He travelled to the *Studium generale* in Lund where he obtained his baccalaureate in 1476. In 1478 he became lector in Stockholm, and between 1482–1485 was *custos* for the custody of Stockholm (the custody being a geographical division of the Franciscan province of Dacia (Scandinavia)). In 1484–1488 and again in 1490–1495 he was the guardian of the Stockholm Friary. In 1491 he studied at the University of Uppsala for his doctorate. In 1495 he was named at the provincial chapter meeting at the Franciscan friary at Kungälv as *lector principalis* at the *studium generale* in Lund. In 1483 Kanutus Johannis was responsible for bringing in the printer Johann Snell to the Stockholm Friary and while there he printed the first book in the Kingdom of Sweden, the *Dialogus creaturarum*.⁴⁵ This was a moralizing text that also dealt with the attributes of animals. Kanutus Johannis is also known to have bought or bound seventeen manuscripts and four incunabula in the Stockholm Franciscan library.⁴⁶

Of the former extensive library at the Stockholm friary, twenty-five manuscripts and fourteen incunabula survive, of which four contain scientific material and three are works of computus. The manuscript Uppsala UUB C 636, written somewhere in the Kingdom of Sweden in the first half of the fifteenth century contained astrological, astronomical, and computistical texts and diagrams. It belonged to Birgerus Magni,

---

43 For this see Schmid, 'Sigtunabrödernas böcker och böner', pp. 77–79.
44 A good introduction into the use of medieval science in Dominican learning can be found in Mulchaney, *"First the Bow is Bent in Study..."*, pp. 219–32.
45 For books in the Stockholm friary see Roelvink, *Riddarholmens kyrka och kloster*, pp. 119–27, for an edition of the earliest printed book in the Kingdom of Sweden see *Dyalogus creaturarum moralizatus*, trans. by Hedlund.
46 For the Stockholm friary library see Roelvink, *Riddarholmens kyrka och kloster*, pp. 125–28, for more on Franciscan libraries and books in the Kingdom of Sweden in general see Roelvink, *Franciscans in Sweden*, pp. 122–51, with a manuscript list pp. 172–80.

bishop of Västerås, until 1464 before entering the friary library. Uppsala UUB C 654, another manuscript written in the Kingdom of Sweden, this time from the second half of the thirteenth century, contains at least parts of the *Termini naturales*, of Johannes Garisdale, the *De proprietatibus rerum* of Bartholomeus Anglicus and the *Prognostica temporum* of Pseudo-Bede. Finally, the manuscript Uppsala UUB C 627 consists of various works of logic and a commentary on Aristotle's *De anima*. It has been dated to 1462–1464 and was written at the University of Erfurt. Laurentius Nicolai, a Franciscan friar from Stockholm bought it back with him from Erfurt.[47] The Stockholm friary was finally dissolved in 1527 along with its library, but the church itself was spared due to its use as a royal mausoleum.[48]

Vadstena Birgittine Abbey Library

Founded in 1370, the Birgittine double abbey of Vadstena contained the greatest library in medieval Sweden and was also home to a short-lived printing press in 1495.[49] This library actually consisted of three libraries, these being the brothers' library, the sisters' library, and the liturgical library. Together these totalled over 1500 volumes. In 1595 the last nun left the abbey and the remaining manuscripts were taken to Vadstena Castle. In 1621 these manuscripts were taken by King Gustav Adolf II (1594–1632), who donated the Latin manuscripts to Uppsala University and the Old Swedish manuscripts to the royal collection at Tre Kroner in Stockholm.

Today 367 volumes survive of the brothers' library, of which forty-nine contain scientific content. These works were either written in Vadstena on paper in a fast cursive script or imported. The library consisted of bookcases *c*. 2 m × 1 m with six shelves and lettered A-O. Most books with scientific content were to be found in bookcases H and I. Bookcase H also included law books while bookcase I included theological works and sermons. There is not space here to detail all the scientific treatises held in Vadstena so instead I will focus here on one example. The manuscript Uppsala UUB C 28 was given the Vadstena shelfmark I 2 7. So, it was to be found in bookcase I on the second shelf and seven books in. The Latin manuscript was a collection of medical, scientific, and theological texts that dated from *c*. 1200–1400. The first section, dating from 1200–1300 was part of the *Articella*, containing the *Aphorisms* and *Prognostics* of Hippocrates along with the Isagoge of Johannitus. The later part of the manuscript from 1300–1400 contains *De coitu* (On Sexual Intercourse) from Constantine the African. This concludes with the

---

47 Described in *Mittelalterliche Handschriften der Universitätsbibliothek Uppsala: Bd. 6. C 551–935*, ed. by Andersson-Schmitt, Hallberg, and Hedlund, pp. 157–59.
48 For more on the dissolution of the Stockholm friary see Roelvink, *Riddarholmens Kryka och klostre*, pp. 125–28.
49 The most comprehensive work on Vadstena's library is Walta, *Libraries, Manuscripts, and Book Culture in Vadstena Abbey*.

De viribus herbarum of Macer and the *Liber lapidum* of Marbod of Rennes. It is unclear where this manuscript originates from and how it arrived in Vadstena.[50]

Uppsala University

The final repository of books in this section is that belonging to the University of Uppsala. This was the first university in Scandinavia, with the papal bull that granted the corporate rights issued by Pope Sixtus IV in 1477. The university began with between ten and fifteen students and five known magisters. Science was taught from the beginning utilizing among others the works of Albertus Magnus, Aristotle, Alfraganus, Avicenna, Ptolemy, and Sacrobosco. Of the students we are fortunate to have the detailed lecture notes surviving of Olaus Johannis from Gotland who studied at Uppsala from 1477 until 1486. In the spring of 1486, he was taught Aristotle's *Physica* by a magister Petrus Olai. Olaus was also taught Euclid's *Elements* and the *De Sphaera* of Sacrobosco, he was taught Aristotelian logic as well as *De anima*.[51] Olaus Johannis became a magister around 1490 and in 1506 became a monk in Vadstena where he took his seven volumes of lecture notes which were deposited in the abbey library.

The German Vadstena monk Petrus Astronomus taught Sacrobosco's *De sphaera* in 1508 at Uppsala University. A detailed astronomical manuscript written by Petrus Astronomus in 1506 survives and is now Stockholm, Royal Library X770.[52] The following year, he designed the great astronomical clock of Uppsala cathedral, which rivalled that of Lund until it was destroyed in the cathedral fire of 1702.[53] Uppsala University had a small number of students in the late Middle Ages, but it was able to teach the full curriculum of the faculty of the arts, especially Aristotelian treatises and the astronomical works of Sacrobosco. One must consider that Petrus Astronomus in particular would most likely have been a highly effective teacher of astronomy.

Conclusion

The medieval fragment collection held in the archives of Sweden and Finland is a rich and until relatively recently a much under-utilized resource. Contained within the fragments is a noteworthy collection of scientific treatises that have been subjected to little scholarly analysis. This chapter has endeavoured to show that the variety of treatises that survive in the fragment collection is remarkable and covers almost all genres of medieval science. Future work on both the manuscript fragments themselves

---

50 Manuscript catalogued in, *Mittelalterliche Handschriften der Universitätsbibliothek Uppsala: Bd. 1. C I–IV, 1–50*, ed. by Andersson-Schmitt, Hallberg, and Hedlund, pp. 257–59.
51 For the lecture notes of Olaus Johannis, many of them edited, see Piltz, *Studium Upsalense*.
52 For a description of this still unedited manuscript see Lindberg, 'Petrus Astronomus' astronomiska tabeller', 250–56.
53 Described in Götlind, *Technology and Religion in Medieval Sweden*, p. 150.

and the sixteenth and seventeenth-century government records that contain them will undoubtably shed light on their origins.

It is possible to see from the surviving source material detailed above that all aspects of the sciences were well known in the Kingdom of Sweden. Certainly, scientific works were to be found in the great cathedral libraries and those of the most important religious houses. Swedish and Finnish magisters taught scientific subjects in the University of Paris from the late thirteenth century onwards, in Leipzig and Vienna from the fifteenth century and finally in the new university of Uppsala from 1477 onwards. We can see that the student Olaus Johannis at Uppsala was taught the works of Aristotle on natural philosophy as well as geometry and astronomy. Undoubtedly future research in the archives of other universities will bring forward more magisters and the subjects that they taught.

The manuscript fragment collection adds to our knowledge of medieval science in the Kingdom of Sweden that has already been previously ascertained from surviving booklists, testaments, and manuscripts. There is of course the distinct possibility that some of the manuscripts fragments that survive are those belonging to one of the books mentioned in the aforementioned booklists or testaments. Toni Schmid proposed this in 1954. Her theory was that the Arzachel fragments were the remains of the volume listed as the *Tabulae Toletanae* in Canon Hemming's testament.[54] Following from Schmid's theory I would propose that more work needs to be done on this account. As I have endeavoured to show, there were many scholars from the Kingdom of Sweden that taught scientific subjects at university level and owned books with scientific treatises. These scholars would be likely to have brought the manuscripts that the fragments represent to Scandinavia.

The work of collecting manuscript fragments is not over. Lehtinen has alerted us to the substantial number of fragments held in London.[55] There needs to be a detailed catalogue made of the George Stephens collection held in the British Library to the same standard as in Sweden and in Finland and then all three collections can be catalogued together, and a final number of manuscripts can be collated. This can then be cross-indexed with the other source material to get a better understanding of the reception and transmission of science in the Kingdom of Sweden during the Middle Ages.

---

54 Schmid, 'Al Zarkali in Schweden'.
55 Lehtinen, 'From Fragments into Codices' pp. 111–15.

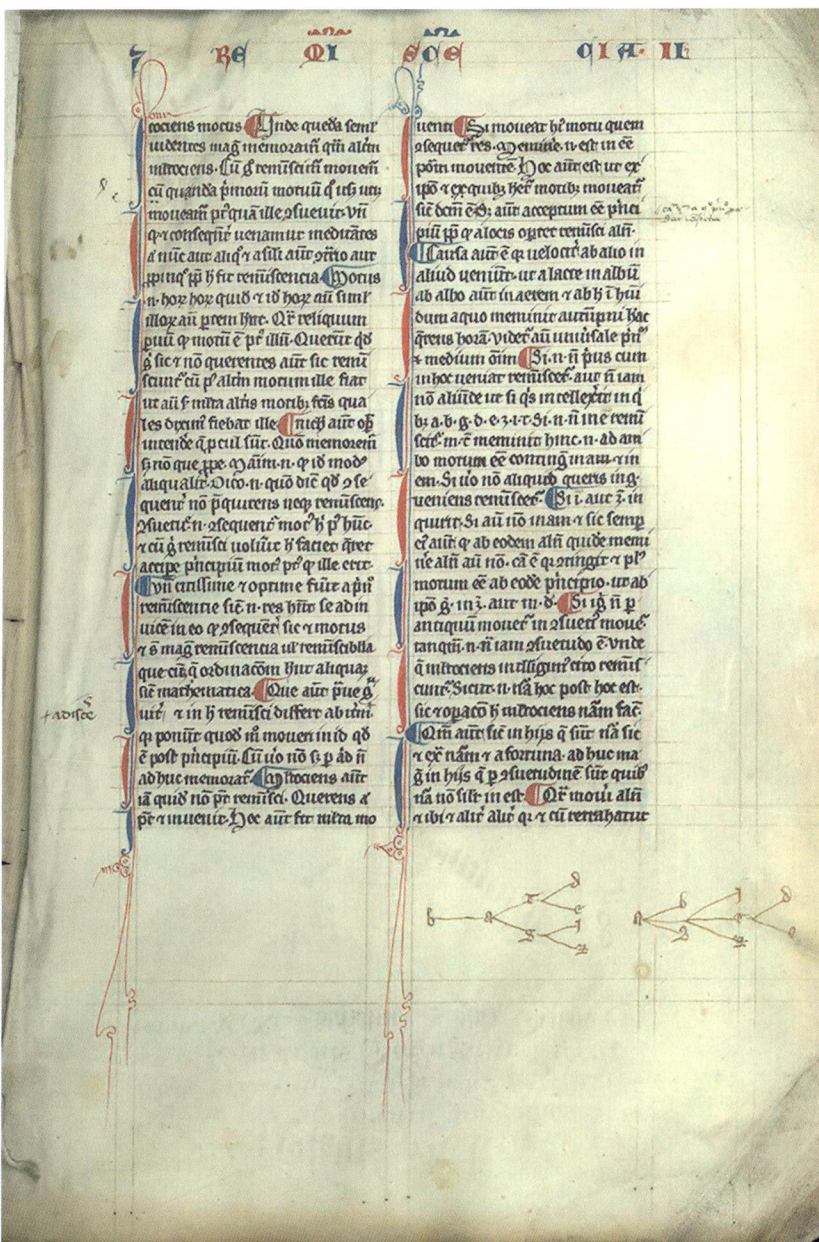

Figure 5.1. Stockholm, Riksarkivet Var 2, Aristotle, *De memoria et reminiscentia* (On Memory and Reminiscence), thirteenth century. Photo: The Swedish National Archives, Stockholm.

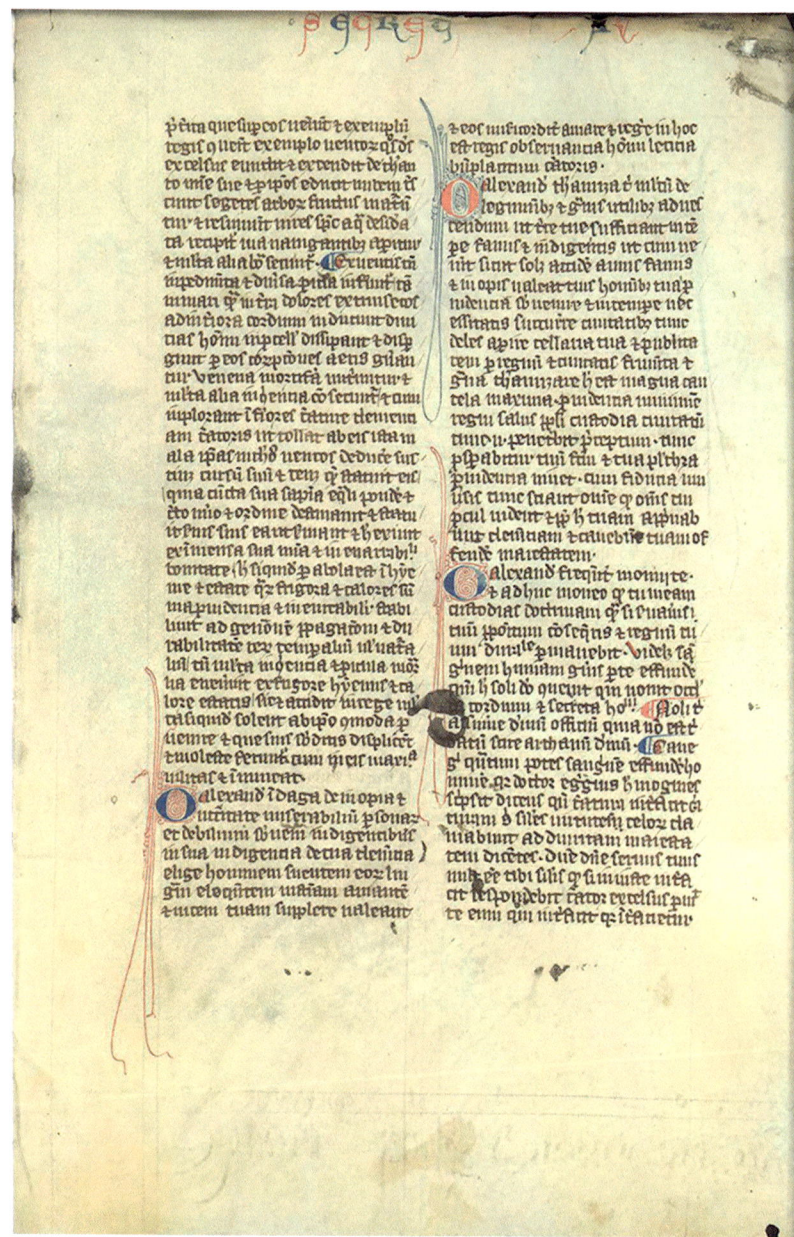

Figure 5.2. Stockholm, Riksarkivet Var 33, Pseudo-Aristotle, *Secretum secretorum* (The Secret of Secrets), fourteenth century. Photo: The Swedish National Archives, Stockholm.

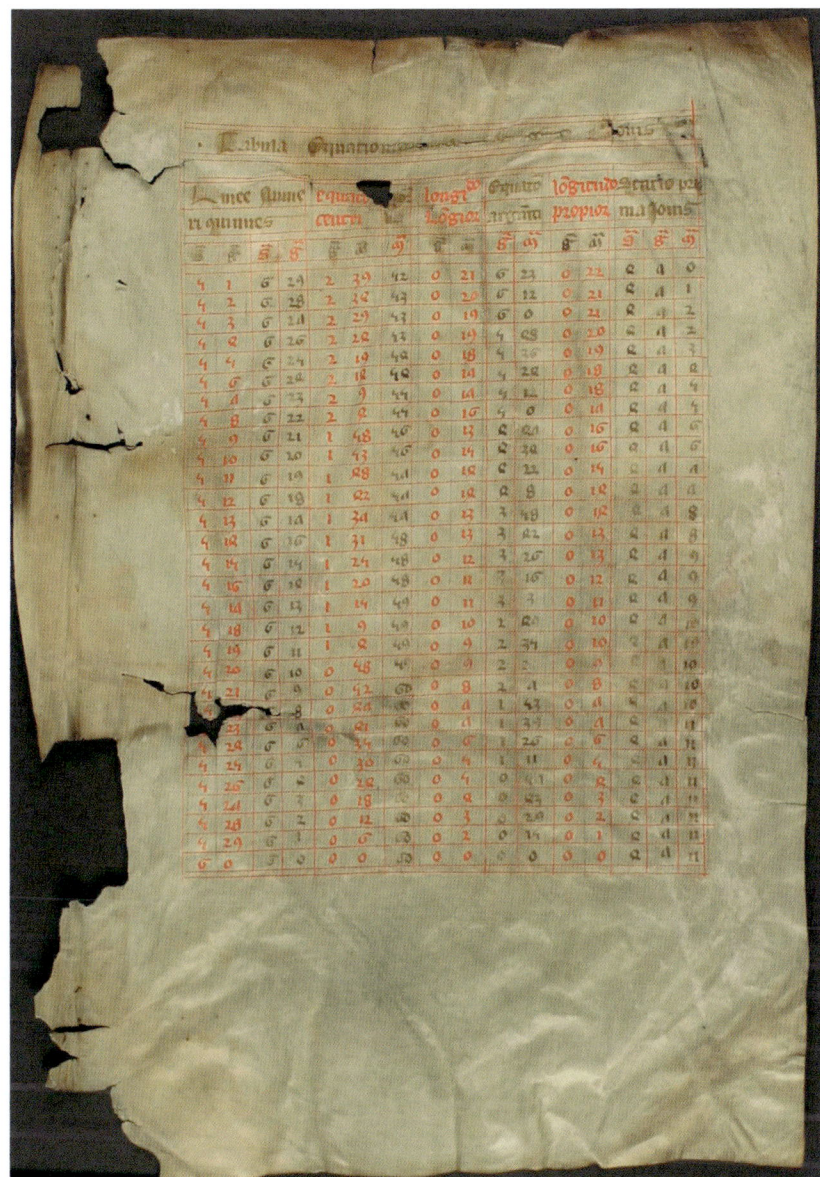

Figure 5.3. Stockholm, Riksarkivet Var 41, Arzachel, *Canones in motibus celestium corporum* (Rules for the Movements of the Heavenly Bodies), thirteenth century. Photo: The Swedish National Archives, Stockholm.

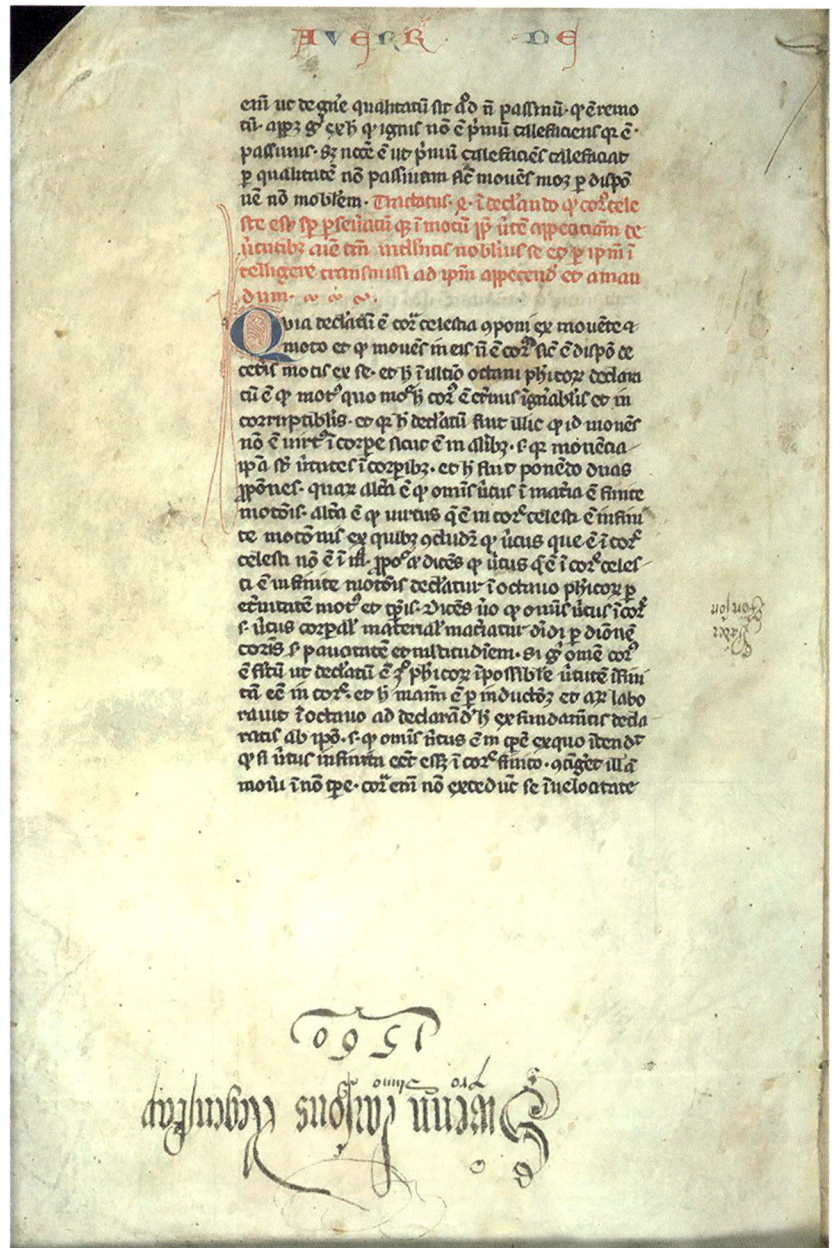

Figure 5.4. Stockholm, Riksarkivet SRA Fr 9399, Averroes, *De substantia orbis* (Concerning the Substance of the Celestial Sphere), fourteenth century. Photo: The Swedish National Archives, Stockholm.

Appendix

Works of Aristotle

De anima (On the Soul)
1. SRA Var 22 13th century *Book III* (William of Moerbeke translation)
2. SRA Var 53 13th–14th centuries *Book III*
3. SRA Var 3 14th century *Book III*

De caelo et mundo (On the Heavens)
1. SRA Var 21 13th century *Books II and IV*
2. SRA Fr 7714 15th century *Books I–IV* (Owned by Henrici de Ansbeke)[56]

De generatione et corruptione (On Generation and Corruption)
1. SRA Fr 7714 15th century (Owned by Henrici de Ansbeke)

De memoria et reminiscentia (On Memory and Reminiscence)
1. SRA Fr 9728 13th century
2. SRA Var 2 13th century (see Fig. 5.1)

De motu animalium (On the Movement of Animals)
1. SRA Var 2 13th century

De sensu et sensato (On Sense and the Sensible)
1. SRA Fr 9728 13th century
2. SRA Var 2 13th century
3. F.m.V.Var.4 14th century (From France in a William of Moerbeke translation)

De somnis et vigiliis (On Sleep and Sleeplessness)
1. SRA Var 30 13th–14th centuries

Metaphysica (Metaphysics)
1. SRA Var 2 13th century *Books VIII, IX, X, XI, XIII and XIV*[57]
2. SRA Var 3 14th century *Books I and VII*

---

56 A Henrici de Ansbeke of the Picardy Nation wrote this manuscript, presumably at the University of Paris in June 1449. Henrici probably came from Hansbeke in Flanders.

57 There are fourteen books in the William of Moerbeke translation from the Greek which he undertook sometime in the 1260s. Earlier translations were made from Greek by James of Venice (fl. twelfth century) and two anonymous translators in the twelfth century but they only translated thirteen books. An early thirteenth-century translation by Michael Scot (*c.* 1175–1232) translated eleven books from Arabic. Therefore this must be the version of William of Moerbeke and so dates the manuscript *c.* 1270–1300. See Borgo, 'Latin Medieval Translations of Aristotle's Metaphysics' pp. 19–30.

Physica (Physics)
1. F.m.V.Var.3 15th century[58]

Pseudo-Aristotelian Works

De coloribus (On Colours)
1. SRA Var 33 14th century

De lineis insecabilibus (On Indivisible Lines)
1. SRA Var 33 14th century

De physionomia (Physiognomics)
1. SRA Var 33 14th century

Liber de causis (Book of Causes)
1. SRA Var 2 13th century

Secretum secretorum (The Secret of Secrets)
1. SRA Var 33 14th century (see Fig. 5.2)

Aristotelian Commentaries

Albertus Magnus

De animalibus (On Animals)
1. SRA Var 1 and F.m.V.Var.2. 14th–15th century *Books II, III, IV, VI, XV, XIX, XXI, XXII,* and *XXIII*[59]

Physicorum libri cum commentario (Book of Physics with Commentary)
1. SRA Var 4 13th century
2. SRA Var 31 14th century *Books I* and *II*

Giles of Rome

Super De anima (Concerning 'On the Soul')
1. SRA Var 17 14th century

---

58 This version goes by the title *Physica novae translationis cum notis*, and so would seem to be one of the medieval commentators on Aristotle. F.m.V.Var.3 has been dated from 1434–1466 and coming from either Austria or Germany, which would indicate a student at one of the universities there such as Vienna or Leipzig as the likely owner of the work.

59 The appearance of this work in the manuscript fragment collection was first published in a brief description by Toni Schmid in 1955 Schmid, 'Albertus-Magnus-Fragmente in Schweden'. Both fragment collections are part of the same manuscript containing forty-five leaves in total. The manuscript has been dated to 1367–1433 and originates from France.

Thomas Aquinas

Commentarius ad libros de caelo et mundo (Commentary of On the Heavens and the Earth)
1. SRA Fr 7702 13th–14th centuries *Book II*

Expositio in Posteriorum (Exposition on Posterior Analytics)
1. F.m.V.Var.21 14th century

Astronomy and Computus

Arzachel

Canones in motibus celestium corporum (Rules for the Movements of the Heavenly Bodies)
1. SRA Var 41 13th century (see Fig. 5.3)[60]

Averroes

De substantia orbis (Concerning the Substance of the Celestial Sphere)
1. SRA Fr 9399 14th century Chapters three and four (see Fig. 5.4)

Anonymous Computus Collection

1. SRA Kal 19 11th–12th century[61]

---

60 The appearance of this work in the manuscript fragment collection was first published in a brief description by Toni Schmid in 1954. Schmid, 'Al Zarkali in Schweden'. The fragment contains the following tables: *De theorica motus circulorum Veneris et Mercurii* (On the Theory of the Circular Motion of Venus and Mercury), *Tabula equationis solis* (Table on the Equation of the Sun); *Tabula medii motus in annis Arabum ad meridiem civitatis toleti* (Table of the Medium Motion in the Arabic Years to the Meridian of the City of Toledo); *Tabula equationis, ad inueniendum eclipsim lune* (Table of Equations to find the Eclipses of the Moon); *Tabula equation[um] Iovis* (Table of the Equation of Jupiter); and *Tabula medii motus Martis in annis Arabum ad meridiem tholeti* (Table of the Medium Motion in the Arabic Years to the Meridian of Toledo). See Kennedy, 'A Survey of Islamic Astronomical Tables'.

61 Dated from the eleventh century in the collection, it may in fact date from the twelfth century. The recto side contains a calendar for the month of December and an illustration of the constellation Capricorn. The saints listed on this calendar are standard apart from the English saint Egwin. The interpretation of the calendar is of it coming from Germany but the inclusion of Egwin does create the possibility of an English origin. The verso side contains several computus texts. The left-hand column contains details of the *lunar regulares* and *ferial regulares*. It also details the epacts and concurrantes. The first two texts on the right-hand side are *None Aprilis* and *De epistolis Graecorum*. These are to be found in the *Bobbio Computus* Chapter five. See Mosshammer, *The Easter Computus and the Origins of the Christian Era*, pp. 206–07 and Jones, 'A Legend of St Pachomius'. My thanks to Immo Warntjes for help with the identification of these passages. The third is a passage on advent

Medicine

Alexander of Tralles

Practica (Practice)
1. F.m.V.Var.28 11th–12th centuries chapter on *De podagra* (On Gout)[62]

Articella

1. SRA Var 23 13th century (the *Prognostica* of Hippocrates and the *De urinis* of Theophilus)

Avicenna

Canon medicinae (The Canon of Medicine)
1. SRA Var 24 14th century *Book III*
2. SRA Var 28 14th century *Book III*
3. SRA Var 5 14th century *Book I*

Guillelmus de Saliceto

Summa conservationis et curationis (Concerning Preservation and Treatment)
1. F.m.V.Var.7, SRA Var 15 and London, British Library, Add. 34390 14th century[63]

Rhazes

Liber medicinalis Almansaris (Almansor's Book of Medicine)
1. SRA Var 40 13th century *Book IX*

---

that appears in the *Liber ordinarius* of St Arnulf in Metz. See *Der Liber ordinarius*, ed. by Ondermatt, pp. 53–54 and *Der karolingische*, ed. by Borst, p. 1542. The calendar needs further analysis to discover its origin but is certainly the oldest testament to the use of computus in Sweden.

62   Alexander of Tralles (c. 525–605) was a Byzantine physician who wrote his important work, the *Therapeutica* on medicine. His work was translated into Latin as *Practica*. There are two leaves which make up the fragment F.m.V.Var.28 which are dated to the eleventh to the twelfth century from Italy. The fragment has been described incorrectly in the catalogue as the Salernitan tract the *Antidotarium Nicolai*. The leaves are instead from the chapters on *De podagra* (On Gout). Langslow, *The Latin Alexander Trallianus*. I must thank Valerie Knight and Monica Green for this observation. See Knight, 'The De podagra (On Gout)'.

63   This is most likely from Italy. Swedish students studied at the University of Bologna in the fourteenth century, and *Summa conservationis et curationis* is written by Guillelmus de Saliceto, a Bolognese scholar. It would therefore be likely that Bologna would be the source of this manuscript. See Siraisi, *Medicine and the Italian Universities 1250–1600*, p. 39 n. 6. Siraisi, *Medicine and the Italian Universities 1250–1600*, pp. 41–44.

Encyclopaedias and Compilations

Bartholomaeus Anglicus

De proprietatibus rerum (On the Properties of Things)
1. SRA Var 25 14th century *Books XVII* and *XVIII*
2. SRA Var 26 14th century *Book XVIII*
3. SRA Var 27 and F.m.V.Var.6 14th century *Books VIII, XVII* and *XVIII* (England)
4. SRA Var 6 and F.m.V.Var.5 14th century *Books VI* and *VII* (England)
5. SRA Var 29 14th–15th century *Book XVI*
6. SRA Var 67 14th–15th centuries *Books XVII* and *XVIII*

Gaius Plinius Secundus

Naturalis historia (Natural History)
1. SRA Var 34 14th–15th century *Book XIII*

Pseudo-Burley

De vita et moribus philosophorum (On the Lives and Manners of the Philosophers)
1. SRA Fr 11713 15th century[64]

Works Cited

Manuscripts and Archival Sources

Bamberg, Staatsbibliothek, MS Bibl. 94
Helsinki, National Library, F.m.V.Var.2
—, F.m.V.Var.3
—, F.m.V.Var.4
—, F.m.V.Var.5
—, F.m.V.Var.6
—, F.m.V.Var.7
—, F.m.V.Var.21
—, F.m.V.Var.28
London, British Library, MS Add. 34390
Stockholm, Riksarkivet (SRA), SDHK 1903
—, SDHK 25006 and fragments Bi 67

---

64 This work was long misattributed to the English scholastic philosopher and logician, Walter Burley (c. 1275–1345). This same misattribution is recorded in the catalogue entry. It is now believed to be the work of an anonymous fourteenth-century Italian author: see Vittorini, 'Life and Works', pp. 41–42. Marenbon, *Pagans and Philosophers*, p. 107. Lutz, 'Walter Burley's De Vita et Moribus Philosophorum', p. 249. Stigall, 'The Manuscript Tradition', p. 45.

—, Fr 7702
—, Fr 7714
—, Fr 9399
—, Fr 9728
—, Fr 11713
—, Kal 19
—, Mi 1
—, Var 1
—, Var 2
—, Var 3
—, Var 4
—, Var 5
—, Var 6
—, Var 15
—, Var 17
—, Var 21
—, Var 22
—, Var 23
—, Var 24
—, Var 25
—, Var 26
—, Var 27
—, Var 28
—, Var 29
—, Var 30
—, Var 31
—, Var 33
—, Var 34
—, Var 40
—, Var 41
—, Var 53
—, Var 67
Stockholm, Royal Library, X770
Stockholm, Statens historiska museum, MS 21288
Uppsala, Universitetsbibliotek, C 28
—, C 381
—, C 579
—, C 620
—, C 627
—, C 629

Primary Sources

Bero Magni de Ludosia, *Disputata super libros De anima*, ed. by Robert Andrews (2017) <https://www.medeltid.su.se/Nedladdningar/Bero%20Magni%20Disputata%20De%20anima.pdf>

Chartularium Universitatis Parisiensis Sub auspiciis consilii generalis facultatum Parisiensium, ed. by Heinrich Denifle and Emile Chatelaine (Cambridge: Cambridge University Press, 2014)

Der karolingische Reichskalender und seine Überlieferung bis ins 12. Jahrhundert, ed. by Arno Borst (Hannover: Hahnsche Buchhandlung, 2001)

Der Liber ordinarius der Abtei St Arnulf vor Metz, ed. by Alois Ondermatt (Fribourg: Academic Press Fribourg, 1987)

Dyalogus creaturarum moralizatus, trans. by Monica Hedlund (Stockholm: Bra böcker, 1983)

Finnish Fragment Collection <http://fragmenta.kansalliskirjasto.fi/>

Mittelalterliche Handschriften der Universitätsbibliothek Uppsala: Katalog über die C-Sammlung: Bd. 1. C I–IV, 1–50, ed. by Margarete Andersson-Schmitt, Håkan Hallberg, and Monica Hedlund (Stockholm: Almqvist & Wiksell International, 1988)

Mittelalterliche Handschriften der Universitätsbibliothek Uppsala: Katalog über die C-Sammlung: Bd. 4. C 301–400, ed. by Margarete Andersson-Schmitt and Monica Hedlund (Stockholm: Almqvist & Wiksell International, 1991)

Mittelalterliche Handschriften der Universitätsbibliothek Uppsala: Katalog über die C-Sammlung: Bd. 6. C 551–935, ed. by Margarete Andersson-Schmitt, Håkan Hallberg, and Monica Hedlund (Stockholm: Almqvist & Wiksell International, 1993)

Studium Upsalense. Specimens of the Oldest Lecture Notes taken in the Mediaeval University of Uppsala, ed. by Anders Piltz, Skrifter rörande Uppsala universitet. C. Organisation och historia, 36 (Uppsala: Acta Universitatis Upsaliensis, 1977)

Svenskt Diplomatariums huvudkartotek över medeltidsbreven SDHK-nr: 1903 <https://sok.riksarkivet.se/sdhk>

Swedish Fragment Collection <https://sok.riksarkivet.se/MPO>

Secondary Studies

Abukhanfusa, Kerstin, *Mutilated Books: Wondrous Leaves from Swedish Bibliographical History* (Uppsala: Almqvist & Wiksell International, 2004)

Andrews, Robert, *Bero Magni de Ludosia Questions on the Soul: A Medieval Swedish Philosopher on Life* (Stockholm: Stockholm University Runica et Mediævalia, 2016)

—, 'Tideman of Närke's Termini naturales' in *Swedish Students at the Universities of Cambridge and Oxford in the Middle Ages*, ed. by Robert Andrews and Olle Ferm (Stockholm: Centre for Medieval Studies, University of Stockholm, 2017), pp. 39–138

Borgo, Marta, 'Latin Medieval Translations of Aristotle's Metaphysics' in *A Companion to Aristotle's Metaphysics* ed. by Fabrizio Amerini and Gabriele Galluzzo (Leiden: Brill, 2014), pp. 19–58

Brunius, Jan, *From Manuscripts to Wrappers: Medieval Book Fragments in the Swedish National Archives* (Växjö, Davidsons Tryckeri, 2013)

—, 'The Recycling of Manuscripts in Sixteenth-Century Sweden' in *Nordic Latin Manuscript Fragments: The Destruction and Reconstruction of Medieval Books*, ed. by Åslaug Ommundsen and Tuomas Heikkilä (London: Routledge, 2017), pp. 66–81

—, *Vasatidens samhälle: en vägledning till arkiven 1520–1620 i Riksarkivet* (Stockholm: Riksarkivet, 2010)

Ferm, Olle, 'Swedish Students and Teachers', in *Swedish Students at the University of Leipzig in the Middle Ages*, ed. by Olle Ferm and Sara Risberg (Stockholm: Stockholm University Runica et Mediævalia, 2014), pp. 11–66

—, 'Swedish Students in Cambridge and Oxford' in *Swedish Students at the Universities of Cambridge and Oxford in the Middle Ages*, ed. by Robert Andrews and Olle Ferm (Stockholm: University of Stockholm, 2017), pp. 11–25

—, 'Swedish Students in Vienna 1389–1491' in *Swedish Students at the University of Vienna in the Middle Ages*, ed. by Olle Ferm and Erika Kihlman (Stockholm: Stockholm University Runica et Mediævalia, 2011), pp. 11–88

Gullick, Michael, 'Preliminary Observations on Romanesque Manuscript Fragments of English, Norman and Swedish Origin in the Riksarkivet (Stockholm)', in *Medieval Book Fragments in Sweden*, ed. by Jan Brunius (Stockholm: Almqvist & Wiksell International, 2005), pp. 31–82

Götlind, Anna, *Technology and Religion in Medieval Sweden* (Falun: Sahlanders Grafiska AB, 1993)

Heikkilä, Tuomas, *The Fragmenta membranea collection* <http://fragmenta.kansalliskirjasto.fi/esittely/>

—, 'From Fragments Towards the Big Picture: Reconstructing Medieval Book Culture in Finland', in *Nordic Latin Manuscript Fragments: The Destruction and Reconstruction of Medieval Books*, ed. by Åslaug Ommundsen and Tuomas Heikkilä (London: Routledge, 2017), pp. 82–111

Heininen, Simo, 'Finland: Medeltiden', in *Ur Nordisk Kulturhistoria: Universitetsbesöken i Utlandet Före 1660, XVIII Nordiska Historikermötet, Jyväskylä 1981, Mötesrapport I*, ed. by Mauno Jokipii and Ilkka Nummela (Jyväskylä: Jyväskylän Yliopisto, 1981), pp. 67–75

Helmfrid, Staffan, *Vallentuna anno domini 1198: Vallentunakalendariet och dess tid* (Vallentuna: Vallentuna kulturnämnd, 1998)

Heyman, Harald J., 'Suno Karoli de Suecia och de Första Föreläsningarna i Astronomi vid Paris Universitet', *Lychnos: Lärdomshistoriska samfundets årsbok* (1950), 212–16

Jones, Charles W., 'A Legend of St Pachomius', *Speculum*, 18.2 (1943), 198–210

Kennedy, Edward Stewart, 'A Survey of Islamic Astronomical Tables', *Transactions of the American Philosophical Society*, 46.2 (1956), 123–77

Kihlman, Erika, 'Bero Magni de Ludosia: Student and Teacher', in *Swedish Students at the University of Vienna in the Middle Ages*, ed. by Olle Ferm and Erika Kihlman (Stockholm: Stockholm University Runica et Mediævalia, 2011), pp. 89–134

—, 'The Inventory of Bero's Library: An Edition with Analysis', in *Swedish Students at the University of Vienna in the Middle Ages*, ed. by Olle Ferm and Erika Kihlman (Stockholm: Stockholm University Runica et Mediævalia, 2011), pp. 135–74

Kleberg, Tönnes, *Medeltida Uppsalabibliotek, 1: Biskop Siward av Uppsala och hans Bibliotek* (Uppsala: Almqvist & Wiksells International, 1972)

—, *Medeltida Uppsalabibliotek*, II:*Bidrag till deras historia fram till år 1389* (Uppsala: Almqvist & Wiksells International, 1972)

Knight, Valerie, 'The De podagra (On Gout): A Pre-Gariopontean Treatise Excerpted from the Latin Translation of the Greek Therapeutica by Alexander of Tralles' (Unpubl. PhD thesis, University of Manchester, 2015) <https://www.research.manchester.ac.uk/portal/files/32297884/pdf>

Knorr, Wilbur R, 'The Latin Sources of Quadrans Vetus, and what they Imply for its Authorship and Date', in *Texts and Contexts in Ancient and Medieval Science Studies on the Occasion of John E. Murdoch's Seventieth Birthday*, ed. by Edith Sylla and Michael McVaugh (Leiden: Brill, 1997), pp. 23–67

Langslow, D. R., *The Latin Alexander Trallianus: The Text and Transmission of a Late Latin Medical Book*, Journal of Roman Studies Monograph, 10 (London: Society for the Promotion of Roman Studies, 2006)

Lehtinen, Anja Inkeri, 'From Fragments into Codices: On Reconstitution of Theological and Philosophical Works', in *Medieval Book Fragments in Sweden*, ed. by Jan Brunius (Stockholm: Almqvist & Wiksell International, 2005), pp. 109–31

—, 'Suomalaisia teologeja Sorbonnen kollegion kirjastossa 1400-luvulla', *Opusculum*, 7 (1987), 147–78

Lindberg, Sten G., 'Petrus Astronomus' astronomiska tabeller', in *Lychnos* (1973–1974) 250–56

Lutz, Cora E., 'Walter Burley's De Vita et Moribus Philosophorum', *The Yale University Library Gazette*, 46.4 (1972), 247–52

Marenbon, John, *Pagans, and Philosophers: The Problem of Paganism from Augustine to Leibniz* (Princeton: Princeton University Press, 2015)

Mosshammer, Alden A., *The Easter Computus and the Origins of the Christian Era* (Oxford: Oxford University Press, 2008)

Mulchaney, Marian Michèle, *"First the Bow is Bent in Study…": Dominican Education Before 1350*, Studies and Texts, 132 (Toronto: Pontifical Institute of Mediaeval Studies, 1998)

Niblaeus, Erik Gunnar, 'German Influence on Religious Practice in Scandinavia, c. 1050–1150' (Unpubl. PhD thesis, King's College London, 2010) <https://kclpure.kcl.ac.uk/portal/files/2932017/539896.pdf>

Ommundsen, Åslaug and Tuomas Heikkilä, eds, *Nordic Latin Manuscript Fragments: The Destruction and Reconstruction of Medieval Books* (London: Routledge, 2017)

Pedersen, Olaf, 'The Origins of the Theorica Planetarum' in *Journal for the History of Astronomy*, 12 (1981), 113–23

Roelvink, Henrik, *Franciscans in Sweden: Medieval Remnants of Franciscan Activities* (Assen: Van Gorcum, 1998)

—, *Riddarholmens Kryka och klostre: Varför är Sveriges kungar begravda hos franciskanerna?* (Stockholm: Veritas Förlag, 2008)

Sällström, Åke, *Bologna och Norden intill Avignonpåvedömets tid* (Gothenburg: University of Gothenburg, 1957)

Schmid, Toni, 'Albertus-Magnus-Fragmente in Schweden', *Beiträge zur Geschichte der Philosophie und Theologie des Mittelalters*, 4 (1955), 30–31

—, 'Al Zarkali in Schweden. Ein beitrag zur kenntnis der mittelalterlichen Sternkunde', *Classica et mediaevalia*, 15 (1954), 252–57

—, 'Om Sigtunabrödernas böcker och böner', in *Sigtuna Mariakyrka, 1247–1947*, ed. by H. Arbman (Sigtuna: Sigtuna fornhems förlag, 1947), pp. 45–82

—, 'Undersökningen av medeltida svenska bokfragment', *Scandia: Tidskrift för Historisk Forskning*, 6.1 (1933), 103–15

Schück, H., 'Svenska Pariserstudier under medeltiden', in *Kyrkohistorisk Årsskrift*, ed. by Herman Lundström (Stockholm: Norstedt & Söner, 1902), pp. 118–77

Siraisi, Nancy G., *Medicine and the Italian Universities 1250–1600* (Leiden: Brill, 2001)

Stigall, John O., 'The Manuscript Tradition of the De Vita et Moribus Philosophorum of Walter Burley', *Medievalia et Humanistica*, 11 (1955), 44–49

Vittorini, Marta, 'Life and Works', in *A Companion to Walter Burley: Late Medieval Logician and Metaphysician*, ed. by Alessandro Conti (Leiden: Brill, 2013), pp. 17–48

Walta, Ville, *Libraries, Manuscripts, and Book Culture in Vadstena Abbey* (Helsinki: University of Helsinki, 2014)

JOHNNY GRANDJEAN GØGSIG JAKOBSEN

Friars of Science

Dominican Transmission and Usage of Scientific Knowledge in Medieval Scandinavia

The Friars Preachers of the Dominican Order were learned men. Not only were they supposed to know the Bible by heart, but they were also expected to be well-acquainted with literature on how to fully interpret and explain the Good Book to others. As a means of achieving the latter, even the average Friar Preacher was therefore expected to master a basic knowledge of sciences beyond theology — such as astronomy, geography, natural sciences, physics, and medicine — since features from these schools could be used comparatively in sermons to explain the word and the will of God. Occasionally, however, some Friars Preachers took their scientific studies way beyond what was needed for the purpose of preaching. This chapter will demonstrate how such additional fields of science were transmitted by the Friars Preachers into Scandinavia and Northern Europe; how they were incorporated and used in Dominican sermons; how some of these topics were taught at the conventual schools; how on-going astronomical studies were recorded in Scandinavian-Dominican annals; and how some of the friars excelled so much in these additional studies that their theses became the leading scientific works in their respective fields.

The Dominican Order and its *raison d'être*

The Dominican Order or, more correctly, the Order of Preachers (*ordo predicatorum*) was founded in 1216. The basic idea of the new order was to form a mobile corps of elite preachers, all specialists in theology as well as in communication. Furthermore, the members were to live in accordance with what they were preaching, being good examples of Christian humility themselves, and for this purpose they were to be mendicants, that is beggars, only to live on alms and donations from the rest of society. The more profound aim of the Friars Preachers, as the members were called, was to fight heresy and paganism, and spread basic Christian knowledge among laypeople everywhere, to help improve their behaviour and increase their chances of salvation. Geographically, the Dominican Order took off in the south of France and north

---

Johnny Grandjean Gøgsig Jakobsen • is Associate professor at the Department of Nordic Studies and Linguistics at the University of Copenhagen. jggj@hum.ku.dk

of Italy, and by the beginning of the 1220s, only a few years after the foundation of the order, Friars Preachers were already being sent out to establish convents in all regions of Europe, including the more peripheral areas of the north and the east. Before 1228, Dominican convents in these frontier regions of Christianity were, just as in the rest of Europe, organized into provinces with a significant degree of semi-autonomy. For the region around the Baltic Sea, the convents were organized in three provinces (see Fig. 6.1): Dacia (for the kingdoms of Denmark, Norway, and Sweden with adjacent duchies), Polonia (Prussia, Pomerania, Poland, Silesia, and Bohemia, the latter segregated into its own province in 1301), and Teutonia (initially the entire German-speaking area, but in 1303 divided into Saxonia in the north and Teutonia in the south).[1]

The Order of Preachers was, as its name suggests, first and foremost an order of *preachers*. Preaching constituted the essential axis in the entire structure of the Order and life of the individual friars; it was the fundamental Dominican *raison d'être*. This is not to say that Dominican life was exclusively about preaching, nor that the Friars Preachers were the sole clergy practising this pastoral occupation. Within the secular church, preaching does not seem to have been considered a key assignment of the parish clergy, who were busy administering the Sacraments, but rather was something left for the prelates. Most canons secular and bishops, however, only had a limited theological insight, as their education and interests were usually focused on canon law, and it was based on personal gift rather than deliberate training if a prelate showed any communicative talent for giving sermons. Some other monastic orders — like the Cistercians, Augustinians, and Franciscans — also devoted serious attention to the pastoral task of preaching, but none of these facilitated equal conditions for studying the Bible and theological literature to that of the Order of Preachers. Right from its beginning, the Dominican Order depended less on the chances of inborn talents and divine interception, and instead built up a highly systematic and academic school system to nurture this knowledge, so that its friars could become the best possible preachers.[2] This school system worked both at an average level for the rank-and-file friars and at an elevated stage for the most advanced intellectual minds. Likewise, no other monastic order devoted more attention than the Dominicans to the practice of 'communication skills' in the form of preaching and disputing, since all this collected knowledge was considered worthless if it were not passed on effectively to others for the benefit of everyone. Thus, studying and understanding theology, and learning the practical skills to communicate theological knowledge to the rest of society, were the two basic pillars of medieval Dominican training. However, as will become evident in the following, most Friars Preachers *kunne mere end deres Fadervor* (knew more than their Pater Noster), to cite an old Danish saying that describes skills going beyond the

---

1   On the general history of the Dominican Order, see Hinnebusch, *The History of the Dominican Order*, I.
2   On the educational system of the Dominican Order, see Hinnebusch, *The History of the Dominican Order*, II, 3–280; and Mulchahey *"First the Bow is Bent in Study"*; on the educational system of the Franciscan Order, see Roest, *Franciscan Learning, Preaching and Mission*.

expected. Partly to increase reception of the key pastoral messages of their sermons among the intended target audiences along with what seems to be often personally motivated interests, numerous Friars Preachers also engaged in other academic fields of science outside their normal pastures of pure theology.

The Content of Dominican Convent Libraries

Even if Friars Preachers were, according to Dominican ideology, helped by the Holy Spirit to preach eloquently and convincingly,[3] the brethren could also rely on more temporal tools of assistance. In fact, the single largest element in the life of a Dominican friar was studying the Bible together with the extensive amount of related literature. For this purpose, every single convent was to be equipped with an adequate collection of books for friars to study, an activity that was far from limited to the young friars alone: as stated by Fr Humbertus de Romanis,[4] healthy preaching relied on theology and, thus, on studies of the Gospels, and no preacher should ever become so self-assured that he stopped repeatedly re-reading the Scriptures and educating himself with related thoughts and interpretations by others.[5]

All Dominican convent libraries in Dacia were dissolved at the time of the Lutheran Reformation. For a couple of them (Sigtuna and Tallinn) it is, however, possible to reconstruct what must be a rather good impression of their collective content, partly from surviving manuscripts later included in other libraries, partly from written references to now-lost material.[6] Such references also exist referring to a small number of books belonging in particular to other Dominican convents in Sweden. In total, we know of almost 200 literary works which at some point were part

---

3   St Dominic himself, allegedly, sent out young friars to preach even before they had studied theology with these words of comfort: 'Go in trust, because the Lord will be with you and he will put the words of preaching in your mouth'. Tugwell, *The Way of the Preacher*, pp. 31–35.
4   Apart from in the case of St Dominic himself, all names of Dominican and Franciscan friars will be given in the Latin form. *Fr* is in this article used as abbreviated honorific for the Latin title *Frater*, the standard prefix used in medieval texts for men of the regular clergy, here always used on mendicant clergy, corresponding to *Friar* in English and *broder* in Danish. *Dr* is used in its common meaning *Doctor*, although in this chapter always referring to doctors of theology. The orderly abbreviation *OP* after a name states an affiliation to the Dominican Order (*ordo predicatorum*), while *OFM* refers to the Franciscan Order (*ordo fratrum minorum*).
5   See Brett, *Humbert of Romans*, p. 156.
6   A number of manuscripts from the Dominican convent library in *Sigtuna* have survived as they were first transferred to the extensive Brigittine library of Vadstena Abbey, from where they eventually came to the University Library of Uppsala (as part of the so-called 'C-collection'). The Dominican connection is apparent from personal inscriptions in the books, but in addition to these, several other books from Vadstena without inscriptions may also have originated from Sigtuna. Furthermore, several references to named books that were donated or otherwise related to the convent in Sigtuna are preserved in diplomas. From the convent in *Tallinn*, a collection of sixteen medieval manuscripts and twenty-six incunabula and early printed works is preserved in the City Archives of Tallinn under the name of 'The Dominican Library'.

of a Dominican convent library in the province of Dacia. The bulk of this collective bibliography consists of theological literature and sermon collections. Among the standard theological classics recommended by the Order, we know for sure that Gratian's *Decretum*,[7] Gregory IX's *Decretales*,[8] Petrus Lombardus's *Libri sententiarum*,[9] Petrus Comestor's *Historia scholastica*,[10] and Fr Raymundus de Peñaforte OP's *Summa de casibus* were to be found in Dacian convent libraries.[11] The more challenging works by Augustine (*De civitate Dei*) and Fr Thomas Aquinas OP (*Summa theologiae*) were also present within the province,[12] along with several works by Fr Albertus Magnus OP,[13] famous Franciscan works like Fr Bonaventura's *Pharetra*,[14] and the versified Latin grammar book *Doctrinale puerorum* by Fr Alexander de Villa Dei.[15] Dominican libraries in Dacia also contained several works within neighbouring academic fields, likely to help the preachers deduct logical conclusions, and to enrich their sermons with philosophical and physical argumentation and examples. These include Petrus Hispanus's *Summulae logicales*,[16] and various commentaries on Plato and Aristotle.[17] Among the more curious finds in Dacian-Dominican libraries are a manual for sign language,[18] a compilation on medicine,[19] and a small treatise on the theological symbolism of chess.[20]

A locally produced example of such a literary aid from outside the normal sphere of theology, but most likely used for Dominican sermon preparation in Dacia, is a Danish

---

7   Sigtuna: *Diplomatarium Suecanum* (hereafter: DS) no. 2672 / *Diplomatarium OP Dacie* (hereafter: DOPD), accessible at <www.jggj.dk/DOPD.htm>, 1328 10/8; Västerås: DS no. 4716 / DOPD 1351 10/5). All published source instances referred to in this chapter concerning the Dominican province of Dacia have also been collectively web-published by the author in the DOPD.
8   Sigtuna: DS no. 2672 / DOPD 1328 10/8.
9   Visby: DS no. 9637 / DOPD 1377 27/12; Sigtuna: *Svenskt Diplomatariums Huvudkartotek* (hereafter: SDHK), accessible at <http://sok.riksarkivet.se/SDHK>, no. 25006. Tallinn: *Die Dominikaner in Livland*, ed. by von Walther-Wittenheim, pp. 33–34.
10 Stockholm: University Library of Uppsala (hereafter: UUB) 34:28; Sigtuna: Schmid 'Om Sigtunabrödernas böcker', p. 79.
11 Sigtuna: SDHK no. 25006.
12 Visby: DS no. 9637 / DOPD 1377 27/12; Schleswig: Schnabel, 'Mittelalterliches', p. 202.
13 e.g. Sigtuna: *Compendium theologice veritatis* and *De veris virtutibus* (see Schmid, 'Om Sigtunabrödernas böcker', pp. 75 and 78–79); Stockholm: *Mariale seu De Laudibus beatæ virginis Mariæ* (UUB 34:28, see Collijn, 'Svenska boksamlingar', pts 1 and 2 (esp. pt. 2, pp. 128–29).
14 Sigtuna: UUB C 254, cf. Schmid, 'Om Sigtunabrödernas böcker', p. 81.
15 Tallinn: *Die Dominikaner*, ed. by von Walther-Wittenheim, pp. 33–34.
16 Sigtuna: SDHK no. 25006; UUB C 620, see Schmid, 'Om Sigtunabrödernas böcker', p. 51.
17 SDHK no. 25006. Lindroth, *Svensk lärdomshistoria I*, p. 52.
18 UUB C 222, see Schmid, 'Om Sigtunabrödernas böcker', p. 52.
19 UUB C 579, see Schmid, 'Om Sigtunabrödernas böcker', p. 81. It includes a part of *Commentum super tabulas magistri Salerni* by Bernardus Provincialis de Salerno and *Summa de modo medendi* by Gerardus de Cremona.
20 *Die Dominikaner*, ed. by von Walther-Wittenheim, p. 30. The manuscript is even said to be of local Dominican origin, but as it was lost during World War Two, the exact content and authorship of the treatise is uncertain. If it was indeed Dominican and about chess, it may have been inspired by a similar treatise, popular in England, that was written by the Italian Fr Jacobus de Cessolis OP in the thirteenth century. Greatrex, 'Monks and Mendicants in English Cathedral Cities', p. 105.

collection of proverbs from the mid-thirteenth century called *Parabolae*. The oldest extant complete version of the *Parabolae* is a printed book from 1506, which is known to have been in the possession of an otherwise unknown Fr Gedeon Gedeonis of the Friars Preachers in Ribe in the early sixteenth century, as he has inscribed his name in it.[21] The collection, which had been compiled by a certain Petrus Laale, possibly a secular priest originating from the island of Lolland,[22] consists of 1200 proverbs in Latin with parallel translations in Danish. Several of the included proverbs are still commonly known phrases in the Danish language, such as 'A terra fugias vel morum convena fias: Man scal seedh følghe eller landh fly' (Follow the customs or leave the country); or 'Ampnen parvorum facit unda frequens fluviorum: Monghe becke oc smaa gøre een stoor aa' (Many small streams make a big river).[23] Its main medieval function seems to have been as schoolbook for students who were just beginning to learn Latin,[24] but this was hardly its purpose for Fr Gedeon, who rather seems to have used it the other way around, as a tool to understand and learn Danish proverbs that might be useful in his sermons. To judge from his name, Fr Gedeon was not a native Dane or Scandinavian, but was more likely from the Low German regions, which fits well with the fact that the convent in Ribe at this time was a well-integrated part of the Dutch Congregation, within which brethren were frequently transferred between nations. The collection cannot by any means be termed particularly religious in content; the bulk of the proverbs included relate to universal and more temporal life matters, most likely to originate from a lay peasant culture. However, whereas some phrases are positively vulgar and — while perhaps good for laughs between the friars inside the priory — hardly serious sermon material, numerous others could easily be included by a preacher to stress his point to the audience, and indeed, even to form a core element of the sermon.

Whereas treatises on sign language, medicine, and the 'common knowledge' of Petrus Laale may be attributed with some direct usefulness to a convent of Friars Preachers, works on less obvious topics could also be found in Dominican libraries around the Baltic Sea. The Friars Preachers in Lübeck are known to have possessed a rare two-volume set of Pliny's *Naturalis historia*, apparently transcribed in northern France or northern Germany in the early thirteenth century, and illustrated by a Danish artist, Petrus de Slaulosia. Already by the fifteenth century, this classical work was so difficult to obtain that the Medici family in Florence, after long negotiations, persuaded the Dominicans in Lübeck to send the two books to Italy to be copied in return for a safety deposit of 100 Rhine guilders. When it arrived in the 1430s, Cosimo de Medici was apparently so pleased with the Dominican manuscript that he gladly gave up the deposit and simply kept the books, which can still be seen at the Biblioteca Medicea Laurenziana in Florence.[25]

---

21 Petri Laale, *Parabolae*. The Royal Library, Copenhagen, N 1170 4°. *Peder Låles ordspråk*, ed. by Kock and Petersen, I, 6.
22 *Peder Låles ordspråk*, ed. by Kock and Petersen, II, 732–35.
23 Petri Laale, *Parabolae* nos 7 and 38.
24 *Peder Låles ordspråk*, ed. by Kock and Petersen, II, 736–40.
25 Schnabel, 'Mittelalterliches', pp. 177–78.

Dominican Theological and Theology-Related Literature

Friars Preachers not only indefatigably read and transcribed the works of others, probably more than any other religious order of their time, they also produced a solid stream of theological treatises and various kinds of manuals for pastoral care in general and preaching in particular. Such manuscripts were widely transcribed (and later printed) and distributed within the entire Order, including its northernmost province of Dacia.

The best-known Dominican work of them all is the already-mentioned *Summa theologiae* by the outstanding theologian and professor at the University of Paris, Fr Thomas Aquinas OP (d. 1274), a compendium for all the main theological teachings of the Catholic Church and probably one of the most influential works of medieval Western literature. The *Summa* was most likely a standard work in every Dominican convent library, perhaps in several copies. The Friars Preachers in Visby had even loaned two volumes of it to Archbishop Birger of Uppsala in 1377, who also had Aquinas's commentaries on the Gospels at his castle on Arnö.[26] A preserved version of one volume of the *Summa* at the Uppsala University Library is believed to originate from the Franciscan convent in Lund, from where it was later transferred to the Brigittine library in Vadstena, illustrating that this Dominican classic had an audience in Scandinavia outside the Order as well.[27] However, although the *Summa theologiae* seems to have been intended by Aquinas himself as a manual for beginners in theology, it apparently never became recommended reading by the Order for its own preachers, unless they were considered lector material.

The most prominent Dacian contribution to Dominican theological literature was authored by Fr Augustinus de Dacia (d. 1285), who was prior provincial of Dacia in the periods 1261–1266 and 1272–1285. At some point in his career he wrote the *Rotulus pugillaris*, a short manual for beginners in the study of theology, offering a condensed presentation of all useful and necessary theological knowledge that a friar needed to acquire before being allowed by his prior to preach. Fr Augustinus de Dacia's *Rotulus pugillaris* was apparently meant for the education of young Friars Preachers in his own province. We know that a copy existed at the Dominican convent in Helsingborg, as one of the two extant versions according to an inscription belonged to Fr Nicolaus Lagonis from this house;[28] he is most likely identical to a lector of that same name known in the convent in the 1370s.[29] This educational text book from Dacia may, however, also have played a role in the Order's teaching elsewhere

---

26 DS XI no. 9637 / DOPD 1377 27/12. The archbishop had borrowed the first two volumes of the *Summa*, which he was obliged to return to the friars or alternatively buy from them along with two other books for a total of 50 marks. The commentary, which belonged to the cathedral chapter of Uppsala, probably was *Catena aurea in quatuor evangelia*.
27 Roelvink, *Franciscans in Sweden*, p. 122.
28 'Liber fratris Nicholai Laghonis de conventu Helsingburgensi ordinis predicatorum provincie Dacie'. UUB C 647.
29 *Diplomatarium Danicum* (hereafter: DD), IX no. 35 / DOPD 1371–1380.

in Europe, especially in the German provinces, where the title appears in several library lists, and the only other extant copy was found at the Universitätsbibliothek in Basel.[30] A final indication of an international recognition of Fr Augustinus's *Rotulus* is given in the fact that it is included in a list of important authors and works of the Dominican Order.[31]

Friars Preachers also produced literary works outside actual theology and pastoral caretaking, although undoubtedly meant for theological considerations or as examples in sermons. Indeed, two of the most widely distributed Dominican works of medieval Europe were not really theological, but nevertheless considered some of the most prominent works of their time in these supplementary fields of science.

The French Fr Vincent of Beauvais OP (d. *c.* 1264) took it upon himself to compile a compendium of no less than all of the knowledge available to man at his time! This he did in three books dealing respectively with physics and nature (*Speculum naturale*), human skills and behaviour (*Speculum doctrinale*), and history (*Speculum historiale*), collectively known as the *Speculum maius* (The Great Mirror). To this was later added a *Speculum morale* on moral issues, mainly based on Fr Thomas Aquinas and Fr Stephanus de Borbone, and the treatise *De morali principis institutione* on the proper behaviour of princes. All the books are known to have been in Dacian-Dominican possession.[32] The *S. historiale* was given to the convent in Sigtuna in 1328, while the *S. naturale*, *-doctrinale*, *-historiale*, *-morale* and *De morali principis* were at the convent in Stockholm in the late fifteenth century.[33]

Probably the most famous of all Dominican works in the Middle Ages was the *Legenda aurea* by the Italian Fr Jacobus de Voragine OP (d. 1298). It is a *legendarium*, a collection of hagiographic legends concerning saints' lives structured according to the saintly calendar. Originally published around 1260 under the less showy title *Legenda sanctorum*, it soon earned its golden byname by becoming the undisputed 'bestseller' of the Late Middle Ages.[34] The overwhelming popularity of the *Legenda aurea* probably had to do with the narrative skills of Fr Jacobus, who was not only an excellent storyteller, writing in a simple Latin, but also a formidable compiler,

---

30 Universitätsbibliothek Basel, MS B X 9, fols 37ʳ–69ʳ.
31 '*Frater Augustinus, provincialis Dacie, scripsit libellum pro informatione predicantium, quem pugillarem rotulum nuncupavit*'. Schück, 'Svenska Medeltidsförfattare', 1, 161.
32 *De morali principis institutione* was sold by the Dominican convent in Lübeck (prov. Saxonia) to Fr Gregorius of the Friars Preachers in Stockholm in 1407. Both this and the *Speculum doctrinale* are still extant, the latter with an inscription stating that it was donated by Fr Laurentius Magni of the convent in Stockholm in 1482 from his private collection to the common convent library. '*Ad jubileum fratris Laurencii Magni sacre theologie magistri post eius absolucionem ab officio prioratus conventus Stokholmensis. Anno Domini MCDLXXXII*'. UUB 33:30 (*Speculum doctrinale*) and C 616 (*De morali principis*). Schnabel, 'Mittelalterliches', p. 178.
33 DS no. 2672 / DOPD 1328 10/8; Collijn, 'Svenska boksamlinger', pt. 1, p. 127, and pt. 2, pp. 128–36. In the period 1483–1485, Fr Clemens Ryting, lector of the convent in Stockholm, had lent out his copy of *S. historiale* to a Dominican colleague in Skara, Fr Gudmundus Benedicti, and subsequently to the cathedral chapter in Skara.
34 Over 800 manuscript copies have survived, far outnumbering any other literary work of the time, and in the period 1450–1500 it was even printed in more editions than the Bible.

smoothly bringing together different versions of the individual legends.[35] To several of the legends he added tales of miracles alleged to have occurred recently in connection to people praying for saintly help, often somehow including a Dominican involvement. However, as a continued reading for the layman the tales must have been quite monotonous, and Fr Jacobus clearly had not thought of his legendary as a work for popular entertainment, but rather as a compendium of saintly lore meant for the use of preachers in their sermons. *Legenda aurea* is also preserved in numerous editions in Scandinavian libraries, and although none of these or other references can be linked to Dominican convents, several Dacian-Dominican works, as will be shown in the following, were clearly inspired or even based upon the work of Jacobus Voragine, who must have been known to every Friar Preacher in the province, as well as most preachers outside the Order.

Friars Preachers of Northern Europe were able to contribute to this additional literary field as well. Late medieval translations of the *Legenda aurea* were made in most major languages, some of them by Friars Preachers, such as the *Anonymous Bohemus* (anonymous Bohemian friar) who in the fourteenth century wrote a Czech translation called *Passionál*.[36] Also widely distributed on a national scale was an independent legendary written in the vernacular around 1260, the *South English Legendary*, which contains ninety-two hagiographies written in verse and in a popular style that can be said to place narrative over theological concerns. This legendary may have been authored by English Friars Preachers,[37] but its monastic provenance is disputed. Dominican authorship is, however, plausible for another North European legendary in the vernacular produced in the province of Dacia, the *Fornsvenska legendariet* (Old Swedish legendarium), recognized as one of the most prominent literary works to come out of high medieval Sweden. It is, however, not really an independent work as much as an edited compilation of existing legendaria and chronicles from abroad that had been translated into Swedish, among them the above-mentioned *Legenda aurea* and the *Chronicon pontificum et imperatorum* by Fr Martinus Polonus OP, discussed further below. The extant and, as it seems, original version of the *Fornsvenska legendariet*, which has been dated to the 1290s, belonged to the convent of Dominican sisters in Skänninge, and its anonymous author/editor/translator was most likely a friar from the same town's male Dominican convent.[38] Closely related to this genre are two extant works made by two known Swedish Dominicans, Fr Israel Erlandi of Sigtuna and Fr Gregorius of Stockholm, who in the

---

35   More than 130 different sources have been identified for the *Legenda aurea*, although a great part of it is especially based upon two preceding collections of hagiographies of fellow Dominican origin: *Abbreviatio in gestis miraculis sanctorum* by Fr Johannes de Mailly OP, and *Epilogum in gesta sanctorum* by Fr Bartholomeus de Trent OP.

36   Machilek, 'Reformorden und Ordensreformen', p. 71; Koudelka 'Zur Geschichte der böhmischen Dominikanerprovinz' III, 57–59.

37   Hinnebusch, *The Early English Friars Preachers*, pp. 310–12.

38   Jansson, 'Fornsvenska legendariet', cols 518–22; Lindroth, *Svensk lärdomshistoria I*, pp. 70–71. The dialect of the Swedish text certainly points to a provenance from Östergötland, in which Skänninge was the sole Dominican base.

early fourteenth and early fifteenth centuries respectively wrote a hagiography on the king saint of Sweden, Erik, and a miracle collection connected to a picture of the *Defixionis Domini* kept in the Dominican priory church in Stockholm.[39]

Dominican Non-Theological Literature — and the Case of Dr Nicolaus de Dacia

Nonetheless Dominican authorship in Northern Europe extended way beyond pastoral caretaking and hagiographies. A world-historical chronicle, widely distributed all over Europe, was authored by Fr Martinus Polonus OP (Martin de Opava/Troppau, d. 1278), who was of Silesian origin, but seems to have spent most of his career in Rome as papal chaplain and confessor for a series of popes for about a quarter of a century. His *Chronicon pontificum et imperatorum* was novel in the way that it presented papal/ecclesiastical and imperial/secular history in two parallel sections making it highly useful for teaching.

Greatest international fame in the non-theological literary field for a Dacian Dominican was earned by Fr Nicolaus Johannis (a.k.a. 'Nicolaus de Dacia' or 'Nicolaus Lundensis'), who originated from the convent in Lund and was a doctor of theology, probably from the University of Paris. His more spectacular talents and interests were, however, more in the line of natural science, especially astronomy. He is first heard of in the late 1460s, when he was based at the Dominican convent in Leipzig, and here he calculated the exact time of the vernal equinoxes (that is, the two occasions per year when the sun is situated exactly above the equator). Shortly after, around 1470, he authored a considerable work on astrology and astronomy, *Libri tres anaglypharum*, which became fairly widespread and recognized throughout Europe.[40] The somewhat curious title is explained in the preface by the author himself, whose book aims to let important issues stand out as a relief (Lat. *anaglyphe*) at the expense of the superfluous. The work is solidly founded in traditional views on astronomy, taking no notice of the revolutionary new discoveries and interpretations of its time. Fr Nicolaus diligently refers to all his sources, of whom Fr Albertus Magnus OP is especially accredited, but his work also has close similarities to what was going on in contemporary Cracow. In addition to the main section on astronomy, the book also contains a geographical description of the world with special focus on Northern

---

39 The *Miracula sancte crucis Stockholmie* was written by Fr Gregorius of the Dominican convent in Stockholm in the fifteenth century. It is preserved at the Diocese Library of Linköping (Kh 27) and published as *Miracula defixionis Domini*, ed. by Lundén.
40 Author and work are included by the contemporary French astrologer Simon de Phares in a list of famous scholars in his field. The best-preserved version of *Libri tres anaglypharum* is found at the British Museum (Sloane MS, 1680 fols 48–130; the transcript was made in Wroclaw), incomplete versions at Bibliothèque Nationale (MS lat. 7336 and MS lat. 72923) and the University Library of Cracow (MS 18504 and MS 5455), minor extracts and references in Munich, Berlin, and Prague, and a now-lost copy has been in Venice. Jørgensen, 'Bidrag til ældre nordisk Kirke- og Litteraturhistorie', p. 197.

Europe, including Dr Nicolaus's own home province of Skåne, where he states that his work was written. More exotically, he also informs that the Christians in Greenland by now all had been extorted by pygmies![41]

It is worth noting that Fr Nicolaus de Dacia was not just some peculiar 'nerdy' friar, who had gone off on some academic side-track and completely lost touch with everyday Dominican life around him. On the contrary: since 1455, the convent in Leipzig had been reformed according to the ongoing Observant movement within the Dominican Order,[42] and his stay in Leipzig seems to have made the Danish Dr Nicolaus a convinced Observant as well, long before such ideas were introduced in his home province. In 1474, the master general of the Order complied with a request from the convent in Tallinn, Estonia, which was part of the province of Dacia, to have the house reformed as the first of the province, and it was Dr Nicolaus Johannis Lundensis who was put in charge of implementing the reform.[43] Apparently, however, Nicolaus soon fell out with the local assistants appointed for him in Tallinn; they complained about him to the master general, who finally had him removed from office the following year.[44] The reason for his removal is not known, but as he was replaced by a Dominican reformer from the so-called Dutch Congregation, which followed a slightly different line than that of Leipzig, it was most likely due to a disagreement about the exact nature of the reforms (unless maybe it was just due to personal issues with the Danish Dominican doctor). Nicolaus de Dacia certainly did not lose his esteemed position in the Order as such — at least not at this point — as he was then transferred to the convent of Greifswald, were he was matriculated at the university in 1477, and in 1480 it was even he who was chosen to promote the rector of the university with the doctoral degree in theology.[45] During his stay in Greifswald, Nicolaus Johannis was entrusted with an amount of eighty Rhine guilders with which he was to go to Rome and make the mandatory annual payment of his home province's contributions to the master general and the Order's procurator at the Curia. On arrival, his travel expenses were considered to have made an excessive dent into the amount, but after some controversies, he was finally acquitted for any further claims in the case in early 1480.[46] Money again became an issue in 1481, when the — until then well-esteemed — astronomer was suddenly expelled from the university in Greifswald and sent home in dishonour because he had earned a nice sum of money by practising medicine — obviously not considered a suitable *Nebengeschäft* for a doctor of theology. Dr Nicolaus had not done this out of financial concern for himself, as he is said to have sent the money back home to his 'native convent' in Lund; nonetheless, he was now sent the same way in grave shame.[47] However, Nicolaus de Dacia was apparently not exactly a 'stay-at-home' kind of friar,

---

41 Jørgensen, 'Om nogle middelalderlige Forfattere', pp. 248–53 and Jørgensen, 'Bidrag', pp. 197–98.
42 Löhr, 'Die Kapitel der Provinz Saxonia', pp. 9*-10*.
43 *Handlingar rörande Dominikaner-Provinsen Dacia* ed. Karlsson, Ch. IV:1 no. 6 / DOPD 1474 9/6.
44 *Handlingar* ch. IV:1 no. 19 / DOPD 1475 3/7. Walther-Wittenheim, *Die Dominikaner*, pp. 109–10.
45 *Kirkehistoriske Samlinger*, IV, ser. VI, pp. 359–61 / DOPD 1477 25/10 and 1480 13/11.
46 *Handlingar* ch. IV:2 no. 45 / DOPD 1480 25/1.
47 *La Congrégation de Hollande*, ed. De Meyer, pp. 114–15 / DOPD 1481 29/8.

since in 1484 he was transferred outside the province again, this time to Hungary, a province which, just like Dacia, was often used by the Order as a location for deporting friars who had somehow drawn unfortunate attention to themselves.[48] After that, he is never heard of again.

Other Signs of Dominican Science Practice — and Regulations Against it

It was not just Fr Nicolaus de Dacia, who had an astronomical interest within the Dominican province of Dacia. Several of the preserved Dacian-Dominican annals or yearbooks contain entries on solar eclipses and similar phenomena.[49] In particular, the chronicler of the Annales 980–1286, supposedly written at some Danish convent in the 1280s, had a profound liking for the topic. Not only did he note solar eclipses for 1229 and 1262, but for the year of 1240 he even noted a simultaneous eclipse of the moon and the sun (*sic*!): *Ecclipsis lunæ facta est et solis circa meridiem, et apparuit stella juxta solem* (an eclipse of the moon and the sun occurred around noon, and a star appeared next to the sun).[50] For 1282 it is stated that *Signum mirabile visum est quasi draco in multis terris* (a wonderful sign like a Dragon was seen in many lands).[51] And finally, in 1286, we are told that *In vigilia omnium sanctorum visum est signum mirabile quasi tres soles, et qui erat ultimus ascendit in medium coeli et factus quasi iris* (on the night before All Saints Day, there was a magnificent sign like three Suns, and the one which was the last rose up into the middle of Heaven and became like rainbow).[52]

Dr Nicolaus's medical interest was also shared by brethren elsewhere in his home province. From the convent in Skänninge, extant sets of notes taken by two students during classes provide us with a rare insight into what was taught at a Dominican convent school in the mid-fifteenth century. Whereas one of these classes was apparently on penitence, a rather predicatable topic for a Dominican school, the other was from a lesson given by Fr Olavus Johannis in 1461 on human anatomy.[53] More than an actual medical lecture, however, the notes instead give the impression

---

48 *Monumenta Ordinis Fratrum Praedicatorum Historica* (hereafter: MOPH) VIII, p. 383 / DOPD 1484 10/10.
49 For instance, the *Annales Dano-Suecani (916–1263)*, supposedly made at the Dominican convent in either Roskilde or Lund, briefly notes a solar eclipse in 1230, *Eclipsis solis facta*, which is probably the same event that another Dominican chronicle, *Annales Skeningenses*, dates to 1229. A more detailed entry in the latter chronicle, which most likely originates from Skänninge, informs of a solar eclipse in 1270 on Laetare Sunday (23 March) around 1 o'clock: *MCCLXX. Eclipsis solis contigit in dominica letare circa horam primam*. *Annales Ordinis Predicatorum de Provincia Dacie* (hereafter: *Annales*), accessible at <www.jggj.dk/annales.htm>.
50 *Annales*, Ann. 980–1286.
51 *Annales*, Ann. 980–1286.
52 *Annales*, Ann. 980–1286.
53 Schück, 'Ur Skänninges medeltidshävder', p. 109. Both small writs were later included in a mixed compilation made at the Brigittine Vadstena Abbey. UUB C 223.

of a philosophical introduction to physiology and natural science — possibly with the aim of being used in a pastoral context.

This is certainly how Fr Olavus might best have explained his lecture, if asked by any superior visitor of his Order. Since 1293 the Friars Preachers were formally forbidden to practise medicine by the Order itself, as the General Chapter this year apparently felt it necessary to strictly prohibit any such practical involvement with the art of medicine; the friars were only admitted to learn the bare minimum about its practice.[54] Indeed, it was a continuous and delicate balance for the Order of Preachers to endorse its brethren to study potentially relevant topics outside theology with the limitation of using such acquired knowledge in sermons only. Already at the General Chapter in 1243 the friars were prohibited from reading 'the books of the philosophers' or to 'make curious writings on them' themselves.[55] In the new edition of the Dominican constitutions of 1256, this restriction was slightly eased as the friars generally were not allowed to study 'the books of pagans and philosophers', nor to acquire any knowledge on the 'secular sciences' or 'liberal arts', unless they had been granted dispensation from this prohibition beforehand by the master general of the Order or the General Chapter.[56] From 1306, a formal dispensation could also be given by the prior provincial or the provincial chapter.[57]

One of the great Dominican masters, who were allowed a deeper engagement with science, was of course Fr Albertus Magnus ('Albert the Great'). Several of his works take up scientific questions on a philosophical level as well as more concrete descriptions on various topics, not least astronomy, minerals, plants, and animals; indeed, the long-standing division of physical bodies into 'the kingdoms of minerals, plants and animals' can be ascribed to Albertus Magnus.[58] Mention of the latter almost inevitably leads on to one particular area of the more dubious fields of medieval science often associated with the Order of Preachers': alchemy, something that no less than Fr Albertus Magnus and Fr Thomas Aquinas are said to have been devoted to. In his treatise on minerals, *Liber mineralium*, Albertus Magnus did critically discuss both the theories and experiments performed by alchemists since ancient time to his own days, but without expressing much trust in its hope for success. Rather than performing alchemy himself, he focused on the scientific arguments and approaches launched by alchemists, which he largely dismissed as

---

54 'Item, inhibemus districte, ne aliquis frater artem medicine exerceat, nec de medicina se aliquatenus intromittat, nisi prius in seculo audiverit et fuerit sufficienter instructus'. *Acta Capitulorum Generalium* (hereafter: ACG) 1293, in: MOPH III, p. 268.
55 'Item, fratres non studeant in libris philosophicis, nisi secundum quod scriptum est in constitutionibus, nec etiam scripta curiosa faciant'. ACG 1243, in: MOPH III, p. 26.
56 'In libris gentilium et phylosophorum non studeant, et si ad horam inspiciant. Seculares scientias non addiscant, nec artes quas liberales vocant, nisi aliquando circa aliquos magister ordinis vel capitulum generale volverit aliter dispensare: sed tantum libros theologicos tam juvenes quam alii legant'. *Constitutiones OP 1256*, in *Analecta sacri Ordinis*, pp. 172–73.
57 'In capitulo de studentibus, ubi dicitur: seculares sciencias non addiscant, post illud: vel prior provincialis, addatur: vel capitulum provinciale. Et hec habet ·iii· capitula'. ACG 1307, in: MOPH IV, p. 15.
58 Führer, 'Albert the Great', Ch. 6.

erroneous.[59] More specific alchemical treatises like *Libellus de alchimia* and *Secreta Alberti*, commonly attributed to the great Dominican scholar, were almost certainly not written by Albertus.[60] His pupil, Thomas Aquinas, also took up the question of alchemy, but only to engage in a theoretical and moral discourse on whether gold and silver produced by alchemy were to be considered as valuable as gold and silver found in nature.[61]

However, some Friars Preachers must have taken up the more practical side of alchemy as well, since the General Chapter of the Order in 1273 strongly admonished its friars not to be involved with the study of alchemy, let alone to teach it, write about it, or practise it.[62] And when all kinds of diabolical sorcery had entered the inquisitorial agenda in the 1310–20s, any Dominican involvement with alchemy was strictly forbidden at the General Chapter in 1323.[63]

Science used in Dominican Sermons: The Case of Fr Mathias Ripensis

Thus, the main purpose for friars of the Order of Preachers to engage with sciences outside theology in the first place was to improve their preaching skills. By taking

---

59 Kibre, 'Albertus Magnus on Alchemy'.
60 Kibre, 'Albertus Magnus on Alchemy', pp. 196–202; Eamon, *Science and the Secrets of Nature*, pp. 71–72.
61 'Ad primum ergo dicendum quod aurum et argentum non solum cara sunt propter utilitatem vasorum quae ex eis fabricantur, aut aliorum huiusmodi, sed etiam propter dignitatem et puritatem substantiae ipsorum. Et ideo si aurum vel argentum ab alchimicis factum veram speciem non habeat auri et argenti, est fraudulenta et iniusta venditio. Praesertim cum sint aliquae utilitates auri et argenti veri, secundum naturalem operationem ipsorum, quae non conveniunt auro per alchimiam sophisticato, sicut quod habet proprietatem laetificandi, et contra quasdam infirmitates medicinaliter iuvat. Frequentius etiam potest poni in operatione, et diutius in sua puritate permanet aurum verum quam aurum sophisticatum. Si autem per alchimiam fieret aurum verum, non esset illicitum ipsum pro vero vendere, quia nihil prohibet artem uti aliquibus naturalibus causis ad producendum naturales et veros effectus; sicut Augustinus dicit, in III de Trin., de his quae arte Daemonum fiunt'. *Summa theologiae*, II.77.2,1.
62 'Item, magister ordinis, de voluntate et consilio diffinitorum precipit districte, in virtute obediencie fratribus universis, quod in alchimia non studeant, nec doceant, nec aliquatenus operentur, nec aliqua scripta de sciencia illa teneant, sed prioribus suis restituant quam cito poterunt, bona fide per eosdem priores prioribus provincialibus assignanda'. ACG 1273, in: MOPH III, p. 170.
63 'Item, cum ars, que alchimia vocatur, sit in pluribus capitulis generalibus districte et sub gravioribus penis prohibita, et adhuc ex hoc in diversis locis ordinis pericula scandalosa surrexerint, precipit magister ordinis de diffinitorum consilio et assensu in virtute spiritus sancti et sub pena excommunicacionis late sentencie, fratribus universis, quod nullus in dicta arte studeat vel discat, operetur vel faciat operari, et scripta de ea, si que habet, nulla teneat, sed infra octo dierum spacium a noticia presencium ea destruat et comburat; in secus autem facientes sentenciam excommunicacionis incurrant, quam magister ordinis in capitulo tulit publice in scriptis et eos nichilominus, de quibus legitime constiterit, ex tunc adiudicat custodie carcerali. Si qui tamen tales aliquos sciverint operari et prelatis suis non manifestaverint, pene gravioris culpe debita subiacebunt'. ACG 1323, in: MOPH IV, p. 147.

in cases known from the physical world, the often rather intangible aspects of Christianity, and how to behave in a proper Christian way, could be more easily explained to common laypeople. For both the Friars Preachers and Friars Minor it was considered crucial to make sure that the core theological moral of their sermons was understood by the audience, and for this reason mendicant preachers were systematically trained in various communicative techniques. One particularly widespread method for doing this was the inclusion of *similitudines* in mendicant sermons, where difficult theological matters and moral issues were compared with everyday phenomena well known to the audience.[64] This could, for instance, take a mercantile approach, as in a Franciscan sermon by Fr Guilbertus de Tournai OFM, who compared sinning with taking unredeemable loans from an usurer, as the Devil took eternal mortgage in the sinner's soul, whereas the Dominican Fr Guillelmus Peraldus compared Lent with a seasonal market, since this was a particular good time to make a fine bargain (a shorter way to salvation) with the merchants (confessors) for a fair price (confession of your sins); only people with a childish mind, who do not understand the mechanisms of market economy, stated Guillelmus, can also believe that they will receive God's grace without giving anything in return.[65] A creative comparison of more regional character was launched by Fr Stephanus de Bourbon OP, who compared the value of indulgence with the Flemish tradition of pole-vaulting over the canals: in a similar way, he claimed, indulgence made it possible to 'pole-vault' one's way through purgatory.[66]

It was in this *similitudines* context of sermons that essentially all Dominican science studies were meant to adhere. Such semi-scientific *similitudines* can also be found in a sermon collection produced in the Dominican province Dacia, written by Fr Mathias Ripensis of the convent in Ribe in the early fourteenth century.[67] Fr Mathias Ripensis in several of his sermons makes use of physics and features of nature to explain biblical events or as comparisons for theological issues. This includes the sparseness of the desert, the currents of the ocean, and the movement of stars. In a more philosophical approach, in one place he compares the miracle of a blind man, who for a moment regains visual capability without losing his blindness, with the way that Mary kept her virginity after giving birth to Christ. Likewise, when God took human form, it is as if a piece of wood was turned into stone, meat, or a horse — something which could not happen in a natural way, due to the differences of their species. Fr Mathias also includes medical knowledge in his sermons, when the healing of leprosy allegorically is compared to the absolution of sins, just as both sickness and sin may end up ruining the entire body. And confessing ones heaviest sins during Lent and again just before Easter is according to Mathias just like when the winegrower prunes the vines in the

---

64 d'Avray, *The Preaching of the Friars*, pp. 252–53.
65 d'Avray, *The Preaching of the Friars*, pp. 208–11.
66 Maier, *Preaching the Crusades*, p. 118.
67 The *Sermones de tempore* by Fr Mathias Ripensis is transmitted in three different transcripts, all preserved at the Uppsala University Library: UUB C 342, C 343 and C 356. A full transcript edited by Johannes Schütz for his doctoral dissertation in 2013 is accessible at <www.jggj.dk/Mathias.htm>.

Figure 6.1. Dominican convents in the Baltic Sea region with provincial borders around 1500. Map by the author.

early spring and rakes up dead leaves in the fall.[68] One cannot, though, rightfully claim that Fr Mathias engaged in any natural-scientific, medical, or agricultural teaching of his audience, as all such elements only are used to visualize his theological points with comparisons to phenomena from the temporal world that he expected his audience to be familiar with.[69] In one sermon, Fr Mathias Ripensis even takes the opposite approach and uses the well-known biblical phenomenon of God's vineyard to explain the structure of the Holy Church. Both were in fact founded by Christ, the fence around the yard is like the guardian angel of the Church, the wine press is like the Cross of the Lord, and the tower in the middle of the vineyard is the Mother of God, connecting Heaven and Earth. The preachers, and here not least the Friars Preachers, are the cultivators of this allegorical vineyard, weeding and ditching the fields to keep Church and Christianity clear of sins.[70]

---

68 Mathias Ripensis, *Sermones de tempore*. UUB C 343, fols 34^r, 38^r, 49^r–49^v, 57^v, 59^r, 62^r–62^v. Schütz, *Hüter der Wirklichkeit*, pp. 178–79 and 196–97.
69 Schütz, *Hüter der Wirklichkeit*, pp. 180–81.
70 Mathias Ripensis, *Sermones de tempore*. UBB C 343, fol. 61^v. Schütz, *Hüter der Wirklichkeit*, pp. 174–75.

Concluding Remarks

Despite widespread rumours of an involvement with alchemy, medieval Dominican scholarship only appears to have struck gold in a more figurative sense with the medieval bestseller *Legenda aurea*. Even so, scholars of the Order did produce a solid stream of literature that was more properly grouped on the shelves of medieval science in the terms of natural philosophy, medicine, and astronomy, than on the more common Dominican-dominated shelves of theology and pastoral care. Even the ordinary rank-and-file friar of the Order was not only supposed to know the Bible by heart, but was also expected to master a basic knowledge on sciences beyond theology — such as astronomy, geography, natural sciences, physics, and medicine — since features from these schools could be used comparatively in sermons to explain the word and will of God. Occasionally, however, some Friars Preachers took their scientific studies way beyond what was needed for the purpose of preaching, such as the multi-talented Danish Dominican of the fifteenth century, Dr Nicolaus de Dacia, who — besides being a doctor of theology — also practised as monastic reformer, renowned astronomer, geographer — and university quack.

Works Cited

Manuscripts and Archival Sources

Basel, Universitätsbibliothek Basel, MS B X 9
Petri Laale, Parabolae. The Royal Library, Copenhagen, NKS 1170 40
Krakow, Jagiellonian Library, MS 5455
––, MS 18504
London, British Museum, Sloane MS 1680
Paris, Bibliothèque Nationale, MS lat. 7336
—, MS lat. 72923
Uppsala, Universitetsbibliotek (UUB), C 222
—, C 223
—, C 342
—, C 343
—, C 356
—, C 579
—, C 616
—, C 620
—, C 647

Primary Sources

Analecta sacri Ordinis Praedicatorum, vol. 3 (Rome: Istituto Storico Domenicano, 1897–1898)
Annales Ordinis Predicatorum de Provincia Dacie, Centre for Dominican Studies of Dacia (2010–2012), accessible at <www.jggj.dk/annales.htm>

La Congrégation de Hollande, ed. by A. De Meyer (Liège: Soledi, 1946)
Diplomatarium Danicum (Copenhagen: Det danske Sprog- og Litteraturselskab, 1938–)
Diplomatarium OP Dacie: Centre for Dominican Studies of Dacia (2005–), accessible at <www.jggj.dk/DOPD.htm>
Diplomatarium Suecanum (Stockholm: Riksarkivet, 1829–)
Die Dominikaner in Livland – Die natio Livoniae, ed. by Gertrud von Walther-Wittenheim (Rome: Istituto Storico Domenicano, 1938)
Handlingar rörande Dominikaner-Provinsen Dacia ('Historiska Handlingar' 18:1), ed. by K. H. Karlsson (Stockholm: Kungl. samfundet för utgifvande af handskrifter rörande Skandinaviens historie, 1901)
Kirkehistoriske Samlinger (Copenhagen: Selskabet for Danmarks Kirkehistorie, 1849–)
Miracula defixionis Domini. En mirakelsamling från Stockholms dominikankloster efter Kh 27 i Linköpings stifts- og landsbibliotek utgiven med inledning, översättning och register av Tryggve Lundén, ed. Tryggve Lundén (Gothenburg: Göteborgs högskola årsskrift 55:4, 1949)
Monumenta Ordinis Fratrum Praedicatorum Historica (Rome: Istituto Storico Domenicano, 1897–)
Peder Låles ordspråk, ed. by A. Kock and C. Peterson, 2 vols (Copenhagen: Samfund til Udgivelse af Gammel Nordisk Litteratur, 1889–1894)
Svenskt Diplomatariums Huvudkartotek (Stockholm: Riksarkivet), accessible at <http://sok.riksarkivet.se/SDHK>

Secondary Studies

d'Avray, David L., *The Preaching of the Friars – Sermons Diffused from Paris Before 1300* (Oxford: Clarendon Press, 1985)
Brett, Edward Tracy, *Humbert of Romans: His Life and Views of Thirteenth-Century Society* (Toronto: Pontifical Institute of Mediaeval Studies, 1984)
Collijn, Isak, 'Svenska boksamlingar under medeltiden och deras ägare', *Samlaren*, 23 (1902), 125–30, *Samlaren*, 24 (1903), 125–40
Eamon, William, *Science and the Secrets of Nature* (Princeton: Princeton University Press, 1994)
Führer, Markus, 'Albert the Great', in *The Stanford Encyclopedia of Philosophy* (Fall 2016 Edition), accessible at <https://plato.stanford.edu/archives/fall2016/entries/albert-great/>
Greatrex, Joan, 'Monks and Mendicants in English Cathedral Cities', in *The Friars in Medieval Britain: Proceedings of the 2007 Harlaxton Symposium*, ed. by Nicholas Rogers (Donington: Shaun Tyas, 2010), pp. 97–106
Hinnebusch, William A., *The Early English Friars Preachers* (Rome: Istituto Storico Domenicano, 1951)
—, *The History of the Dominican Order*, 2 vols (New York: Alba House, 1966–1973)
Jansson, Valter, 'Fornsvenska legendariet', in *Kulturhistorisk Leksikon for nordisk middelalder fra vikingetid til reformationstid*, 4 (Copenhagen: Rosenkilde og Bagger, 1959), pp. 518–22
Jørgensen, Ellen, 'Bidrag til ældre nordisk Kirke- og Litteraturhistorie', *Nordisk tidskrift för bok- och biblioteksväsen*, 20 (1933), 186–98

—, 'Om nogle middelalderlige Forfattere, der nævnes som hjemmehørende i "Dacia"', *Historisk Tidsskrift*, 8, 3 (1910–1912), 234–62

Kibre, Peal, 'Albertus Magnus on Alchemy', in *Albertus Magnus and the Sciences*, ed. by James A. Weisheipl (Toronto: Pontifical Institute of Medieval Studies, 1980), pp. 187–202

Koudelka, Vladimír J. 'Zur Geschichte der böhmischen Dominikanerprovinz im Mittelalter', *Archivum Fratrum Praedicatorum*, 27 (1957), 39–119

Lindroth, Sten, *Svensk lärdomshistoria I: 'Medeltiden – Reformationstiden'* (Stockholm: Nordstedt, 1975)

Löhr, Gabriel M., 'Die Kapitel der Provinz Saxonia im Zeitalter der Kirchenspaltung 1513–1540 (Einleitung)', *Quellen und Forschungen zur Geschichte des Dominikanerordens in Deutschland*, 26 (1930), 1–79

Machilek, Franz, 'Reformorden und Ordensreformen in den böhmischen Ländern vom 10. bis 18. Jahrhundert', in *Bohemia Sacra: Das Christentum in Böhmen 973–1973*, ed. by Ferdinand Seibt (Düsseldorf: Schwann, 1974), pp. 63–80

Maier, Christoph T., *Preaching the Crusades – Mendicant Friars and the Cross in the Thirteenth Century* (Cambridge: Cambridge University Press, 1994)

Mulchahey, Michèle, *"First the Bow is Bent in Study..." - Dominican Education Before 1350* (Toronto: Pontifical Institute of Medieval Studies, 1998)

Roelvink, Henrik, *Franciscans in Sweden* (Assen: Van Gorcum, 1998)

Roest, Bert, *Franciscan Learning, Preaching and Mission c. 1220–1650* (Leiden: Brill, 2014)

Schmid, Toni, 'Om Sigtunabrödernas böcker och böner', in *Sigtuna Mariakyrka, 1247–1947*, ed. by Holger Arbman (Sigtuna: Sigtuna fornhems förlag, 1947), pp. 45–82

Schnabel, Kerstin, 'Mittelalterliches Buch- und Bibliothekswesen geistlicher Gemeinschaften in Schleswig und Holstein', in *Klöster, Stifte und Konvente nördlich der Elbe*, ed. by Oliver Auge and Katja Hillebrand (Neumünster: Wachholtz Verlag, 2013), pp. 165–216

Schück, Adolf, 'Ur Skänninges medeltidshävder', in *Skänninge stads historia*, ed. by Anders Lindahl (Skänninge: Stadsfullmäktige i Skänninge, 1970), pp. 68–124

Schück, Henrik, 'Svenska Medeltidsförfattare', *Samlaren*, 12 (1891), 154–70

Schütz, Johannes, *Hüter der Wirklichkeit – Der Dominikanerorden in der mittelalterlichen Gesellschaft Skandinaviens* (Göttingen: V&R unipress, 2014)

Tugwell, Simon, *The Way of the Preacher* (London: Darton, Longman & Todd Ltd, 1979)

MARTEINN HELGI SIGURÐSSON

Master Perus of Arabia

An Exemplary Magician in Medieval Iceland

Meistari Perus, an illustrious man of learning and wisdom, plays a central role in the Old Icelandic *riddarasaga* (tale of knights) or courtly romance known as *Clári saga keisarasonar* (The Saga of Clarus the Emperor's son).[1] Invited from Arabia to teach Prince Clarus, the wondrously fair and learned son of Emperor Tiburcius of Saxony (Saxland), Perus becomes inadvertently (or so we are led to believe) the initiator of his pupil's quest for Serena, the supremely beautiful but haughty and cruel princess who governs France with her father, King Alexander. She is hostile to all suitors and far superior to the Saxon prince in her knowledge of science, wisdom, and guile, and after humiliating and painful attempts at wooing her in the French capital, Clarus finally succeeds with the help of his crafty tutor. Perus disguises Clarus as Eskilvarð, an extravagantly rich prince from Ethiopia, and by means of her avaricious longing for three marvellous pavilions belonging to Eskilvarð (and made by Master Perus), Serena is lured to lose her virginity and marry the visiting prince in the French capital. The newly wedded Serena thereafter undergoes a year-long and punitive test of her conjugal love at the hands of Perus in the guise of a black, hideous, and abusive vagabond-buffoon to whom the princess believes she has been tricked into marriage. Subjecting her to the rough life of vagrancy and begging, Perus brings about a dramatic reform of the French princess, who demonstrates long-suffering compassion in the course of her ordeal with her tramp-husband. Serena is ultimately united to her true husband, Clarus, at the imperial court in Saxony, and the saga ends with a moral exposition of the *ljós dæmi* (clear examples) of wifely virtues exhibited

---

1   In the following discussion, references are to Cederschiöld's 1907 edition, *Clári saga*. The saga was first edited by Cederschiöld in 1879 with the title *Clarus saga* and a Latin translation by Cavallin. For a German rendering by Maußer see 'Prinz Clarus'. Commentators and editors spell the title variously as *Clári, Clárus, Clarus, Klári, Klarus* or *Kláruss saga*. Nearly a decade before my paper at the Odense conference on 'Travelling Wisdom', I presented a shorter study of Perus at the Saga Conference of 2006 in Durham. For an abstract of this older paper, see Marteinn Helgi Sigurðsson, 'The Fantastic Feats of Master Perus'.

Marteinn Helgi Sigurðsson • holds a PhD in Old Norse from the University of Cambridge. His fields of research include medieval Scandinavian history and Old Norse literature. He is co-editor of the recent edition of *Jómsvíkinga saga* in the *Íslenzk fornrit* series (vol. 33). marteinnhs@gmail.com

by Princess Serena. The vacillation in the saga between the uppermost and basest levels of society is strikingly reflected in its eloquent style and colorful lexicon,[2] where Middle Low German loanwords are curiously frequent and the place-name Paris is conspicuously absent.

Besides his role in *Clári saga*, Perus appears as the chief protagonist or moralizing hero in a set of three far shorter tales that are likewise only attested in Old Icelandic.[3] His name has the same Latin declension as in the saga, and he is again entitled *meistari*, that is 'master' in the scholarly or sapiential sense of the term (Latin *magister* is, however, never used in these tales or the saga). Perus is not in this context presented as a hired tutor, and neither is he associated with Arabia or the Orient. He is, however, similarly depicted as an itinerant sage of high learning and a master-magician, and each of the three tales concern his dealings with powerful but demonstrably unwise and unjust noblemen who suffer great humiliation while Perus remains unscathed and free to wander on from place to place (none of these places are, however, given a name). The earliest extant manuscript of both *Clári saga* and the three tales devoted to Perus bears the shelfmark AM 657 a-b 4to. It dates from around 1350 and is housed in Copenhagen at Den Arnamagnæanske Samling.

Marred by many illegible places, defective leaves, and lacunae, the Copenhagen manuscript consists largely of translations or adaptations of Latin sermon-tales, that is, edifying and amusing 'exempla', of foreign origin. AM 657 4to contains the largest compilation of such exemplary tales in Old Icelandic,[4] and the three exempla about Perus are recorded together on fols 30^v–33^r under the heading *Af meistara Pero ok hans leikum* (Of Master Perus and his Feats).[5] The beginning of *Clári saga* is lost in 657 on account of a lacuna, and the surviving text occupies fols 83^r–88^v and, following another lacuna, fols 89^r–90^v. The saga's beginning is preserved in its second oldest manuscript, Cod. Holm. Perg. 6 4to, which is dated to *c*. 1400–1425 and housed in Stockholm at Kungliga Biblioteket. *Clári saga* occupies the final place (on fols 128^v–137^v) in this Icelandic anthology of thirteen romances (*riddarasögur*), where it immediately follows *Mǫttuls saga*, a (prose) translation of the Old French fabliau *Le lai du cort mantel*, which concerns a chastity-testing cloak at King Arthur's court.[6]

---

2   For a detailed study of the saga's wealth of terms derived from Middle Low German see Kalinke, '*Clári saga*: A Case of Low German Infiltration'.
3   The three tales in question are edited by Gering as nr. LXXXI in *Islendzk æventyri* I, 217–31. Gering provides a German translation in *Islendzk æventyri* II, 159–65. These tales were first edited in 1860 by Konráð Gíslason in his *Sýnisbók*, pp. 419–27.
4   An overview of the manuscript's contents is found in Hjalti Snær Ægisson, *Þýdd ævintýri*, pp. 203–06.
5   The first two tales about Perus in 657 are also preserved (in the same order) in AM 586 4to from *c*. 1450–1500, where they appear in a collection of several other exempla. AM 343 a 4to, from *c*. 1450–1475, contains all three tales (in the same order) at the end of another miscellany that contains mostly Norse sagas of the legendary past (*fornaldarsögur*). Gering uses the latter manuscript in the variant apparatus in *Islendzk æventyri* I, 217–31, and AM 586 in the apparatus on pp. 217–27.
6   The romances are all written in prose as is in fact the case with all surviving Old-Norse Icelandic romances or *riddarasögur*. On the contents of the Stockholm manuscript see Slay, 'The Original State', and Slay, 'Introduction'. The continued popularity of *Clári saga* from *c*. 1400 until the

Clári saga opens with a statement concerning its provenance in the form of a written Latin poem in rhyming and rhythmical verse that was 'found' in France by Jón Halldórsson, bishop of Skálholt in Iceland (1322–1339), and transmitted by him through his retelling or translation:

> Þar byrjum vér upp þessa frásǫgn, sem sagði virðuligr herra Jón byskup Halldórsson, ágætrar áminningar, – en hann fann hana skrifaða með látínu í Frannz í þat form, er þeir kalla 'rithmos', en vér kǫllum hendingum, – ok byrjar svá: At Tíburcíús, Saxlands keisari, stýrði sínu ríki með miklum heiðri ok sóma.[7]

> (There we begin this story, as it was told by the venerable Lord Bishop Jón Halldórsson of excellent memory — he found it written in Latin in France in the form they call 'rithmos', but which we call rhyme, — and it begins thus: Tiburcius, Emperor of Saxony, governed his realm with great honour and glory.)[8]

Jón Halldórsson of Skálholt (Johannes Skalholtensis), the bishopric of southern Iceland, was a *prédikarabróðir*, that is a Friar Preacher (i.e. Dominican), renowned for telling delightful moral stories, 'adventures' or exempla, *æfintýr* or *dæmi(sögur)*, that he had encountered abroad, and Jón is known to have studied in the French capital, Paris, around 1300 when Philip the Fair was king of France.[9] No Latin source for *Clári saga* is, however, known to exist, and neither is any version attested in another language although the storyline bears similarities with international tale-types such as the 'Taming of the Shrew', 'Patient Griselda', and the Brothers Grimm *märchen* known as 'König Drosselbart' (King Thrushbeard).[10] Although Serena is presented as a co-ruler and never called a king (*kóngr/konungr*), *Clári saga* is commonly held to be the oldest *riddarasaga* to involve a 'maiden king' (*meykóngr* or *meykonungr*) figure, and since this type of female royal ruler — a maiden monarch cruel and antagonistic to all suitors — appears to be unknown outside Old Icelandic fiction, there is reason to suspect that the saga was in this respect influenced by indigenous narrative tradition.[11]

None of the personae in *Clári saga* appear in other narratives with the exception of Perus, and he is otherwise only known from the three Old Icelandic exempla concerning his magic feats. In view of his part in *Clári saga*, Alfred Jakobsen has tentatively suggested that Bishop Jón might also be the source of the three exempla about Perus.[12] Parallels in terms of motifs and plotline demonstrate that the three

---

nineteenth century is indicated by the fact that versions of it are extant in over twenty known manuscripts. For a list of known surviving manuscripts see Kalinke and Mitchell, *Bibliography*, pp. 72–73.

7   *Clári saga*, p. 1.
8   Translation mine, as elsewhere below.
9   On Jón's educational and Dominican background see further below.
10 On the last-named tale-type, see the monograph of Philippson, *Der Märchentypus von König Drosselbart*. The folktale connections of *Clári saga* are briefly treated in Düwel, 'Clárus saga'.
11 See further Kalinke, '*Clári saga, Hrólfs saga*', pp. 276–77.
12 Jakobsen, 'Ævintýri', p. 616.

shorter tales are partly inspired by (or modelled on) foreign narratives.[13] This is most evident in the third exemplum, where Perus becomes a kingmaker by means of his powers of illusion. It relates how a certain (unnamed) duke, *hertogi*, possessed many warships but no domain on land, and he is said to have been regarded as a righteous (*réttlátr*) man of his *stétt* or social standing. At one point, when the duke lies at anchor and has his cooks prepare a cock for dinner, Perus appears before him and proposes to make him a king if he so wishes. The duke accepts the stranger's offer, and Perus soon arranges for this advancement in power and dignity to happen in a kingdom nearby. Perus subsequently visits the newly made king to collect the ten marks of gold they had agreed he would pay Perus annually for the kingship (the duke had eagerly offered 200 marks, but Perus was content with the ten he initially requested). The erstwhile duke pays the sum gladly at first, but he does so with great reluctance the year thereafter. When Perus returns for the third time, the king refuses with anger to pay any further tribute, threatening to have Perus imprisoned or killed. Reminding the king of the circumstances of their first meeting, Perus therewith reveals, at the very moment the cock is fully cooked, that all the events subsequent to that first encounter were but a magic illusion, *sjónhverfing*, created by Perus in order to put the duke's integrity to the test. With Perus having thereby exposed that the duke was in fact unjust and had lusted for power, the tale concludes with a moral on the corruption, and ingratitude, that so often follows the gaining of power:

> Þetta sama snertr marga ofmjök; þvíat þat er ofgjarnt valldinu, at metnaðrinn spillir ok vill einn öllu ráða; virða ok stundum oflítils þat sem vel var áðr við hann gjört en hann fekk valldit. Lýkr hèr at segja af Pero at sinni.[14]

> (The same affects many a man overmuch; for power is greatly inclined to make the ambition of a man corrupt him so that he wants to rule everything alone. Sometimes he also appreciates too little favours he received before he gained power. Tales of Perus end here for the time being.)

The concluding phrase, 'at sinni' (for the time being), evidently refers to Perus's reappearance in *Clári saga* much later in 657, where his Arabian background must have been mentioned in the now-lost beginning of the saga. In the beginning preserved in the Stockholm manuscript, Perus is introduced as a famous *meistari* living 'út í Árábía' (out in Arabia), and his summoning from Arabia is in any case recalled later on in the saga according to both 657 a-b 40 and Cod. Holm. Perg. 6 4to. (see below).[15] These

---

13 See the comments in *Islendzk æventyri*, ed. by Gering, II, 165–69; Krappe, 'The Delusions of Master Perus'; *Ævintýri frá miðöldum*, ed. by Bragi Halldórsson, p. 168, and Hjalti Snær Ægisson, *Þýdd ævintýri*, pp. 62–63.
14 *Islendzk æventyri*, ed. by Gering, I, 231.
15 *Clári saga*, pp. 3–4. For the context of this depiction of Perus in the saga see below. AM 586 4to, which does not contain this final tale in the Perus trilogy, concludes the second tale with the comment 'ok kemr hann víða við sögur. Þar fór núna' ('and he appears in many tales. So it now went'). See *Islendzk æventyri*, ed. by Gering, I, 227 (apparatus). In AM 343 4to, which contains all of the three tales in 657, but not *Clári saga*, the final tale ends: 'Höfum vèr nú eigi meira heyrt sagt af meistara P. at sinni, ok lýkz hèr þersu æventýri' (Now we have not heard more said about master P. for the time being,

circumstances concerning Perus in AM 657 may be significant to the question of his ethnic identity. Perus is in fact nowhere expressly said to be an Arab (or Saracen), and neither does he show any signs of allegiance to Islam, the religion of Muslims.[16] Judging from the situation in 657, the reader might gather that Perus had at some point travelled to Arabia and, having there acquired further wisdom, science, or magical skills, returned to the West when summoned to Saxony. Speculation along such lines remains limited by the fact that it is left unclear where his previous adventures took place, although the personal names involved (Vilhjálmr, Eiríkr, Ingibjǫrg and Prinz) give the impression that the events are supposed to be set somewhere in Europe.[17]

The tale about Perus and the deluded duke has numerous analogues in medieval European literature. Variants appear in such works as Don Juan Manuel's *El Conde Lucanor*, a Castilian collection of exemplary tales completed *c.* 1335, and two influential Dominican collections of exempla intended for use in sermons, Stephanus de Borbone's (d. 1261) *Tractatus de diversis materiis praedicabilibus* and Humbertus de Romanis (d. 1277) *De dono timoris*.[18] AM 657 contains another Icelandic variant of the selfsame tale-type within a lengthier and composite exemplum (on fols 44^v–49^r) concerning the feats of a wise, just, and most powerful schoolmaster in Paris who assumes (in the first part of this piece) a magical role corresponding to that of Perus whereas a poor student-protégé takes the place of the landless duke.[19] This exemplum set in Paris might well stem from Bishop Jón Halldórsson's storytelling, and the ultimate origin of the didactic tale-type in question is perhaps Oriental as Alexander Haggerty Krappe has suggested.[20]

The name Perus is never explained in the sources, and Shaun Hughes has proposed that it 'may be a variant of Porus, the King of India, who was one of the main opponents of Alexander the Great'.[21] The Indian warrior-king Por(r)us, who was famously subdued by Alexander, was a widely known figure in the days of Bishop Jón, and he appears, for example, in Walter of Châtillon's twelfth-century epic poem *Alexandreis*, which was (as Hughes notes) translated into Icelandic by Brandr Jónsson, who was bishop of Hólar, Iceland's northern diocese, in 1263–1264.[22] While the names Perus and Por(r)us sound similar and share the same Latin declension (gen. -*i*, dat. -*o*, acc. -*um*), the two figures are quite dissimilar, and as an alternative conjecture it

---

and here this adventure ends). Oddly, although 586 does record the second tale, the first tale there concludes with these words: 'ok lýkr hèr at segja frá meistara Perus, en guð geymi vár allra, amen' (and here ends the account of Master Perus, and may god protect us all, Amen).

16   On Perus's ties to Arabia see further below.
17   These personal names appear only in the first two tales. The third tale likewise contains no place-name, and its only personal name is Perus. See further below.
18   See Hjalti Snær Ægisson, *Þýdd ævintíri*, pp. 61–63.
19   This tale is edited in *Íslendsk æventyri*, ed. by Gering, I, 256–67. Gering notes this as a variant of the third exemplum about Perus in *Íslendsk æventyri* II, 169 and 202–03. For further discussion see Hjalti Snær Ægisson, *Þýdd ævintýri*, pp. 61–62.
20   Krappe, 'The Delusions of Master Perus', pp. 220–22.
21   Hughes, '*Klári saga* as an Indigenous Romance', p. 53 (fn. 69).
22   For further information on Brandr and his prose translation or retelling of the *Alexandreis*, see Wolf, '*Gyðinga saga*, *Alexanders saga*, and Bishop Brandr Jónsson'.

may be suggested that the name-form Perus is derived from Latin Petrus, i.e. Peter, a name that often loses the intervocalic *t* in vernacular variants (such as Old Provençal Peire or Middle English Piers). Such an etymology might not seem far-fetched when one considers Perus's qualifications as a champion of the distinctive clerical class of 'vagabond' scholars that were on the loose in Western Europe during the twelfth and thirteenth centuries (see further below).

Shortly after *Clári saga*'s incipit, we find the commonplace notion or literary topos of the transfer of learning, *translatio studii*, from East to West in a description of the education of the young Prince Clarus of Saxony, whose name is said to mean *bjartr* (bright) on account of his great beauty.[23] Endowed with a wondrous *skilningr* (understanding, discernment, or intellect), Clarus rapidly masters the *sjǫfaldar listir* (sevenfold arts), i.e. the seven *artes liberales*, so well that he could be called the supreme learned master, *yfirmeistari*, 'allra þeirra, sem í þessum þriðjungi váru, er Erópa heitir' (of all men in this third part of the world named Europe).[24] Eager to enhance the education of his precocious son, Emperor Tiburcius is then said to have searched for a *typtunarmeistari* (disciplinary master (of courtly conduct))[25] who could teach Clarus arts, *listir*, that were *ósénar ok fáheyrðar* (unseen and rarely heard of) in those lands, and Perus is introduced at this early point in the saga as follows:

> Í þenna tíma spurðiz af einum mektugum meistara út í Árábía, sem Pérús hét at nafni, frábærrar speki ok visku yfir fram alla menn í verǫldinni, af hverjum víða er lesit í bókum ok mǫrg æfintýr við snertr af sínum listum ok klókskap. Svá mikit berz keisarinn fyrir, at hann gørir sendiboða svá langan veg lands ok lagar, at hann lokkar til þenna meistara sik heim að sœkja með fǫgrum féboðum ok blíðum fyrirheitum, til þess at hans son megi hluttakari verða hans margfróða meistaradóms.[26]
>
> > (At this time there came news of a magnificent master out in Arabia who bore the name Perus, who in his outstanding sagacity and wisdom surpassed all men in the world and of whom one reads widely in books and on account of whose arts and cunning appears in many adventures. The emperor therewith became so eager that he sent emissaries on a long journey by land and sea in order to entice this master to his residence with fair offers of wealth and kind promises so that his son might partake in the multifarious learning of his mastership.)

---

23 *Clári saga*, p. 2.
24 *Clári saga*, p. 3. Europe is here conceived of as one third (*þriðjungr*) of the world along with Asia and Africa in accordance to the conventional cosmography of the time. Africa has a (albeit fake) representative in *Clári saga* when Clarus is disguised by Perus as a wealthy prince from Bláland). See further below.
25 On the precise meaning and coutly connotations of this term, a Middle Low German loanword, see Kalinke, '*Clári saga*: A Case of Middle Low German Infiltration', p. 19.
26 *Clári saga*, pp. 3–4.

Clarus finds his previous education to have been mere child's play in comparison with the great *klerkdómr* ('(scholarly) learning', but literally 'clerical learning') of his new master, and after a long period of study under Perus, and while the two are one day taking a walk (*spázérandi*), the master presents his pupil with a *lítit æfintýr* (little adventure). In this educational 'adventure' Perus describes how the most beautiful Princess Serena of France resides with grandeur in a great tower overlooking the capital of the king (*í konungsins hásætisborg*) and from where she governs the realm along with her father (King Alexander) on account of her great wisdom. She is attended by a lioness and sixty high-born ladies-in-waiting, and her tower is guarded by a thousand knights who let no man enter except for male servants and the king. Perus concludes his account of Serena by stating that she is so wise that she would deem Clarus's learning as being no more than that of a country bumpkin (*akrkarl*, literally 'field-worker'). The prince is then given the task of composing five verses (*fimm versa*) on this subject-matter before the following morning at prime, but Clarus becomes so captivated by the words of his master that he can neither versify nor sleep and feels compelled to see Serena and marry her. Perus seeks in vain to dissuade Clarus by informing him of Serena's great cunning and maltreatment of all previous suitors, and the master refuses to join Clarus on a wooing mission. With his father's consent, Clarus nevertheless sails with great pomp, but without Master Perus, to the capital of the French king, *hásætisborg Frakkakonungs*.[27] King Alexander receives Clarus well, and they entertain each other to splendid feasts, but the prince beholds Serena first at a banquet to which she thereafter invites him in her tower. Serena's wisdom is illustrated in this context by a description of how the course of the stars (*stjǫrnugangr*) and an array of multifarious learning (*margfræði*) decorate her chambers:

> Ok aldri kom enn svá mikill meistari inn um þær dyrr, at eigi mætti nema enn meira, en hann kunni áðr, af þeim meistaradóm, sem þar mátti líta.[28]
>
> (And never once did any great master enter through those doors without learning still more than he knew before on account of the magisterial wisdom that one could there behold.)

Near the end of their opulent dinner, Clarus asks Serena to marry him, and she reacts amiably, offering to share with him a soft-boiled egg (the narrator notes at this point that Clarus had then reached the age of eighteen). Feigning to take the first sip, she hands the egg to Clarus but subtly makes him lose grip so that he spills the contents onto his chest and all the way down to his belt. Serena thereupon openly scolds Clarus by calling him vulgar names (including *fúll farri* 'foul vagabond'),

---

27 *Clári saga*, p. 9. Curiously, the city of Paris is not once mentioned by name in the saga. In Cederschiöld's 1907 edition, the index of place-names (p. 75) is short: Arábía, Arábíaland, Bláland, Erópa, Frakkland, Frannz and Saxland. Neither are the saga's personal names (listed on the same page) many: Alexander ('Frakkakonungr'), Clárus ('keisarason'), Eskilvarð ('konungsson'), Jón Halldórsson ('byskup'), Pérús ('meistari'), Séréna ('konungsdóttir'), "Severa" (a pun on Serena's name), Tecla ('konungsdóttir ok þjónustumær'), and finally, Tiburciús ('keisari').

28 *Clári saga*, p. 18.

indicating thereby that he is beneath her status and an unworthy suiter.[29] Courtly attitudes toward social rank are ironically brought into focus later on when Princess Serena becomes a destitute vagrant begging barefooted for alms in order to feed her (supposed) vagabond husband.

Serena is, we are told, overjoyed with this outcome of the feast, and having been shamefully expelled from her court, Clarus returns to Saxony, where Perus warns his pupil against seeking any revenge and returning to France. The master's Arabic background is in this context recalled when the frustrated prince reproaches his tutor in rage:

> Til hvers kom þat mínum feðr keisaranum at láta gøra eptir þér, bannsettum manni, allt út í Arábíam, svá sem mér til nǫkkurs styrks ok auka mannanar ok vizku, ef þú komt hér til enskis nema spá mér illspár? Ok aldri steigt þú enn lengra fram með þitt vit en einn nautreki eða rotit laukshǫfuð.[30]
>
> > (Why did it occur to my father the emperor to summon you, an accursed man, all the way from Arabia, that I might gain some support and be advanced in my upbringing and wisdom, if you came here only to offer me prophecies of misfortune? Never once did you take a step forward with your wit rather than a common cow-herd or a rotten onion-head.)

The adjective *bannsettr* could mean 'excommunicate' as well as more generally 'accursed', and so its use here might be intended to suggest that Perus was a man who had, possibly through his pursuit of non-Christian Arabic learning, somehow transgressed orthodox Christianity. However, his ties to Arabia are only mentioned in this one place apart from his first appearance in *Clári saga*, and it might be added that Perus appears to be simply invoking the Christian God when he previously expresses dismay at Clarus's sudden obsession with Serena by exclaiming: 'Guð fyrirláti mér, at ek gaf þér svá úþarft efni!' (May God forgive me for having given you such a troublesome subject-matter!).[31]

Faced with his pupil's threats of beheading (his head is then called *bannsett* by the prince), Perus finally agrees to assist Clarus on condition that he be given control of the Saxon empire for three years. This is quickly granted by Emperor Tiburcius, who appears throughout, much like King Alexander, to be a rather weak figure of authority. During this period, Perus employs the vast resources at his disposal to make three spectacular land-tents (*landtjǫld*), each of which is drawn by a (seemingly) mechanical animal: a brown bear, a lion, and, most magnificently, a golden vulture (*gammr*).[32] The elaborate accounts devoted to each of these three costly creations,

---

29 The significance of this scene for the saga's plot has been discussed in detail by Kalinke in her 'Table Decorum'.
30 *Clári saga*, p. 26.
31 *Clári saga*, p. 7. Arabia is only mentioned in one other place in the saga, namely when Clarus (on his first visit to the French capitol), informs Tecla, the Scottish handmaid of Serena, that Arabia is the land richest in gold. See *Clári saga*, p. 16.
32 For further discussion of the animals which draw the tents, see Hughes, '*Klári saga* as Indigenous Romance', pp. 54–55.

whose purpose is intriguingly not stated and only becomes apparent later on in France, illustrates Perus's great powers of engineering, which may be viewed as an aspect of his image as a master-magician. Technical marvels are often attributed to learned magicians in European literature of the period,[33] and as a positive, albeit fictional, representative of Arabic science or learning who temporarily governs the great Saxon empire, Master Perus is without parallel in Old Icelandic literature.[34]

At the end of his three-year rule of the Saxon empire, Perus sails with Clarus to the capital of France (*hásætisborg Frakkakonungs*). Shortly before their arrival in its harbour, Perus takes some ashen powder out of his pouch and rubs this into the face of Clarus, disguising him thereby as Prince Eskilvarð from Bláland ('Blackland', i.e. Ethiopia). The land-tents are then employed to appeal to Serena's insatiable avarice (*ágirni*) and desire to possess everything precious, a moral weakness that Perus evidently intended all along to exploit by means of the three pavilions. In exchange for the bear-drawn tent, which is first brought on land for display, Serena offers to share her bed with Eskilvarð with the understanding that if he attains her sexual favours she will marry him. Drugging him with a sleeping-potion, Serena tricks the prince twice out of a land-tent, first that with the bear and then that with lion, and Eskilvarð is on both occasions severely flogged by Serena's men while lying unconscious by her bed. Once Perus intervenes with his magic and alters the mind of Serena's handmaid Tecla, a Scottish princess (when she steps over the threshold of a tent set up within the vulture-drawn tent), Prince Clarus finally succeeds in bedding Serena. With the princess having lost her virginity to Eskilvarð (Clarus), the two are promptly married, and the couple spend a happy fortnight together in bed.

While one will at this point have gathered that Perus held mastery over technology as well as psychology or magical 'mind-control', he subsequently becomes a moralistic agent in a way reminiscent of his interactions with a few overly proud, avaricious, and wrongful nobles in the three exempla recorded previously in AM 657. On the day when Serena is to sail away with her princely husband from Bláland, she awakes outdoors beside a detestable black man whom she believes to be Eskilvarð. Calling her a wily and wicked whore (*vánd púta*) who surrendered her virginity out of mere avarice, he rejects Serena at once, ordering her to return to the house of her father, and paying no heed to her desperate promises of loving him purely, dressing him finely, and making him the next king of France, this abusive man quickly turns out to be an itinerant buffoon. He swiftly runs off to entertain and beg for a living, but Serena devotedly pursues her tramp and a band of *forumenn* (travellers) or *vagnamenn* (wagon-folk). Descending thereby into the world of vagabonds, the

---

33   For a study on late medieval links between outstanding engineering and magic see Eamon, 'Technology as Magic'.
34   Insofar as Perus can be regarded as an Arabic character, he is a rare, but not entirely unique, figure in Old Icelandic literature for being portrayed in a positive light. For Perus's significance as a representative of Arabic/Islamic science within the context of medieval Icelandic science see Etheridge, 'The Evidence for Islamic Scientific Works in Medieval Iceland', pp. 67–71. Perus's image as a positive representative of Arabia has also been noted by Sverrir Jakobsson. See his 'Íslam ok andstæður', p. 23.

erstwhile princess and co-ruler of France then endures a twelve-month period of utter poverty, homelessness, and begging, and once her supposed husband breaks a leg, she even carries him on her back. Her fortune changes when she reaches a church and, begging there for alms while the injured tramp waits for her elsewhere, receives the kind help of a mysterious stranger. During her travels with the vagabond-buffoon, who is eventually revealed to have been Master Perus in disguise, Serena proves to be a model wife for her patience, generosity, and compassion towards him and, once rejoined with Prince Clarus and restored to royal status, her transformation is highlighted with a concluding moral explication.[35] Whereas Perus serves, in effect, as Serena's 'disciplinary master' in the fields of charity and common humanity (as opposed being the courtly *typtunarmeistari* of Prince Clarus), it is left obscure how Clarus's ordeal would have served his educational programme. The concluding comments indicate, however, that Perus did in actual fact exercise more control over the course of events than one might at first have assumed:

> Hafa þat ok flestir fyrir satt, at meistari Pérús myndi mátt hafa fyrr til sín tekit, ef hann hefði viljat, en þótti engu varða í fyrstu, þótt Clárús keisarason fengi þvílíkt klódrep af henni, með því at hann vildi ekki hans ráðum hlíta.[36]
>
> (And most people hold it to be true that Master Perus could have intervened earlier had he so wished but that he considered it no matter at first that the emperor's son Clarus would receive such a claw-swipe from her since he would not heed his advice.)

Bishop Jón Halldórsson, who is said to have 'found' the tale about Clarus in France, was a Dominican friar from the city of Bergen in Norway.[37] He entered the Bergen convent of St Dominic's Order of Friars Preachers in his childhood,[38] and he went on to study extensively abroad, first at the University of Paris, and eventually that of Bologna in northern Italy. Paris and Bologna were then the foremost centres of Dominican scholarship on the Continent, and Bishop Jón was held in high esteem for his Latinity and scholarly learning. Following his studies abroad, Jón served as a canon at the cathedral chapter of Bergen for more than a decade before Archbishop Elífur korti Árnason of Nidaros (Trondheim) in Norway made him bishop elect of Skálholt in 1321, and Jón was the first member of any mendicant order to attain the office of bishop in Iceland, a rural country where medicant friars never established houses or any permanent kind of presence. The background of this first Preacher-bishop in Iceland is briefly described in an entry on his consecration in 1322 in the

---

35 *Clári saga*, pp. 72–74.
36 *Clári saga*, pp. 72–73.
37 For the main dates of Jón's career see Kolsrud, *Den norske Kirkes Erkebiskoper*, pp. 264–65. Further biographical information is found in *Islendzk æventyri*, ed. by Gering, II, pp. v–xxiii; Paasche, 'Jon Halldórsson'; Jakobsen, *Studier i Clarus saga*, pp. 16–22; *Biskupa sögur*, ed. by Guðrún Ása Grímsdóttir, pp. cii–cxi; Marteinn Helgi Sigurðsson, *The Life and Literary Legacy of Jón Halldórsson*, and Etheridge, 'Canon, Dominican and Brother'.
38 *Biskupa sögur*, ed. by Guðrún Ása Grímsdóttir, p. 454.

annals of Einar Hafliðason (1307–1393), a prominent Icelandic cleric whom Jón had ordained to the priesthood in 1331: 'vigdr Jo*n* Halldors *s*on brod*ir* af *p*redikara lifnade til b*y*sko*p*s j Ska*l*ho*ll*te. var h*ann* michilshat*t*ar klerkr *ok* stadit leingi utlendiss j Bononia *ok* Paris ad studi*um*'.[39] (Jón Halldórsson [was] consecrated bishop of Skálholt, a friar of the Order of Preachers; he was a cleric of great standing and had for a long time attended *studium* abroad in Bologna and Paris).

Bishop Jón came to be especially remembered as a great preacher and storyteller who used foreign tales to entertain and edify his audience. This aspect of Jón's episcopacy was in keeping with his training as a Friar Preacher since exemplary tales were a conventional part of Dominican preaching, and *Clári saga* is an exceptional *riddarasaga* in that it concludes with an explication of the story's moral import in a manner which one might expect at the end of a sermon-tale. As previously noted, *Clári saga* lies embedded within a large collection of exempla recorded in AM 657 4to, which also preserves three exempla devoted to Master Perus ('Af meistara Pero ok hans leikum'). AM 657 includes moreover, on loose leaves numbered 98–100, a rather short, anecdotal and hagiographical account of Jón now known as *Jóns þáttr Halldórssonar biskups* or *Söguþáttr Jóns Halldórssonar biskups*.[40] The bishop's storytelling and preaching are there commemorated with affection and even presented as the chief reason for his posthumous (though uncanonized) status as a miracle-working saint. The anonymous author begins *Jóns þáttr* as follows:

> Nú skal nefna virðulegan mann er heitir herra Jón Halldórsson, hinn þrettándi byskup í Skálaholti at Íslandi. Hann var hinn sæmilegsti maðr í sinni stétt sem lengi man lifa á Norðrlöndum, því at sú var hans æfi lengst at hann fór, síðan hann hafði gjörz predikari í Nóregskonungs ríki, at studium mjök ungr allt út í París ok um síðir út í Bononiam. Kom hann svá aftr af skolis fullkominn at aldri, at hann var sá vísasti klerkr er komit hefir í Nóreg, því var hann vígðr ok kosinn byskup Skalholtensis af Eilífi erkibyskupi. En hverr man greina mega hverr hans góðvili var at gleðja nærverandis menn með fáheyrðum dæmisögum er hann hafði tekit í útlöndum, bæði með letrum ok eigin raun, ok til vitnis þar um harðla smátt ok lítit man setjaz í þenna bækling af því stóra efni, því at sumir menn á Íslandi semsettu hans frásagnir sér til gleði ok öðrum. Manum vér í fyrsta setja sín æfintýr af hvárum skóla, París ok Bolon, er gjörðuz í hans náveru.[41]

> (Mention shall now be made of a venerable man named Lord Jón Halldórsson, the thirteenth bishop of Skálholt in Iceland. He was the most distinguished man of his station as will long be remembered in northern lands, for he spent most of his life at *studium* once he had become a Preacher in the realm of the Norwegian king, going very young all the way to Paris and eventually Bologna.

---

39 *Islandske Annaler*, ed. by Storm, p. 267.
40 The (*sögu*)*þáttr* on Jón Halldórsson is edited in *Biskupa sögur*, ed. by Guðrún Ása Grímsdóttir, pp. 445–56 and *Islendzk æventyri*, ed. by Gering, I, 84–94.
41 *Biskupa sögur*, ed. by Guðrún Ása Grímsdóttir, p. 445.

He returned from *scolis* at a mature age as the wisest cleric to have come to Norway, and he was therefore consecrated and elected bishop of Skálholt by Archbishop Eilífur. Anyone can relate with what goodwill he would delight people in his presence with novel exemplary tales he had acquired abroad, both from letters and his own experience, and as witness thereof a very little and small part of that large matter shall be included in this booklet since some men in Iceland compiled his stories for their own and other people's delight. We shall first recount one event from each school, Paris and Bologna, that took place in his presence.)

Since no Latin source is attested for *Clári saga*, its purported origin has been doubted by commentators who suspect that Jón's association with the saga is a fiction intended to promote its legitimacy or prestige as entertainment.[42] In a recent discussion of the saga, Marianne Kalinke expresses a different view: 'I see no reason to doubt the authenticity of *Clári saga*'s incipit, which informs the reader that Jón Halldórsson told the story'.[43] Kalinke does not, however, give credence to the statement that Jón came upon the story while in France, and she is instead inclined to endorse Sean Hughes's theory that the bishop composed the saga himself in Iceland, and that the claim that it is based on a Latin *rythmus* discovered by Jón in France is to be viewed as a 'modesty topos'.[44] The Latinate and learned style of the saga certainly accords with Jón's renowned eloquence and mastery of the Latin language. He is said to have spoken Latin as if it were his mother tongue (*móðurtunga*) in *Lárentíus saga biskups*,[45] a biography of Bishop Lárentíus Kálfsson of Hólar (1324–1331) written by the aforementioned annalist Einar Hafliðason. More importantly, *Clári saga* contains many Norwegicisms along with a striking amount of Middle Low German loanwords that make it seem entirely credible that this is a work composed by a learned Norwegian cleric who had grown up in Bergen in the late thirteenth century, a milieu that was heavily influenced by the presence of German merchants as well as the royal court of Norway.[46] In addition to these linguistic and stylistic considerations, it might be noted that the romance's great emphasis on learning and its exemplum-like end could well reflect a Dominican authorship, and the same holds true for the importance of mendicant itinerancy in the reform of the arrogant, cruel, and avaricious Princess Serena of France.

The evidence of *Jóns þáttr* should also be taken into account in this connection. Both of the *æfintýr* that follow the passage quoted above from *Jóns þáttr* are in fact variants of older legends. In the former, which is set in a Paris schoolroom and reflects an international tale-type or migratory legend commonly known as the 'Sorcerer's

---

42 See, for example, de Vries, *Altnordische Literaturgeschichte*, II, 535; Torfi H. Tulinius, 'Kynjasögur úr fortíð og framandi löndum', p. 196; Barnes, *The Bookish Riddarasögur*, p. 72.
43 Kalinke, '*Clári saga*, *Hrólfs saga*', p. 277 n. 9.
44 Kalinke, '*Clári saga*, *Hrólfs saga*', p. 283 n. 16; Hughes, '*Klári saga* as an Indigenous Romance', p. 158.
45 *Biskupa sögur*, ed. by Guðrún Ása Grímsdóttir, p. 403.
46 On the language of the saga, see Kalinke, '*Clári saga*: A Case of Low German Infiltration', and older secondary literature cited in that study.

Apprentice' (see further below),⁴⁷ Jón stealthily reads from his master's book and raises thereby a terrible tempest. In the tale concerning Bologna, which is set in the lion-portal of the city's cathedral, Friar Jón (Johannes Nordmannus) witnesses the death of a fellow student and namesake from England (Johannes Anglicus) when the latter, prompted by a dream he had the previous night, playfully puts his hand into the maw of a sculpted lion where a deadly asp (*aspis*) lay hidden. The Bolognese *æfintýr* has a multitude of related Eurasian analogues, but the closest known parallel appears to be a legend that is recorded within the context of exempla concerning the virtue of Prudence in the *Rerum memorandum libri* by Francesco Petrarca (1304–1374). Although Petrarch studied in Bologna (1320–1323) like Jón, he presents this event as having taken place in ancient times and in the portal of a pagan temple whose location is unspecified.⁴⁸ The liberal way in which Jón has evidently attached these two older legend-types to the figure of himself might be taken as a sign of his potential or inclination to create fictions about himself abroad and the provenance of his tales.

Without any comment on Perus's origin, the first of the three exempla devoted to him relates how he once served as the counsellor (*ráðgjafi*) of two brothers, Vilhjálmr and Eiríkr, who inherited the duchy (*hertogadómr*) of their father. Perus had asked for the hand of their sister, Lady Ingibjǫrg, in marriage, but the brothers refused because he 'var nafnbótarlauss ok óférikr, en allra manna var hann best menntr' (had no [hereditary] title or wealth, although he was the most learned of all men).⁴⁹ Insisting on being relieved from affairs of state from dusk to dawn, Perus was in the habit of secretly making love to Ingibjǫrg in her quarters while leaving his illusory double to drink in the palace of the dukes. The affair of Perus with their sister is eventually discovered by courtiers envious of Perus's dignity and high rank (*sómi* and *metorð*), and once his seat in the hall is proved to be vacant, the dukes break into Ingibjǫrg's chamber. Just as they rush in, Perus leaps from the lady's bed, takes a seat on her throne, and covers himself with the blue cloak he was wont to wear daily. Claiming precedence for being first-born, Duke Vilhjálmr proceeds to pull Perus off the throne and to tear his legs asunder so that his entrails flow onto the floor. Ingibjǫrg faints as the brothers depart forbidding anyone to clean her chamber, but Perus quickly reappears intact and reveals to his beloved (who believes him at first to be a ghost) that Vilhjálmr had in fact merely split a cleft tree-trunk (*skálmatré*) that lay on floor, which then suddenly appears to be clean again. Once fully dressed, and without heeding Ingibjǫrg's pleadings that he should take flight, Perus enters the dukes' hall, where he is immediately shackled and taken to a grove where he is to be executed before a large gathering. Perus then delivers a speech in which he reprimands the dukes before making a fantastic escape:

> 'Nú hefi ek verit bundinn um stund, en hèðan af skal ek lauss vera; en þit bræðr erut mjök heimskir menn: ek bað systur ykkarrar ok villdu þit eigi gipta mèr hana,

---

47 Marteinn Helgi Sigurðsson, *The Life and Literary Legacy of Jón Halldórsson*, p. 37.
48 See further Marteinn Helgi Sigurðsson, 'Djöfullinn gengur um sem öskrandi ljón' (where there is an English summary of the article).
49 *Islendzk æventyri*, ed. by Gering, I, 218.

því at ykkr þótti ek lítils háttar hjá ykkr, en þóat ek sè ykkr úríkari fyrir penninga sakir, þá em ek ykkr þó vitrari, ok ef várar mægðir hefði tekiz, mætti þit minnar vizku æfinlega notit hafa, því at þá hefði ek alldrigi við ykkr skilit. Mæli ek nú þat um, að öll ykkur framferð snúiz ykkr til úfrægðar ok hamingjuleysis.' Síðan sýndiz þeim sem hann tæki eitt blátt hnoða ór brókabelltispungi sínum ok kastaði upp í loptit ok þat yrði at einum streng. Sýndiz þeim sem hann helldi þar í annan enda ok læsi sig upp eptir, ok hvarf hann þeim svá at alldri sáz hann síðan.[50]

> ('I have now been bound for a while, but I shall henceforth be loose. You brothers are very stupid men. I requested the hand of your sister and you would not let me marry her because you deemed me of little worth in comparison with you. But although I am less mighty on the account of money, I am wiser than you, and had I married into your family, you would always have benefitted from my wisdom, for I would never have abandoned you. I now place a spell to effect that all your conduct will lead to infamy and misfortune.' It then appeared to them that he drew a blue bundle out of the pouch on the belt of his pants and threw this into the air and that it turned into a rope. It then appeared to them that he grasped one end and pulled himself up, and he then vanished so that he was never seen again.)[51]

In the second tale about Perus, we are told how he gets the better of a 'ríkr höfðingi er Prinz var kallaðr' (a mighty magnate who was named Prinz). One day, when Prinz rides through the countryside with his entourage of twelve men, he encounters Perus mounted on a fine steed:

> Hann sýndiz þeim mikill ok vaskligr. Hann hefir hest svá fríðan, at engan hafa þeir sèt þvílíkan, bæði sakir vaxtar ok vænleiks. Klæðabúnaðr þessa mannz sýndiz þeim umfram alla þá sem þeir höfðu fyrr sètt.[52]

> (He appeared to them to be both large and valiant. He had such a fine horse, that they had never seen one like it on account of its build and beauty. The clothes of this man seemed to surpass all those they had seen before.)

Prinz asks the stranger to give him his weapons, clothes, and horse since these are 'eigi ótigins manns eign' (not the possessions of an ignoble), and Prinz offers his own steed, clothes, and a great fee in exchange. When Perus declines, Prinz has his men seize Perus, who allows them to undress him without resistance, and Prinz then carries on with his men leaving Perus behind with Prinz's attire and horse. One of Prinz's men, who happened to carry a magical stone (*náttúrusteinn*) with the property of

---

50 *Islendzk æventyri*, ed. by Gering, I, 222–23.
51 It is noteworthy that in the closest known analogue of this tale, an early modern and German poetic account, the great Dominican scholar Albertus Magnus assumes a role corresponding to that of Perus. When, as a student in Paris, Albert becomes the lover of the daughter of the French king and this is discovered, he similarly makes his escape by means of a bundle of thread that he tosses into the air. See *Islendzk æventyri*, ed. by Gering, II, 166.
52 *Islendzk æventyri*, ed. by Gering, I, 223–24.

dispelling illusions, soon discovers that his lord is actually riding a load of brushwood (*hrísbyrði*) bound together with twigs and tatters. When this man has shared his revelation with his fellows and eventually Prinz himself, Prinz's men are sent back to apprehend this *vándr galdramaðr* (evil magician). Perus is subsequently bound and brought before Prinz, who accuses him of deceit in their dealings. Perus then proclaims his innocence, explaining that Prinz's mind had simply been blinded by avarice (*ágirni*), and Perus is thereafter led to the end of a pier by the sea where he is condemned to be drowned in public for his *galdrar ok gjörningar* (magic and wizardry). Hedged in by the crowd, Perus eventually announces: 'Ek hefi verit bundinn nú um stund, en nú vil ek lauss vera, ok þó mun ek eigi hlaupa út á sjóinn at bana mèr'. (I have now been bound for a while, but I now wish to be free, and yet I shall not leap into the sea to kill myself). He thereupon breaks free and takes out of his pouch a bit of chalk with which he draws a picture of a ship set for sail. Following a mighty din, the people assembled look on as Perus sails off and out of sight.[53] Magic illusion or *sjónhverfing* plays an even larger part in the third and final exemplum about Perus in which the concluding moral (see above) on the corruption and tyranny that so often attends worldly power may also apply to the two preceding adventures.

Besides the first two *æfintýr* recounted in *Jóns þáttr*, i.e. those pertaining to Paris and Bologna, this semi-biographical work includes a much longer tale which the bishop is said to have used as an exemplum (*dæmi/æfintýr*) in a sermon on the summer feast-day (20th July) of his predecessor in Skálholt, St Þorlákr Þórhallsson (d. 1193).[54] This sermon-tale, which promotes the idea of ultimate and divine judgement in terms of true merit rather than kinship or wealth, is clearly based on some Latin version of the same tale, which is widely attested in earlier and contemporary exempla-collections.[55] It remains unclear what further Old Icelandic exempla might or should be attributed to Jón although a fair number of those found in AM 657 and other Icelandic manuscripts have no doubt reached the country by way of his retelling or books imported by the bishop while some might in fact be, more or less, his original compositions.[56]

As for the reception of *dæmi* or *æfintýr* related by Bishop Jón, it is interesting to note how we are told in *Jóns þáttr* that his tales sometimes caused offence:

> enginn maðr þvílíkrar stéttar mátti framar fella sik til að vera mönnum til hugléttis ok gleði, ok fyrir þá grein at optlega vóru eigi allir með einum hug er hann heyrðu þá samði hann sik eptir því at allir mætti gleðjaz af hans orðum, því vóru frásagnir hans sumar bæði veraldlegar ok stórorðar svá at sumir menn lögðu honum til lýtis.
>
> (no man of such station could prostrate himself further so as to bring people delight and joy, but because his audience was oftentimes not of one mind, he

---

53 For remote and younger analogues for this tale in terms of a few motifs, see *Islendzk æventyri*, ed. by Gering, II, 167.
54 *Biskupa sögur*, ed. by Guðrún Ása Grímsdóttir, pp. 449–54.
55 For medieval variants of this tale see Hibbard, 'Ekenbald the Belgian'.
56 See Jakobsen, 'Ævintýri', cols 614–16.

would adapt himself so that all might be gladdened by his words. Some of his tales were therefore both worldly and crudely worded so that some people found this to be a fault with him.)[57]

Adaption to the minds and social status of the audience (sermons *ad status*) was a fundamental aspect of Dominican preaching, and although it is unknown whether Jón or any other preacher introduced anecdotes about Perus into actual sermons, it is not difficult to imagine how such matter might have disturbed Icelanders accustomed to old-fashioned patristic homilies from the pulpit.[58] Be that as it may, since notions of social standing are central to the plot and moral import of *Clári saga*, the tales about Perus are remarkable for the way in which they reflect a thoughtworld akin to that found in the Latin poetry (typically rhythmic and rhyming) of the so-called Goliards (*goliardi*) of the twelfth and thirteenth centuries, where the merits of 'wandering scholars' or 'vagabond clerics' such as themselves are celebrated at the expense of other classes, high and low, within the established order of society.[59]

The itinerant Master Perus, who in *Clári saga* takes on the guise of a repulsive vagabond, seems at home within this cultural context. Goliardic poets excelled in parodying Holy Scripture, saintly figures, and the liturgy, and if the name of Perus is (as suggested above) derived from Petrus and he is thus the namesake of St Peter, the Prince of the Apostles who possessed powers to loose and to bind, Perus's narrow escapes might in part be seen as a playful reflection of the immunity from secular jurisdiction that accompanied the clerical and academic status. In fact, the adventures of Perus seem designed to impress upon the reader how mastery of learning (*listir*, *klerkdómr*, *meistaradómr*) provides the true measure of social station and dignity (*stétt* and *tign*), and this potentially subversive kind of meritocracy, which is clearly more clerical and scholarly than courtly or chivalric, has some parallel in the ideal of courtly love, according to which honour and virtue form the true foundation of nobility instead of hereditary rank or wealth. Serena's love for her tramp-husband can be viewed as ennobling along such lines. As Kalinke observes in her article 'Table Decorum and the Quest for a Bride in *Clári saga*': 'Séréna is depicted as having changed from a *grimdarmaðr* to a *dygðarmaðr*. The exact meaning of *grimdarmaðr* is difficult to determine, but in antithesis to *dygðarmaðr* the meaning appears to reside in the moral realm'.[60] The first elements in these two compounds in -*maðr* ('-man') contain the common terms *grim(m)d* and *dyg(g)ð*, which mean, respectively, 'cruelty' and 'virtue'. Kalinke concludes the same study with this comment on the moral import of *Clári saga*:

> Clári saga is both romance and exemplum, neither of which excludes the other. The mishap with the slippery egg provides at first a lesson in courtly decorum,

---

57 *Biskupa sögur*, ed. by Guðrún Ása Grímsdóttir, p. 448.
58 For a detailed discussion of sermon literature in medieval Iceland see Hall, 'Old Norse-Icelandic Sermons'.
59 For a general discussion of the Goliards see Morris, *The Discovery of the Individual*, pp. 128–32 and Le Goff, *Intellectuals in the Middle Ages*, pp. 24–25. A classical treatment of the Goliards and their poetry is to be found in Waddell, *The Wandering Scholars*.
60 Kalinke, 'Table Decorum', p. 66. Kalinke here refers to p. 70 in *Clári saga*.

but finally an even greater and more important lesson: neither table etiquette nor spotless dress, the trappings of the courtly class, reflects or is indicative of inner worth.[61]

In the case of Perus, however, who with his *lítit æfintýr* initiates an eventful lesson on real worthiness with regard to marital love and the role of the wife in particular, it is obviously learning and wisdom, rather than fine love, high rank or riches, that lead a man towards perfection, and Bishop Jón of Skálholt is not an unlikely transmitter of such intellectual attitudes or ideals. Indeed, *Jóns þáttr* presents his outstanding learning as the sole reason for his rise to the rank of bishop when we are told that he returned 'af skolis fullkominn at aldri, at hann var sá vísasti klerkr er komit hefir í Nóreg, því var hann vígðr ok kosinn byskup Skalholtensis af Eilífi erkibyskupi' (he returned from *scolis* at a mature age and was thus the wisest cleric to have come to Norway; he was therefore consecrated and elected bishop of Skálholt by Archbishop Eilífr).[62] If Bishop Jón derived a sense of pride from his Latin learning, a sign thereof might be detected in *Lárentíus saga* when, during a dispute between Jón and Bishop Lárentíus of Hólar, the latter says in Norse (*norræna*) to his fellow-bishop from Skálholt, who had begun by speaking in Latin: 'Vita menn þat, herra Jón, at yðr er svá mjúkt latínu at tala sem móðurtungu yðar, en þó skilr þat ekki alþýða, ok því tölum svá ljóst at allir megi skilja'. (Everyone knows, Lord Jón, that you speak Latin so fluently as if it were your mother tongue, but common people do not understand it, and let us therefore speak clearly that all may understand).[63]

Having introduced the two *æfintýr* concerning Paris and Bologna with the statement that these *gjörðuz í hans náveru* (took place in his presence), a choice of words that recalls the humility topos, the author of *Jóns þáttr* begins the former tale as follows:

> Þann tíma sem hann var í París nýliga kominn, gekk hann inn í skóla þann er æztr til var. Var hann þann tíma ungr ok lítit skiljandi hjá því sem síðar. Byrjaðiz þá þegar þat sem lengi helz síðan um hans ráð at Guð gaf honum mikla mannheill alla götu, einkanlega af þeim er æztir vóru ok vitrastir, því leggr höfuðmeistarinn ok allr samnaðr skólans mikla blíðu ok góðar virðingar á sveininn sem í því lýsiz er eftir fer.[64]
>
> > (Shortly after his arrival in Paris, he entered its most eminent school. He was young at the time and of little understanding compared to later. From the very beginning, and this would long remain so throughout his life, God always gave him great favour among men, especially those most eminent and wise; the headmaster and entire congregation of the school therefore had much affection and high regard for the youth, as will be seen in what follows.)

---

61 Kalinke, 'Table Decorum', p. 72.
62 *Biskupa sögur*, ed. by Guðrún Ása Grímsdóttir, p. 445.
63 *Biskupa sögur*, ed. by Guðrún Ása Grímsdóttir, pp. 403–04. See also pp. 349, 400, 406 for further references in the saga to Jón's great learning.
64 *Biskupa sögur*, ed. by Guðrún Ása Grímsdóttir, pp. 445–46.

Within this prestigious setting where Jón's future advancement is linked to his studies in Paris, we are then given a memorable sample of what kind of learning his enthroned headmaster possessed:

> Svá gekk til á einn dag at yfirmeistarinn sér á sína bók er mjök var stór í sínum vexti. Ok sem hann beiðir at létta sér út af skólanum leggr hann opna bókina niðr í hásætit áðr hann gengr út, ok án dvöl forvitnar hann piltinn, er vér nefndum, hversu greitt hann man lesit fá eitthvert capitulum af bókinni hans meistara; því rennr hann upp gráðurnar er lágu fyrir hásætit ok les þegar þat sem honum bar fyrir augu. Ok sem hann hefir lesit eitt capitulum brestr á húsinu brakandi þytr meðr æðistormi sem allt myndi ór lagi færaz ok jafnbrátt lúkaz upp hurðir. En er Jón heyrir þat ok skynjar at meistari man ganga flýtir hann sér eftir megni aptr til sætis síns. Sem meistari kemr inn farandi segir hann ok sver um við nafn hins hæsta Guðs at ef stormr þessi gengr til aptans man hann þurrka öll þau stöðuvötn sem í eru Franz – 'eðr hvat er', sagði hann, 'hefir enginn glenz við bók mína síðan ek gekk?' Nú var sveinninn Jón svá vel kenndr at eigi ok engi vildi segja eptir honum – 'ok því sá ek nú', sagði hann Jón byskup, 'at setit var meðan sætt var; fell ek þá fram sálugr ok játaði hvat ek hafði gört, en meistari svaraði mér svá: "Líkna man ek þér, Jón", sagði hann, "en þó skaltu hafa augu fyrir hvat þú kannt lesa meðan þú skilr eigi betr"'.[65]

> (It so happened one day as the headmaster looked onto his book, which was of very great size, that he wished to relieve himself outside the school and laid the open book down on his high-seat before walking out. The aforementioned lad was at once curious to know how well he could read a chapter from the book of his master. He therefore ran up the steps leading to the high-seat and read forthwith what met the eye. But when he had read one chapter, a thundering gust struck the house with such a raging storm that everything seemed about to be thrown into disorder, and at that very moment the doors were opened. When Jón heard this and sensed that the master had returned, he hastened as he could back to his seat. As the master came rushing in, he said and swore by the name of the highest God that if this storm continued into the evening it would dry up every lake in France – 'Or how is it', he said, 'did any of you play with my book when I left?' Jón was such a well-liked boy that not one wished to tell on him – 'And so I therewith saw', said Bishop Jón, 'that I had sat as long as I could stay seated. I fell forth contrite, confessing what I had done, and the master gave me this reply: "I'll be merciful with you, Jón", he said, "but you should be wary of what you might read while you do not have better understanding"'.)

With these cautionary words on the difference between *lesa* and *skilja*, the acts of reading and understanding (Latin *legere* and *intellegere*), and the potential dangers of undertaking the former without the latter, the headmaster quickly assumes control of the situation:

---

65 *Biskupa sögur*, ed. by Guðrún Ása Grímsdóttir, p. 446.

Síðan skundar meistari upp til bókarinnar ok veltir henni á aðra hálfu – 'sem þar les hann eitt capitulum', sagði herra Jón byskup, 'þótti mér vonligt at þat væri nærri því langt sem ek hafði áðr lesit, ok án dvöl sem capitulum var úti fellr stormrinn svá skjótt at með öllu varð vindlaust. Má af slíku marka', sagði byskupinn, 'hve listin lifir í bókunum þótt heimrinn gjöriz mjök gamall'.[66]

> (The master thereupon hurried up to the book and turned to another page. 'He then read one chapter', said the Lord Bishop Jón, 'and it seemed to me to be about the same length as the one I had read previously. And without delay, once the chapter was completed, the storm fell so suddenly that there was absolute calm. One can infer from events such as these', said the bishop, 'how much art endures in books although the world grows very old'.)

This tall tale of the bishop can obviously be taken to suggest that Jón studied magic while in Paris or that he at least allowed people to entertain the notion that he had done so. But as with the magic feats of Master Perus, one is hardly meant to see anything damnable or demonic about the bookish art (*list*) whose enduring power is set against the grave *senectus mundi* (world grown old) topos. One might furthermore gather from this Parisian tale that matter about a master-magician like Perus and his adventures in France or elsewhere could have appealed to Bishop Jón and even featured in his repertoire of exempla. Some of his tales may well have been designed to entertain and edify young clerics trained under the auspices of Jón within the diocese of Skálholt.

According to *Jóns þáttr*, Bishop Jón died from an illness on Candlemas (2nd February) in the house of the Bergen Dominicans, where miracles were soon said to occur by his grave. Before a description of his saintly death, we are told in the *þáttr* how the bishop had a dream on the night before he embarked on his final journey to Norway in 1338. He dreamed that he stood in the church of his brotherhood in Bergen and was asked to give a sermon on the theme from Revelation 14:13: *Beati mortui qui in domino moriuntur* (Blessed are the dead who die in the Lord). When Jón completed his sermon, King Magnús Hákonarson (d. 1280) and Archbishop Eilífr (d. 1332) appeared beside the bishop and led him up a ladder from the choir and through the roof where he was to rest on a bed made for him on top of the church. This bed recalls the words about the reward of the blessed dead that follow in Revelation 14:13: *ut requiescant a laboribus suis opera enim illorum sequuntur illos* (they will rest from their labour and their deeds follow them). The ladder is obviously inspired by the one that Dominic, the holy founder of the Friars Preachers, was famously seen to ascend to heaven at the hour of his death in the vision of Friar Guali, the prior of the Preachers in Brescia.[67]

One can assume that Jón's purported dream stood close to his personal notions, as a Friar Preacher, of what constituted spiritual perfection or divine holiness in a

---

66 *Biskupa sögur*, ed. by Guðrún Ása Grímsdóttirm pp. 446–47.
67 For an Old Norse version see *Mariu saga*, ed. by Unger, pp. 811–16. In some early versions of the vision the saint is lifted into heaven between two ladders, one held by Christ and the other by Mary. See, for example, de Mailly's Life of St. Dominic in *Early Dominicans*, ed. by Tugwell, p. 59 and Jacobus de Voragine, *The Golden Legend.*, pp. 54–55.

human. While Perus is certainly a moralizing and exemplary figure, it is less clear how he might be appreciated in terms of saintly merits. As a fictional master-magician, he appears to embody a culmination of wisdom idealized by wandering scholars who felt themselves entitled to be critical of established authority and merit positions of power on the basis of their learning. Perus is not portrayed as a university master, but his relationship to ideas about worldly merit, social status and power does not appear unrelated in perspective to the celebrative reverence for academic learning illustrated in the exemplum about the powerful Paris master and his ungrateful student, a variant of the final tale in the Perus trilogy, where the realm of the university *studium* seems to be envisioned as a separate realm or estate beside the older orders of the (clerical) *sacerdotium* and (monarchical) *regnum*.[68] No-one in the first half of the fourteenth century appears more likely than Bishop Jón Halldórsson to have introduced Icelanders to such attitudes or ideals in exempla, and he doubtless became acquainted with the goliardic tradition during his studies in the French capital. As for the question of Master Perus's possible ties to the figure of St Peter (Petrus), one might well imagine that Jón first composed tales involving Perus in Iceland after his consecration as the bishop of St Peter's cathedral in Skálholt, an event that took place on 1 August 1322, that is on the feast commemorating how the Apostle with the power of loosing and binding was miraculously freed from his chains in prison and could thus resume his ministry of preaching: *festum sancti Petri ad vincula*.[69]

Works Cited

Primary Sources

Ævintýri frá miðöldum, vol. 1, ed. by Bragi Halldórsson (Reykjavík: Skrudda, 2016)
Biskupa sögur, vol. 3, ed. by Guðrún Ása Grímsdóttir, Íslenzk fornrit, 17 (Reykjavík: Hið íslenzka fornritafélag, 1998),
Clarus saga. Clari fabella. Islandice et latine, ed. by Gustav Cederschiöld (Lund: Gleerup, 1879)
Clári saga, ed. by Gustav Cederschiöld, Altnordische Saga-Bibliothek, 12 (Halle: Niemeyer, 1907)
Early Dominicans. Selected Writings, ed. by Simon Tugwell (Ramsey: Paulist Press, 1981)
Islandske Annaler indtil 1578, ed. Gustav Storm (Christiania: Grøndahl, 1888)
Islendzk æventyri: Isländische legenden, novellen und märchen, ed. by Hugo Gering, 2 vols (Halle: Verlag der Buchhandlung des Waisenhauses, 1882–1883)
Jacobus de Voragine, *The Golden Legend. Readings on the Saints*, vol. 2, trans. by William Granger Ryan (Princeton: Princeton University Press, 1993)

---

68 On this development in thirteenth-century notions of polity see Grundmann, 'Sacerdotium—Regnum—Studium'.
69 On Jón's consecration see *Islandske Annaler*, ed. by Storm, pp. 152, 205, 345–46. On the feast of St. Peter in Chains, see Jacobus de Voragine, *The Golden Legend*, pp. 34–39.

Mariu saga, ed. by R. Unger (Oslo: Brögger & Christie, 1868–1871)
'Prinz Clarus und Prinzessin Serena. Ein romantisches Liebesmärchen. Aus dem Altnorwegisch-Isländischen übertragen', trans. by Otto Maußer, in *Walhalla. Deutsche Warte für ein wahres Kultur- und Kunstleben*, 8 (1912), 1–49
Sýnisbók íslenzkrar tungu og íslenzkra bókmennta í fornöld, ed. by Konráð Gíslason (Copenhagen: Gyldendal, 1860)

Secondary Studies

Barnes, Geraldine, *The Bookish Riddarasögur: Writing Romance in Late Mediaeval Iceland* (Odense: University Press of Southern Denmark, 2014)
Düwel, Klaus, 'Clárus saga', in *Enzyklopädie des Märchens. Handwörterbuch zur historischen und vergleichenden Erzählforchung*, vol. 3, ed. by Rudolf Wilhelm Brednich and others (Berlin: De Gruyter, 1981), pp. 64–66
Eamon, William, 'Technology as Magic in the Late Middle Ages and the Renaissance', *Janus*, 70 (1983), 171–212.
Etheridge, Christian, 'Canon, Dominican and Brother: The Life and Times of Jón Halldórsson in Bergen' in *Dominican Resonances in Medieval Iceland: The Legacy of Bishop Jón Halldórsson of Skálholt*, ed. by Gunnar Ágúst Harðarson and Karl-Gunnar A. Johansson (Leiden: Brill, 2021), pp. 7–40
—, 'The Evidence for Islamic Scientific Works in Medieval Iceland', in *Fear and Loathing in the North: Jews and Muslims in Medieval Scandinavia and the Baltic Region*, ed. by Cordelia Heß and Jonathan Adams (Berlin: De Gruyter, 2015), pp. 49–74
Grundmann, Herbert, 'Sacerdotium—Regnum—Studium. Zur Wertung der Wissenschaft im 13. Jahrhundert', *Archiv für Kulturgeschichte* 34 (1951), 5–21
Hall, Thomas N., 'Old Norse-Icelandic Sermons', in *The Sermon*, ed. by Beverly Mayne Kienzle, Typologie des sources du Moyen Âge occidental 81–83 (Turnhout: Brepols, 2000), pp. 661–709
Hibbard, L. A. 'Erkenbald the Belgian: A Study in Medieval Exempla of Justice', *Modern Philology*, 17 (1920), 669–78
Hjalti Snær Ægisson, *Þýdd ævintýri í íslenskum handritum 1350–1500. Uppruni, þróun og kirkjulegt hlutverk* (Háskóli Íslands: Reykjavík, 2019)
Hughes, Sean, '*Klári saga* as an Indigenous Romance', in *Romance and Love in Late Medieval and Early Medieval Iceland: Essays in Honor of Marianne Kalinke*, ed. by Kirsten Wolf and Johanna Denzin (Ithaca: Cornell University Press, 2008), pp. 135–63
Jakobsen, Alfred, 'Ævintýri', in *Kulturhistorisk leksikon for nordisk middelalder fra vikingetid til reformationstid*, 20 (Copenhagen: Rosenkilde og Bagger, 1976), pp. 614–16
—, 'Studier i Clarus saga: Til spørsmålet om sagaens norske proveniens', in *Årbok for Universitetet i Bergen, Humanistisk Serie*, 1963, No. 2 (Bergen: Norwegian University Press, 1964)
Kalinke, Marianne, '*Clári saga*: A Case of Low German Infiltration', *Scripta Islandica*, 59 (2008), 5–25
—, '*Clári saga, Hrólfs saga Gautrekssonar*, and the Evolution of Icelandic Romance', in *Riddarasögur: The Translation of European Court Culture in Medieval Scandinavia*, ed.

by Karl G. Johannsson and Else Mundal, Bibliotheca Nordica, 7 (Oslo: Novus forlag, 2014), pp. 273–92

—, 'Table Decorum and the Quest for a Bride in *Clári saga*', in *At the Table: Metaphorical and Material Cultures of Food in Medieval and Early Modern Europe*, ed. by Timothy Tomasik and Julian Vitullo (Turnhout: Brepols, 2007), pp. 51–72

Kalinke, Marianne, and P. M. Mitchell, *Bibliography of Old Norse-Icelandic Romances*, Islandica, 44 (Ithaca: Cornell University Press, 1985)

Krappe, Alexander Haggerty, 'The Delusions of Master Perus', *Scandinavian Studies*, 19 (1947), 217–24

Kolsrud, Oluf, *Den norske Kirkes Erkebiskoper og Biskoper indtil Reformationen*, Diplomatarium Norvegicum, 17 B (Oslo: Det Mallingske Bogtrykkeri, 1913)

Le Goff, Jacques, *Intellectuals in the Middle Ages*, trans. by Teresa Lavender Fagan (Cambridge, MA: Blackwell, 1993)

Marteinn Helgi Sigurðsson, '"Djöfullinn gengur um sem öskrandi ljón": Af Jóni Halldórssyni Skálholtsbiskupi, Francesco Petrarca og fornu ljónahliði dómkirkjunnar í Bologna', *Skírnir*, 178 (2004), 341–48

—, 'The Fantastic Feats of Master Perus of Arabia (Abstract)', in *The Fantastic in Old Norse/Icelandic Literature: Sagas and the British Isles. Preprint Papers of the Thirteenth International Saga Conference. Durham and York, 6th–12th August, 2006*, vol. 2, ed. by John McKinnell, David Ashurst, and Donata Kick (Durham: Centre for Medieval and Renaissance Studies, Durham University), p. 659

—, 'The Life and Literary Legacy of Jón Halldórsson, Bishop of Skálholt: A Profile of a Preacher in Fourteenth-Century Iceland' (Unpubl. M.Phil. diss., University of St Andrews, 1996)

Morris, Colin, *The Discovery of the Individual 1050–1200* (Toronto: University of Toronto Press, 1987)

Paasche, Frederik, 'Jon Halldorsson', in *Norsk biografisk Leksikon*, vol. 7, ed. by Lars O. Jensen–Krefting (Oslo: Aschehoug, 1936), pp. 101–02

Philippson, Ernst, *Der Märchentypus von König Drosselbart*, FF Communications, 50 (Greifswald: Suomalainen Tiedeakatemia, 1923)

Slay, Desmond, 'Introduction', in *Romances: Perg. 4:0 Nr 6 in the Royal Library, Stockholm*, ed. by Desmond Slay, Early Icelandic Manuscripts in Facsimile, 10 (Copenhagen: Rosenkilde and Bagger, 1972), pp. 9–28

—, 'The Original State of Stockholm Perg. 4:0 nr 6', in *Afmælisrit Jóns Helgasonar 30. Júní 1969*, ed. by Jakob Benediktsson (Reykjavík: Heimskringla, 1969), pp. 270–87

Sverrir Jakobsson, 'Íslam og andstæður í íslensku miðaldasamfélagi', *Saga* 50.2 (2012), 11–33

Torfi H. Tulinius, 'Kynjasögur úr fortíð og framandi löndum', in *Íslensk bókmenntasaga*, vol. 2, ed. by Vésteinn Ólason (Reykjavík: Mál og menning, 1993), pp. 165–246

de Vries, Jan, *Altnordische Literaturgeschichte*, 2 vols, 2nd edn (Berlin: De Gruyter, 1967)

Waddell, Helen, *The Wandering Scholars*, 7th edn (Constable: London, 1990)

Wolf, Kirsten, '*Gyðinga saga*, *Alexanders saga*, and Bishop Brandr Jónsson', *Scandinavian Studies* 60 (1988), 371–400

FLORIAN SCHRECK

Science in Medieval Fiction

The Case of Konráðs saga keisarasonar

When considering the role science and learning played in the intellectual history of the Middle Ages it is fruitful to widen the scope beyond medieval scholars and their treatises and look at the avenues of dissemination and cultural engagement with this knowledge.[1] This is particularly vital as modern academic scholarship is still prone to draw a dividing line between the sciences and humanities even when discussing historical periods for which such a separation would have been clearly anachronistic. Consequently, this chapter follows the approach of literature and science studies, inquiring how the two disciplines cross-fertilized one another in historic forms of interaction.[2]

While scientific topics were incorporated in various forms of literature in the Middle Ages, in the following it will be argued that the development of the romance genre and the mode of fictional storytelling opened up new narrative avenues to stage and speculate about scientific knowledge. This proposition will then be discussed with the help of a case study based on an Old Icelandic romance, *Konráðs saga keisarasonar*, noteworthy for its prominent portrayal of topics ranging from natural history to such areas as astronomy and linguistics. As such, this chapter seeks to contribute a uniquely medieval perspective to the field of literature and science studies, which, since the publication of the hugely influential *Darwin's Plots: Evolutionary Narrative in Darwin, George Eliot and Nineteenth-Century Fiction* by Gillian Beer in 1983, has largely been focussed on the early modern and modern era.[3] Over the last decade this research area has steadily attracted increasing scholarly attention from various

---

1 For a more comprehensive account of many of the thoughts articulated here, see the dissertation Schreck, 'Science in Medieval Fiction'.
2 On the roots of the field literature and science studies in the ambition of overcoming the artificial divide between the sciences and the humanities as a leftover from the *Two Cultures* debate of the past, see Sleigh, *Literature and Science*, and Willis, *Literature and Science*. For a general overview of past research in literature and science studies see also Pethes, 'Literatur- und Wissenschaftsgeschichte'.
3 Beer, *Darwin's Plots*. Note that the *Journal of Literature and Science*, established in 2007, noticeably excludes older manifestations of literature and science from its scope, encouraging submissions only on subjects from periods since the 'Scientific Revolution', *Journal of Literature and Science*, 1, p. 3.

Florian Schreck • graduated with a PhD from the University of Bergen with a research project on the reception of learned writing on natural history in Old Icelandic romances.

disciplines.[4] However, contribution on medieval subjects has only very recently picked up more pace.[5]

The medieval period not only made fundamental contributions to the evolution of science but also to the progress of literature. During the twelfth and thirteenth centuries the systematic study of nature, motivated by the theological desire to increase the understanding of the Christian God's creation and fuelled by the translation of scientific treatises from Arabic, gained institutional support in the creation of the first European universities.[6] And, beginning in the twelfth century, at the same time when learning, science and cultural production surged forward, another new and far-reaching development took place in continental and Northern Europe: the blossoming of fictional literature in the form of the romance genre.[7] This mode of storytelling offered both its authors and audiences a platform to engage with subjects in a way which was freed from many of the constraints of more traditional kinds of storytelling. And it was here that we see the transmission of medieval science make a lateral movement: from the medium of factual writing to fictional literature — from treatises, encyclopaedias, maps, charts, and medical texts to narrative tales. In the following this chapter takes a closer look at this important junction in the history of literature and science.

Medieval romance is generally associated with the exploits of stock characters, typically knights, princesses, and other nobles, and a small subset of recurring themes, such as courtly love and the fulfilment of chivalrous quests. And yet, this chapter's title welcomes associations with modern day science fiction for a good reason.[8] A few brief examples will quickly outline the matter: in *Cligès*, written by Chrétien de Troyes, a young empress wants to feign a fatal illness in order to trick her husband and be together with her secret lover. This is a common enough plot device, yet the text uses the occasion to discuss contemporary medical practices as the empress craftily has her urine sample manipulated so that she can fool the doctors at court and even the physicians from Salerno.[9] Building on Chrétien's *Yvain, ou le Chevalier au Lion*, an Old Icelandic romance, *Sigurðar saga þǫgla*, tells of a knight who frees a lion from the

---

4   See for example the establishment of the *Journal of Literature and Science* in 2007 or the recent multitude of publications such as *On Literature and Science*, ed. by Coleman; Klinkert, *Epistemologische Fiktionen*; *Literatur und Wissen*, ed. by Köppe; LaGrandeur, *Androids and Intelligent Networks*, and Fendt, *Wissenschaft und Imagination*.
5   For some recent studies on the interaction between literature and science in the medieval period, see e.g. van Ruymbeke, *Science and Poetry in Medieval Persia*; Bulang, *Enzyklopädische Dichtungen*; Gabrovsky, *Chaucer the Alchemist*, and *Medieval Science Fiction*, ed. by Kears and Paz.
6   For a more comprehensive overview of the historical development of science in medieval Europe, see the Introduction to this volume. See also Grant, *The Foundation of Modern Science in the Middle Ages*, esp. Ch. eight 'How the Foundations of Early Modern Science were Laid in the Middle Ages', and Hannam, *God's Philosophers*.
7   On the development of fictionality and the romance genre see Green, *The Beginnings of Medieval Romance*.
8   Cf. the recent and likewise provocatively named anthology *Medieval Science Fiction*, ed. by Kears and Paz.
9   Chrétien de Troyes, *Cligès*, ed. and trans. by Kasten, st.5719–63.

clutches of a dragon, upon which the narrator includes a lengthy digression detailing the zoological characteristics of lions and their allegorical interpretations according to the Christian bestiaries.[10] Another story, *Hektors saga*, reports a shining suit of armour that is able to effortlessly repel any blows struck against it.[11] The explanation is delivered promptly: the astonishing defensive properties are not powered by magic, but by mineralogy. An adamas gemstone was crafted into the helmet and the stone's ability to make its wearer invincible is transferred onto the armour. The text follows this claim up with an extensive quote from a lapidary listing the gemstone's qualities according to medieval mineralogical lore.[12]

The romance genre has a strong and deep connection to science and learned knowledge, especially in the forms that developed in the medieval North. The introduction of fictionality offered a multitude of creative options for the popularization of science through the narrative medium. In turn fictional literature provided a new way to discuss scientific topics and make contributions to the scientific discourse.

The Narrative Mode of Medieval Romance

Starting in the twelfth century, the romance genre swept from England and France throughout Europe. These tales of chivalric adventure were quickly translated and adapted into the vernacular traditions of each new country they reached, broadening their circulation to Germany, Italy, The Netherlands, Portugal, Spain, Greece and, of course, Scandinavia.[13] Romance's popularity had such a profound and ongoing impact that these stories comprised 'arguably the most influential and enduring secular literary genre of the European Middle Ages'.[14] Although the romances' specifics changed from country to country, they all shared a stock of common character constellations, motifs, and settings. Besides these similarities in content they also all drew on a shared aesthetic and narrative mode.

Romance scholarship has long since argued that the genre was responsible for the popularization of fictional literature in medieval Europe. While fictionality in the West can look upon a tradition reaching back to classical antiquity, it was romance that made it one of the dominant forms of storytelling, lasting to this day. Much of the research on the early development of fictionality was initially focused on Middle High German romances.[15] Building on the work of Walter Haug and Fritz Peter Knapp, Dennis H. Green defined romance's fiction as follows:

---

10   *Sigurðar saga þǫgla*, in *Late Medieval Icelandic Romances*, vol. 2, ed. by Loth, pp. 141–47.
11   *Hektors saga*, in *Late Medieval Icelandic Romances*, vol. 1, ed. by Loth, p. 85.
12   For an in-depth study of this example, see Schreck, 'Science in Medieval Fiction', pp. 99–104.
13   Krueger, 'Introduction', p. 1.
14   Krueger, 'Introduction', p. 1.
15   See esp. Haug, *Literaturtheorie im deutschen Mittelalter* and Knapp, *Historie und Fiktion in der mittelalterlichen Gattungspoetik*, vol. 1: *Sieben Studien und ein Nachwort*, 1997; vol. 2: *Zehn neue Studien und ein Vorwort*.

> Fiction is a category of literary text which, although it may also include events that were held to have actually taken place, gives an account of events that could not conceivably have taken place and/or of events that, although possible, did not take place, and which, in doing so, invites the intended audience to be willing to make-believe what would otherwise be regarded as untrue.[16]

This definition contains several important points that are central to the study of science in fiction. It addresses some of the concerns of medieval poetics, particularly romance's relation to truth and lying, which is a key issue when talking about scientific facts in fictional narratives. According to the early medieval bishop scholar Isidore of Seville there were three kinds of narratives: *historia*, *fabula*, and *argumentum*.[17] Drawing on classic poetics Isidore describes *historia* as tales that report events that have happened, *fabula* as concerning things that could not happen, and *argumentum* as stories about events which although they did not happen, nevertheless could happen. Green defines fiction as situated between *fabula* and *argumentum*. The romances were thus open to represent scientific knowledge truthfully as well as to play with this factual material in any way they liked.

Green's definition also points out that the audience must be willing participants in the game of fictional literature. Fictionality rests on a 'contract between author and audience in which each consciously plays his allotted role'.[18] This meant that the audience needed to be made aware that if they read or listened to romances, they engaged with fictionality. Accordingly, the medieval romances often featured prologues and epilogues discussing the nature of truth in relation to literature. These together with the much more prominent and sometimes even intrusive presence of the narrator's voice in the romances have been discussed as possible signals to the audience marking the texts' special status as pieces of fiction.[19] Green argues furthermore that romance fictionality was influenced by the Horatian notion that a work of literature should both be entertaining and useful.[20] As Green points out the ideals of *prodesse et delectare* were present both in medieval literary poetics and practice.[21] They can certainly be considered a crucial factor for the presence of science in the romances and their impact will be examined more closely in a moment.[22] But it should also be noted that another of Horace's recommendations might have played an important role for science in medieval fiction. In his *Ars poetica* Horace counselled that authors should aim for verisimilitude in their stories, to present what

---

16 Green, *The Beginnings of Medieval Romance*, p. 4.
17 Isidore of Seville, *Etymologiae*, trans. by Barney, I. xliv.
18 Green, *The Beginnings of Medieval Romance*, p. 3.
19 See the discussion in Knapp, 'Subjektivität des Erzählers und Fiktionalität'. Cf. the discussion of narrator *apologiae* in Old Norse romance by O'Connor, 'History or Fiction?', pp. 101–69.
20 Green, *The Beginnings of Medieval Romance*, p. 28. Cf. Horace *Ars poetica*, , ed. and trans. by Rushton Fairclogh, 333.
21 For a list of medieval sources outlining the reception of Horace's ideas on poetics see the endnotes mentioned there by Green, *The Beginnings of Medieval Romance*, p. 28.
22 For a brief discussion of the Old Norse translated romances instructional function see Barnes, 'Current Issues in *riddarasögur* Research'.

they make up for pleasure in a manner that still keeps close to the truth.²³ Thus any possible inclination the romance writers might have had to temper with the facts would likely have been curbed by this advice.

Fictionality as a key element of romance's narrative mode went hand in hand with another characteristic that shaped the tone and form of the genre. The romances are known for being fond of retelling similar stories in new ways. Scholars in the past, particularly in Old Norse scholarship, have often been very critical of this tendency as they deemed it the mark of 'unoriginality'.²⁴ But repetitiveness was a conscious aesthetic choice and a deliberate feature of the romances. It usually took one of two shapes: either the form of rewriting the same story matter, as for example in the many Middle High German versions of the Chrétien romances, or the consequent reuse of certain plot structures, such as the bridal-quest in the case of Old Icelandic romance.²⁵ Joachim Bumke and Franz-Josef Worstbrock describe the medieval practice of retelling using the differentiation in *materia* and *artificium*.²⁶

Materia is the matter of a story, its content consisting of a more or less fixed constellation of characters, plot developments, and narrative cores. The term *Erzählkern* (narrative core) was coined by Joachim Heinzle and describes important traditional nexuses of the matter that could not be left out in any retextualization processes, as they were integral to the story.²⁷ These narrative cores are made up of certain plot elements and motifs and, as Heinzle stresses, could also contain individual phrases or sentences which belonged to a story's staple inventory. This meant that the point of departure for most romance writers was a given *materia* which they then retold in a new fashion.²⁸ An example of such a *materia* would be the life story of Alexander the Great with its fixed biographical and historical events coupled with a set of motifs, such as Alexander's wisdom, martial prowess, or hubris. This *materia* stayed the same even though it was retold again and again, particularly in Middle High German literature with the three retextualizations of Lambrecht's *Alexander*: the *Vorauer Alexander* c. 1160/70, the *Straßburger Alexander* before 1187, and the *Basler Alexander* in the late thirteenth century, and the other Alexander romances by Rudolf of Ems c. 1235/40, Ulrich of Etzenbach before 1290, Seifrit 1352, the *Großer Alexander* before 1397, and the prose versions by Meister Wichwolt about 1400 and Johannes Hartlieb about 1450.²⁹

The sheer number of texts signals that even though the basic story stayed the same the audiences were still attracted to these retellings. Worstbrock argues that when the 'what' of a romance was usually a given, all the writer's attention was focused on the

---

23 Horace, *Ars poetica*, ed. and trans. by Rushton Fairclogh, 338.
24 See Driscoll, 'Late Prose Fiction (*lygisögur*)', p. 191, see also the references mentioned there.
25 See Kalinke, *Bridal-Quest Romance in Medieval Iceland*.
26 Worstbrock, 'Wiedererzählen und Übersetzen'; Bumke, 'Retextualisierungen in der mittelalterlichen Literatur'.
27 Heinzle, 'Konstanten der Nibelungenrezeption in Mittelalter und Neuzeit', p. 92.
28 Bumke, 'Retextualisierungen in der mittelalterlichen Literatur', pp. 10–11.
29 Brunner, *Geschichte der deutschen Literatur des Mittelalters*, pp. 144–45.

'how', the *artificium*.³⁰ He thus claims that a romance writer's artistic contribution to a narrative rested chiefly on how the story was told rather than in the invention of the plot. While Worstbrock views the issue from the point of the writer and the creation of romances, it might be added that the 'how' question, the focus on the *artificium*, was also important for the audience, as the popularity of these rewritings attest. A look at the transmission of Old Icelandic romances quickly reveals that they were very popular indeed, judging by the number of surviving manuscripts.³¹ Yet all they did was retell new variations on the same matter, the bridal-quest. All of them feature the same stock of high nobility characters, a setting at the royal courts of Europe or the East, and a plot revolving around how a young prince or princess after many quests and obstacles finally gets married and settled in their kingdom. But instead of growing tired of listening to the same story over and over again, the Icelandic audience relished these stories and copied them by hand up to the nineteenth century.

The explanation for this popularity might be fairly straightforward. While an audience experienced a romance — either through reading or oral performance — they did not have any trouble in following the plot, as its narrative direction was inevitably the same as with all other stories in the genre. Thus, the audience could devote their energy entirely to appreciating the creativity that went into how the story was told. Analogues making use of the same narrative mechanism count among the most popular forms of entertainment even today, with crime fiction's 'whodunit'-plot structure closely resembling that of the Old Icelandic romance's bridal-quest. In both cases the plot structures remain the same throughout the genre and the art of how the story is told remains central. Any deviations from the general plot break audience expectations, but the tension these deviations create exist only because of the dominance and presupposition of the genre-typical plot structure. This mechanism meant romance authors needed good ways to make their narratives interesting and compelling besides the plot. One solution was to include something as fascinating as the marvels of nature.

One of the signs testifying to the rising widespread interest in science and particularly natural history from the twelfth century onward, was the significantly increased production of encyclopaedias. Carrying on traditions from antiquity, these collections served to disseminate scientific and learned knowledge, primarily to a clerical userbase, preachers, and university men.³² To that end they covered a wide range of subjects, but always devoted a large part of their text to natural history. While some encyclopaedias added moralizing interpretations to their entries others stuck strictly to listing the known or observed facts about nature in chapters on topics such as botany, mineralogy, or zoology. But no matter how they presented their knowledge, the encyclopaedias' aim from the early Middle Ages was to provide the

---

30 Worstbrock, 'Wiedererzählen und Übersetzen', pp. 135–39.
31 Kalinke, 'Norse Romance', p. 316.
32 For a general overview of the development of medieval encyclopaedias, see Hünemörder, 'Antike und mittelalterliche Enzyklopädien'.

basic natural knowledge needed to correctly interpret biblical allegories and serve as material for sermons.[33] Only towards the later medieval period did this priority shift as a growing lay audience showed interest in education and the study of nature, particularly amongst the growing urban elite and middle class.[34] Consequently, more encyclopaedias were compiled and increasingly written in the vernacular languages.

The encyclopaedias provided their European audience with knowledge about distant foreign lands and the natural marvels which could be found there, like exotic animals, gemstones with extraordinary powers, or hideous monstrous peoples. Enriching a romance story with a selection of these astonishing subjects meant offering the audience a wide spectrum of fascinating experiences, distinct from their own daily lives.[35] For clerical authors it also meant a chance to edify the romances' audience in the breadth and wondrousness of the Christian God's creation.

The Case of *Konráðs saga keisarasonar*

The romance genre emerged in Iceland at a time of great upheaval, when the small island lost its political independence and became a part of the Kingdom of Norway. While the previously dominant narrative genres were chiefly concerned with the past of the island's settlers and the history of domestic struggles, writers and audiences alike now began to turn their interests towards the wider world. While there is some evidence that individual European romances may have found their way to Iceland as early as 1200, the genre of Old Icelandic romance is generally assumed to have its origins in a wave of translations of chiefly French and English and to a lesser degree German romances initiated in the middle of the thirteenth century at the court of Hákon Hákonarson, King of Norway.[36]

These translations inspired Icelandic authors greatly and resulted in a stream of original compositions, which, although initially modelled after the foreign imports, quickly developed their own idiosyncrasies and genre conventions. More than half of the Old Icelandic romances, also called indigenous *riddarasǫgur*, known from the Middle Ages survive in more than forty manuscripts and two in almost eighty. The majority of these were copied in the post-medieval period bearing witness to the genre's lasting impact throughout the ages. As Matthew Driscoll observed, these numbers mean that the Old Icelandic romances were 'arguably the most popular sagas of all time and of any type'.[37] The genre's broad and productive reception in Iceland

---

33 Hünemörder, 'Antike und mittelalterliche Enzyklopädien', p. 351.
34 Hünemörder, Antike und mittelalterliche Enzyklopädien', p. 351 and Clark, A *Medieval Book of Beasts*, p. 115.
35 On wonder and the marvellous and its relation to pre-modern scientific writing, see Campbell, *Wonder and Science*.
36 Scenes from *Yvain, ou le Chevalier au Lion* by Chrétien de Troyes are depicted on the church door of Valþjófsstaðir in Iceland dated to as early as 1200, see Harris, 'The Lion-Knight Legend in Iceland' and Acker, 'Dragons in the Eddas and in Early Nordic Art', p. 65.
37 Driscoll, 'Late Prose Fiction (*lygisögur*)', p. 194.

on the periphery of the medieval Christian world thus strongly echoes romance's popularity throughout Central Europe.

The Old Icelandic romances were generally much more reluctant to depict or interpret the phenomena of natural history in their allegorical sense, as was popular with earlier French and German romances.[38] Old Icelandic romance's more literal approach to nature may be seen as a reaction to a larger pan-European development in natural philosophy. Beginning with the translation of scientific treatises from Arabic, Greek, and Hebrew in the twelfth and thirteenth centuries, intellectuals in Europe began increasingly to value the literal sense in the study of nature and scripture.[39] By strengthening ties to these international communities, the Old Icelandic romances acted against the feeling of growing isolation Iceland experienced under Norwegian rule.[40] This expands and reinforces Geraldine Barnes's suggestion that the genre's occupation with exotic and far-away lands, European nobility and learning were rhetorical strategies 'to counter Iceland's progressive political and economic side-lining from mainland Scandinavia during this period'.[41] The Old Icelandic romances should thus no longer be pejoratively considered a 'literature of escape'.[42] Rather, they represent the desire of Icelanders to connect to the world and join in European literary and intellectual trends.

Konráðs saga keisarasonar is a good representative of the more 'bookish' subset of Old Icelandic romances, to use Geraldine Barnes's term.[43] This sub-genre of romances draws upon a large amount of learned writing and stands close to the narrative mode of *argumentum*.[44] The Old Icelandic romances tell stories that did not happen but are plausible and therefore could have happened. *Konráðs saga keisarasonar*'s oldest extant manuscript is Holm perg. 7 4to, dated to about 1300, making the saga one of the earliest attested works in the genre.[45] Five other copies and rewritings survive from the Middle Ages and more than forty from the post-medieval period, showcasing the popularity of the saga.[46] Furthermore, eight sets of *rímur*, ballads, based on *Konráðs saga keisarasonar* were composed since about 1500.[47]

Jürg Glauser described *Konráðs saga keisarasonar* in his monograph on the genre as a piece of formula literature, unfolding against a rigid system of literarily established norms depicting how socio-economic conditions (richness, luxury, elegance) lead to

---

38 See Schreck, 'Science in Medieval Fiction,' pp. 301–09.
39 See Wegmann, 'Naturwahrnehmung im Mittelalter', p. 26, and the references there.
40 For more on this, see Schreck, 'Science in Medieval Fiction', pp. 279–314.
41 Barnes, *The Bookish Riddarasögur*, p. 23.
42 See Kalinke 'Norse Romance', p. 319 and her discussion of the past treatment of Old Icelandic romance as escapism.
43 See Barnes, *The Bookish Riddarasögur*.
44 Other examples of science in fiction writing in this genre can be found in e.g. *Hektors saga, Kirialax saga, Nítíða saga, Sigurðar saga þǫgla* and *Vilhjalms saga sjóðs*, to name but a few.
45 All references to *Konráðs saga keisarasonar* are made to the edition by Zitzelsberger. All quotes and the textual analysis are based on the 'D' version Copenhagen, AM 180 b fol., *c.* 1500.
46 *Konráðs saga keisarasonar*, p. xvii.
47 Driscoll, 'Late Prose Fiction (*lygisögur*)', p. 195.

moral virtues (wisdom, kindness, superiority).⁴⁸ Glauser stresses the saga's adherence to genre conventions and its reinforcement of the medieval aesthetic belief that outer appearances reflect inner worth. A decade later, Wilhelm Heizmann and Tibor F. Péza on the other hand consider *Konráðs saga keisarasonar* to be a sophisticated piece of literature using the established plot pattern of the bridal-quest as the background against which the text discusses the intellectual problem of real and nominal identity.⁴⁹

Both of these interpretations reflect important aspects of *Konráðs saga keisarasonar* as the saga does follow formulaic genre norms, but as discussed above, this only serves to increase its literary value in the framework of romance aesthetics. Against this conventional backdrop the topics which the saga chooses to discuss stand out even more. Heizmann and Péza recognized identity as one of the saga's main themes, but, as this chapter argues, so are education and science.

Konráðs saga keisarasonar's two main characters are Konráðr, the son of the emperor of Saxland, and Roðbert, a jarl's son and his best friend and advisor. The story begins with the detailed description of their education. Both were taught in their childhood by Roðbert's father, an eminent scholar as the narrator notes.⁵⁰ But besides scholarly knowledge, they were also educated in knightly skills and athletic prowess. It soon became apparent to Roðbert that he could not match Konráðr in the physical exercises, so he decided to concentrate on rhetoric and language skills. The saga remarks that he swiftly became able to understand and talk the languages of all the people in the world and that he became so eloquent and cunning that his proficiency of a language often far surpassed that of the native speakers themselves. When it became time for Konráðr to learn foreign languages as well, Roðbert advised him not to waste his time on such things. Any linguistic knowledge he would acquire would only take up his intellectual capacities and displace the knowledge and skills he had already learned.⁵¹ When his father, the emperor, explicitly asks Konráðr if he did not want to begin learning languages himself, Konráðr replies, that he had no need of such things, as his good friend Roðbert would always be at his side and translate for him.⁵²

Unsurprisingly, Konráðr's trust in Roðbert is entirely misplaced. The two have to leave the royal court, after Roðbert impregnates Konráðr's sister, the emperor's daughter. When they sail with their retinue to Constantinople, it falls to Roðbert to communicate with the people there, as he is the only one in the group who speaks the local language. At this point Roðbert chooses to betray Konráðr as he usurps the latter's name and presents himself as Konráðr and son of the emperor of Saxland, while introducing the actual Konráðr as the jarl's son Roðbert.⁵³ While at court in Constantinople Roðbert tries to arrange to be married to the king's only daughter,

---

48 Glauser, *Isländische Märchensagas*, pp. 166–68.
49 Heizmann and Pézsa, 'Reale und nominelle Identität in der *Konráðs saga keisarasonar*'.
50 *Konráðs saga keisarasonar*, p. 129.
51 *Konráðs saga keisarasonar*, p. 132.
52 For a detailed study on the theme of linguistic education in *Konráðs saga keisarasonar* and other romances see Kalinke, 'The Foreign Language Requirement in Medieval Icelandic Romance'.
53 *Konráðs saga keisarasonar*, p. 136.

Matthilldr. She is described as beautiful and extraordinary well learned and wise. The saga once more stresses education, as the narrator tells us that her father, the king, called upon all the scholars and masters of the country to educate her.[54] Although Matthilldr does not speak the language of these foreigners that have come to Constantinople, she immediately notices that something is wrong, as the man that was introduced to her as Roðbert, the actual Konráðr, had a far nobler bearing than their alleged leader Konráðr/Roðbert. Roðbert now tries to get Konráðr killed by daring him to perform increasingly dangerous feats before the court, but Konráðr prevails and eventually grows wise to Roðbert's deception. In collaboration with Matthilldr he masters a quest to prove his worth to the king of Constantinople and defeats Roðbert to demonstrate to everyone that he is superior and thus the actual Konráðr and son of an emperor. Roðbert is exiled and Konráðr marries Matthilldr.

As is apparent, one of the main themes of the saga is education and the entirely negative consequences of lacking an education. This sentiment is hardly unique to *Konráðs saga keisarasonar*, but the saga explores the theme of education to an exceptional depth and focuses on the practical application of learning. This is evident in the case of Roðbert and his linguistic skills, but it is also true when it comes to medieval science, as can be most clearly seen in the quest that Konráðr must undertake in order to prove his worth to the king of Constantinople.

Konráðr is charged with finding a certain gemstone, yet he is only shown what this gemstone looks like.[55] No other information is provided to him. However, Matthilldr's books allow her to not only identify the stone but present her with all other knowledge needed to successfully master the quest. Matthilldr is able to locate the destination to which Konráðr must travel: 'svo visa bækr til ath fadir mínn muni hafa sent þik á Serkland jth mikla j borg Babilonem' (The books indicate that my father seems to have sent you to Serkland the Great, into the city of Babylon).[56] Her books also outline the geography along Konráðr's route, but when it comes to actual navigation, Matthilldr advises him to apply his astronomical knowledge: *enn þu verdr stefno þinni at hatta eptir hímintunglagang* (And you shall set your course by the movement of the celestial bodies).[57]

Her books also put Matthilldr into the position to further guide Konráðr in how to prepare for the journey. She counsels Konráðr: *þu skalt ok hafa med þer hana ok svín* (You shall also bring with you a rooster and a pig).[58] While this may seem like an eccentric instruction at first, Matthilldr immediately reveals the solid zoological foundation on which she bases her recommendation: 'Þat land er fyrst firir þer er ǫngo er bygt nema leonibus. Þau hrædaz ekci nema þat er hani gelr' (You will first encounter a land where no one has settled but lions. They fear nothing except the crow of a rooster).[59]

---

54 *Konráðs saga keisarasonar*, p. 137.
55 *Konráðs saga keisarasonar*, pp. 156–57.
56 *Konráðs saga keisarasonar*, p. 157.
57 *Konráðs saga keisarasonar*, p. 157.
58 *Konráðs saga keisarasonar*, p. 157.
59 *Konráðs saga keisarasonar*, p. 157.

It becomes apparent that the books on which Matthilldr relies not only inform her about the geography on the way to Babylon but also contain knowledge of the regional characteristics, such as dangerous fauna. This mode of presenting different interconnected fields of knowledge, such as geography, zoology, or mineralogy, as they pertain to specific lands is frequently used in medieval encyclopaedias and can also be found in *Stjórn*. This Old Norse Bible compilation appended the Genesis description of the world's resettling by Noah's sons after the Flood with a long description of the different regions of the world, made up of Old Norse translations from Isidore's *Etymologiae* and Vincent of Beauvais's *Speculum historiale*.[60] A geographical outline of the different parts of the world, taken from Vincent, is supplemented by a list of each region's noteworthy zoological, botanical, mineralogical, or otherwise remarkable characteristics, generally translated from Isidore. While it is unlikely that *Konráðs saga keisarasonar* used *Stjórn* as one of its sources for specific zoological or mineralogical knowledge, other Old Icelandic romances did — or at least utilized the same source text on which the *Stjórn* chapter is based.[61]

That lions fear the crowing of a rooster might strike a modern audience as an odd sentiment but in medieval zoological writing it is a well-documented fact which can be traced back to Pliny the Elder's *Naturalis historia* (lib. 8.9). Given this work's influence on natural history the lion–rooster dynamic soon became a fixed element of most medieval encyclopaedias and can for example be found in Bartholomeus Anglicus's *De proprietatibus rerum* (lib. 18.65) or Vincent's *Speculum naturale* (lib. 19.74), albeit it is notably missing from Isidore's *Etymologiae*.

Konráðr does as Matthilldr instructed him and brings a rooster along on his journey. And it is at this point that the saga narratively explores how book-learning on zoology might be gainfully employed in practice. Soon after he enters the lion-country Konráðr makes camp and sets up his rooster.[62] When a lion shortly afterwards attacks the camp everything goes according to plan. The rooster's crow confuses the lion and the beast takes flight immediately, giving Konráðr the opportunity to easily dispatch the fleeing lion. The episode presents itself as a play on the topos of the fight between the hero and a lion, highlighted by a parallel to earlier in the saga. When Konráðr encountered a lion for the first time in Constantinople, he confronted and killed it by relying on his martial prowess and divine providence.[63] But this time, the mode of overcoming the lion is entirely new. Instead of demonstrating his courage and fighting skills he succeeds by sensible application of zoological knowledge gained through the study of learned books.

The design of a structurally familiar scene — the fight of man against beast — is used to introduce the zoological knowledge into the narrative. This eases the audience's intake of the potentially unknown facts by framing it in a narrative manner for which they have established interpretative strategies in place, thanks

---

60 *Stjórn*, ed. by Astås, pp. 100–52.
61 Cf. for example *Kirialax saga* or *Sigurðar saga þǫgla*.
62 *Konráðs saga keisarasonar*, p. 159.
63 *Konráðs saga keisarasonar*, p. 151.

to the universal genre conventions which they know how to process. But the scene not only imparts knowledge about lions to its audience, it also comments on this knowledge. The zoological treatises all agree on the lion's fear of the rooster, but they merely report this fact. The saga on the other hand considers the implications of this behavioural characteristic and shows one of the uses it could be put to, developing the idea of a natural and rooster-based lion repellent strategy, useful as a safety measure to protect one's camp and also as a possible aid when lion hunting. *Konráðs saga keisarasonar* thus uses its zoological source texts on a narrative level as an aesthetically pleasing twist on a well-established topos, as it corresponds well to the maxim of repetition and variance, while at the same time furthering the saga's theme of a necessary balance between mental and physical education, learned study, and practical application. On an extratextual level, we see the saga as a cog in the machinery of the dissemination and production of knowledge, as it teaches the audience about the zoological features of the lion while also generating new insights into the transmitted knowledge by speculating about the consequences of these features. This way of approaching the known facts in a speculative fashion and engaging in thought experiments is fictional literature's unique contribution to the study of nature.

This idea and pattern repeats itself in the encounter with an elephant. As mentioned earlier, Matthilldr advises Konráðr to bring a pig on his travel to Babylon. Her justification runs parallel to the lion episode: *þar næst kemr þu á þat land er ǫngo er bygt nema filum. Þat dyr ottaz ecki nema hrin svína.*[64] (Then you will come to that land where no one has settled but elephants. That animal fears nothing but the squealing of pigs). As predicted by Matthilldr, Konráðr finds himself travelling through Elephant-Land:

> Kemr hann [Konráðr] nu aa Filaland ok sækir nu hartt ferdína ok vard enn vid ecki varr. Ok er hann hafdí miok sott þetta land kom hann þar ath aptne er honum þotti miok vndarlikt þar voro eikur íj. hinar stærsto þær hǫfdo badar hallaz aa einn veg svo ath lauf þeirra ok límar lago aa jordo enn þar vmkringís tradk mikit. hann þottíz vndirstanda ath þar mundí vera huildarstadr filssíns.
>
> (Konráðr then comes to Elephant-Land and presses the journey ahead greatly, but did not become aware of anything. But when he had travelled far through this land, towards evening he came upon something that he thought very curious. There were two great oak trees both leaning in one direction so that their leaves and limbs lay on the ground and around there was much trampled earth. He thought to understand that this must be an elephant's resting place.)[65]

This scene in *Konráðs saga keisarasonar* is based on further zoological knowledge. By the Middle Ages, the elephant's kneelessness was a long-established fact. Although it is

---

64 *Konráðs saga keisarasonar*, p. 157.
65 *Konráðs saga keisarasonar*, p. 159.

absent from the *Etymologiae* and was thus also not included in the *Stjórn* description, it can be found in other works on natural history, such as in Bartholomaeus Anglicus *De proprietatibus rerum*. Pliny (*Naturalis historia* lib. 8.39) repeats Caesar's (*De bello gallico* 6.27) claim that elks have no knees, have to sleep leaning against a tree, and are unable to stand up once they fall down.⁶⁶ In the *Naturalis historia*, the description of the elk and its characteristics follows immediately after the elephant, easily explaining the ensuing confusion. That elks indeed have knees would furthermore have been perfectly obvious to anyone in Northern Europe as elks were readily observable, unlike elephants.

Konráðr carefully studies his surroundings on his travels looking out for signs of the dangerous elephants he was warned about. Through his knowledge of the zoological characteristics of these animals he is able to identify the odd sight of great oak trees precariously leaning towards the ground as the resting place of an elephant. The text thereby uses the factual source it draws on to narratively speculate on the ecological impact of the elephant's characteristics. Its kneelessness and subsequent habit of leaning against trees would deform the trees it leans on because the elephant is also portrayed as very big and heavy. Shrubbery, smaller trees, and grass would be trampled flat in the vicinity of the elephant's resting place as these would not stand up against the heavy elephant passing through them. The saga speculates that elephants would have established sleeping places to which they returned regularly, presuming that it would be too troublesome for the animals to seek out trees sturdy enough to bear their weight anew each night.

After he spotted the leaning oak trees, the saga goes on to describe that Konráðr decided to make his camp for the night there and prepare for the return of the elephant. He dug out a pit close by the path he expected the elephant to take. In this pit Konráðr placed his pig around which he had fastened a rope and then covered the pit with twigs and moss in order to hide it.

> Ok er stund leíd heyrdí hann braukon ogurliga j skógín ok þui næst sa hann fram koma dyr svo mikit ath eckti hafdí hann fyr annat sied jafn mikit. bak þess bar eigi lægra enn hino hæsto tre j mǫrkinní. Þat hafdí hǫfud mikit ok tǫnn fram or ok var aa krockr. sa hann ath þar med mattí hann drepa dyr til matar ser. fætr hans voro vndarliga hafir ok einn lidr nidrí vid hofhuarf. hann vndirstod ath eckí mundí hann ligia mega ok þetta mundí hann nattbol hafa ath hallaz at eikum þessum ok hann mundí eckí upp mega standa ef hann fellí.

> (And some time later he heard a great noise in the forest and then he saw an animal approaching so big that he had never seen any equally big before. Its back was not lower than the highest tree in the darkness. It had a great head and a tooth out of the front, and it was hooked. He saw that he [the elephant] would kill animals with that to feed itself. Its legs were strangely tall and had only one joint down by the fetlock (top of the hoof). He understood that he [the elephant] could not lie down and because of that must have its place

---

66 *Naturalis historia* lib. 6.39.

to sleep leaning against those oak trees and that he could not stand up once he had fallen.]⁶⁷

The elephant finally appears just as Konráðr expected. While approaching its resting place, the animal does not notice the camouflaged pit with the pig inside. Konráðr pulls the rope fastened to the pig, which begins to squeal loudly. The noise of the squealing pig dazes the elephant, which starts to panic. Konráðr seizes the moment and impales the stunned elephant with his lance. Matthilldr's plan to use the elephant's fear of pig squealing to overcome the dangerous pachyderm is successful. The clever utilization of zoological knowledge proves central in managing this important part of the protagonist's quest.

The issue of which sources *Konráðs saga keisarasonar* used must be addressed. The majority of medieval treatises on zoology, such as Isidore's *Etymologiae*, Bartholomeus Anglicus's *De proprietatibus rerum*, Vincent of Beauvais's *Speculum naturale* or the *Liber Floridus* by Lambert of St Omer, claim that elephants are afraid of mice. The possible sources for the zoological characteristic that elephants cannot tolerate the grunting of pigs are either the *Naturalis historia* (lib. 8.27) by Pliny the Elder or the *Liber de natura rerum* by Thomas of Cantimpré, as already Marianne E. Kalinke pointed out.⁶⁸ Kalinke correctly adds that this fact was also included in the *Sendibréf Alexanders*, the Old Icelandic translation of the *Epistola Alexandri ad Aristotelem*, However, I do not agree with her suggestion that this letter is likely the direct source for the elephant–pig interaction in *Konráðs saga keisarasonar*. The particular way in which the pig is used to repel the elephant in the saga is considerably different from *Sendibréf Alexanders*, where soldiers lead the pigs directly into battle.⁶⁹ Furthermore, *Sendibréf Alexanders* does not contain the other zoological or mineralogical knowledge included in *Konráðs saga keisarasonar* whereas Pliny's *Naturalis historia* does. So, the issue may be reduced to whether one wants to assume multiple sources, when a single one would suffice.

Pliny's encyclopaedia deals with another subject which Old Icelandic romance is very fond of: mineralogy or *náttúrusteinar*, the Old Norse term for gemstones with extraordinary powers.⁷⁰ Konráðr's quest to prove his identity begins with the problem of applying lapidary knowledge in order to find out where he must go to search for the stone he is supposed to bring back. It is only fitting that the final moment of that quest also revolves around mineralogy. Upon reaching Babylon, Konráðr enters a strange palace filled with snakes.⁷¹ In one of its halls he sees a golden tray with precious gemstones on it floating in the air without any support. The narrative set-up against which these events unfold is the far-away ruins of heathen Babylon. Under these circumstances the saga's audience might easily jump to conclusions and presume the presence of some unholy supernatural power at work, keeping the tray aloft. Yet,

---

67 *Konráðs saga keisarasonar*, p. 160.
68 Kalinke, 'Foreign Influence on Old Norse-Icelandic Literature'.
69 For more about pachyderm warfare see Kistler, *War Elephants*.
70 Cf. the lapidary chapter of Copenhagen, AM 194 8vo in *Alfræði Íslenzk I*, ed. by Kålund, pp. 77–83.
71 *Konráðs saga keisarasonar*, pp. 161–62.

Konráðs saga keisarasonar quickly dispels any such considerations and clarifies the situation. It is not magic at work here, but science:

> Aungua hluti sá hann [Konráðr] þessum bordskutlí vpp halda. Enn med þui ath hann var vítr madr feck hann þat skiott skilít ath steinar þeir sem Jarndragar heíta mundo settir vera badí j hallargolfit ok rẹfrit ok vtt j veggína ok mundí huer þeirra til sín heímta ok med þessum hætti fastr vera.
>
> (Konráðr saw nothing that would hold this tray up. Yet, because he was a wise man, he soon discerned that those stones called Jarndragar (lit. 'Iron dragger') must be set into the floor, the roof and out in the walls and each must pull to itself and in this way hold [it] fast.)[72]

Clever manipulation of the artfully decorated magnets, or *jarndragar* as they are called here, then allows Konráðr to lower the tray to the ground so that he may collect the stone he is seeking.

The magnet was one of the centre pieces of medieval mineralogy and included in all lapidaries, such as the one in the Old Icelandic fourteenth-century encyclopaedic compilation Copenhagen, Den Arnamagnæanske Samling, AM 194 8vo.[73] But again it is tempting to point towards Pliny as the source for *Konráðs saga keisarasonar*, as the *Naturalis historia* (lib. 34.42) mentions an architect's plan to use magnets in the vaulting of the temple of Arsinoe in Alexandria in order to hold aloft an iron statue.[74] So, while the idea for the scene in the saga might not have been new, *Konráðs saga keisarasonar* did thoroughly expand upon its source. While Pliny mentions the principle at work, the magnets attracting the iron object so that it is suspended in mid-air, the romance goes into specifics of how this can be architecturally achieved. The narrator describes how the magnets are set up along both the vertical and horizontal axes, making sure that the stones exert equal attraction from all sides so that the tray stays fixed at a particular point in the hall's design. Konráðr is then able to interfere with this magnetic equilibrium and thus able to lower the object.

Again, *Konráðs saga keisarasonar* makes artful use of the learned sources at its disposal to craft a scene that puts the practical application of scientific knowledge at its centre while also furthering the saga's story arc. By using his education and acquired knowledge of nature Konráðr is finally able to fulfil his quest, prove his worth and thereby his own identity. Konráðr's re-establishment in society is thus intrinsically linked to how he addressed the imbalance in his knightly skills. Only by embracing the mental disciplines and intelligently applying his scientific knowledge is Konráðr successful in claiming his own identity and becoming a worthy ruler.

This scene in *Konráðs saga keisarasonar* shares several parallels with a similar episode in the *fornaldarsaga Eiríks saga víðfǫrla*. In both sagas, the quest of the eponymous hero

---

72 *Konráðs saga keisarasonar*, p. 162.
73 For a general introduction to medieval lapidaries see Kitson, 'Lapidary Traditions in Anglo-Saxon England' and Keiser, 'Lapidaries', pp. 306–07.
74 This description later found its way into other encyclopaedias, such as Isidore's *Etymologiae* and Alexander Neckam's *De naturis rerum*.

culminates in an encounter with a floating object. This has led Helle Jensen, editor of *Eiríks saga víðfǫrla*, to claim that the author or authors of *Konráðs saga keisarasonar* had directly modelled their scene on the other saga.[75] This statement might be met with some doubt. After all, the magnet in *Konráðs saga keisarasonar* fits neatly into the saga's consistent programme of incorporating scientific knowledge from Pliny's *Naturalis historia*. It nonetheless remains interesting to compare the two episodes, also because medieval and early modern audiences would have had access to both sagas and thus the chance to contrast them with each other.

The protagonist of *Eiríks saga víðfǫrla* seeks to find Ódainsaker, referred to as paradise by Christians. At the end of his journey, having encountered many natural and supernatural marvels along the way, Eiríkr discovers a place which he takes to be Ódainsaker. Here he sees a tower hanging in the air entirely unsupported.[76] When Eiríkr enters the floating tower, he is met by an angel, who explains that he had guided Eiríkr along his quest and that the splendour of the floating tower is not actually paradise, but rather reflects just a fraction of God's power. Eiríkr asks by what force the tower is held aloft and *eingillinn segir. æinnsaman guds kraft helldr honum upp* (The angel says: 'Only the power of God holds it aloft').[77] This explanation illustrates the substantial difference in outlook separating these two sagas. While both make heavy use of learned writing, *Konráðs saga keisarasonar* approaches the world through the eyes of natural philosophy. It chooses to explain an apparently supernatural effect not as a demonstration of the miraculous power of the Christian God, but as the ingenious application of mineralogy.

Conclusion: Science in Medieval Romance

In the late Middle Ages, science and literature were deeply intertwined in the romance genre. Incorporating scientific facts into fictional narratives had important synergetic effects. On the one hand, the romances' fascination with far-away places and astonishing exploits could be indulged by drawing on the marvels of natural history. At the same time, the learned facts equipped the fictional narratives with an anchor to the tangible world by pointing out that all the wondrous things they talked about were not idle fabrications of the imagination but actually represented accepted knowledge of the natural world. It also offered them an alternative to the supernatural by explaining the protagonist's outstanding feats or extraordinary abilities as the ingenious application of scientific knowledge or technology rather than magic. On the other hand, the romances provided a stage for science and a place to discuss it. The stories presented education and the practical use of scientific facts in a favourable light and thus furthered the popularization and dissemination of science. But the romances also supplied a forum to view scientific knowledge from

---

75 *Eiríks saga víðfǫrla*, ed. by Jensen, pp. xxvii–xxviii.
76 *Eiríks saga víðfǫrla*, ed. by Jensen, pp. 77–82.
77 *Eiríks saga víðfǫrla*, ed. by Jensen, p. 96.

a new and more dynamic angle. The narratives' fictionality gave an opportunity to act out thought experiments and speculate about the ramifications of scientific facts in a more open and unrestrained fashion than would have been possible within the confines of strictly factual writing.

Works Cited

Manuscripts and Archival Sources

Copenhagen, Den Arnamagnæanske Samling, AM 180 b fol
—, AM 194 8vo

Primary Sources

Alfræði Íslenzk I, ed. by K. Kålund, Samfund til Udgivelse af Gammel Nordisk Litteratur, 37 (Copenhagen: Møllers Bogtrykkeri, 1908)
Caesar, *Commentarii rerum in Gallia gestarum VII: A. Hirti commentarius VIII*, ed. by T. Rice Holmes (Oxford: Clarendon Press, 1914)
Cligès, ed. and trans. by Ingrid Kasten, *Chrétien de Troyes: Cligès* (Berlin: De Gruyter, 2006)
Eiríks saga víðforla, ed. by Helle Jensen (Copenhagen: Reitzels Forlag, 1983)
Horace, *Satires, Epistles and Ars Poetica*, ed. and trans. by H. Rushton Fairclough (Cambridge, MA: Harvard University Press, 1942)
Isidore of Seville, *Etymologiae. The Etymologies of Isidore of Seville*, trans. by Stephan A. Barney (Cambridge: Cambridge University Press, 2006)
—, *Isidori Hispaliensis Episcopi: Etymologiarum Sive Originum. Libri XX*, ed. by W. M. Lindsay (Oxford: Oxford University Press, 1911)
Konráðs saga keisarasonar, ed. by Otto J. Zitzelsberger, Germanic Languages and Literature, 63 (New York: Peter Lang, 1987)
Late Medieval Icelandic Romances, vol. 1: *Victors saga ok Blávus; Valdimars saga; Ectors saga*, ed. by Agnete Loth (Copenhagen: Munksgaard 1962)
Late Medieval Icelandic Romances, vol. 2: *Saulus saga ok Nikanors; Sigurðar saga þǫgla*, ed. by Agnete Loth (Copenhagen: Munksgaard, 1963)
Pliny, *Natural History*, ed. and trans. by H. Rackham and D. E. Eichholz (Cambridge, MA: Harvard University Press, 1952–1967)
Stjórn, vol. 2, ed. by Reidar Astås (Oslo: Riksarkivet, 2009)

Secondary Studies

Acker, Paul, 'Dragons in the Eddas and in Early Nordic Art', in *Revisiting the Poetic Edda: Essay on Old Norse Heroic Legend*, ed. by Paul Acker and Carolyne Larrington (New York: Routledge, 2013), pp. 53–75
Barnes, Geraldine, *The Bookish Riddarasögur* (Odense: University Press of Southern Denmark, 2014)
—, 'Current Issues in Riddarasögur Research', *Arkiv för Nordisk Filologi*, 104 (1989), 73–88

Beer, Gillian, *Darwin's Plots: Evolutionary Narrative in Darwin, George Eliot and Nineteenth-Century Fiction* (Cambridge: Cambridge University Press, 1983)

Brunner, Horst, *Geschichte der deutschen Literatur des Mittelalters und der Frühen Neuzeit* (Stuttgart: Reclam, 2010)

Bulang, Tobias, *Enzyklopädische Dichtungen. Fallstudien zu Wissen und Literatur in Spätmittelalter und früher Neuzeit* (Berlin: Akademie Verlag, 2011)

Bumke, Joachim: 'Retextualisierungen in der mittelalterlichen Literatur, besonders in der höfischen Epik. Ein Überblick', *Zeitschrift für deutsche Philologie*, 124 (2005), 6–46

Campbell, Mary Baine, *Wonder and Science. Imagining Worlds in Early Modern Europe* (Ithaca: Cornell University Press, 1999)

Clark, Willene B., *A Medieval Book of Beasts. The Second-Family Bestiary. Commentary, Art, Text and Translation* (Woodbridge: Boydell, 2006)

Driscoll, Matthew, 'Late Prose Fiction (*lygisögur*)', in *A Companion to Old Norse-Icelandic Literature and Culture*, ed. by Rory McTurk (Maldon: Blackwell, 2007), pp. 190–204

Fendt, Julia, *Wissenschaft und Imagination in der Literatur. Kulturökologische Analysen zeitgenössischer Romane* (Würzburg: Königshausen & Neumann, 2015)

Gabrovsky, Alexander N., *Chaucer the Alchemist. Physics, Mutability, and the Medieval Imagination* (Basingstoke: Palgrave Macmillan, 2015)

Glauser, Jürg, *Isländische Märchensagas. Studien zur Prosaliteratur im spätmittelalterlichen Island*, Beiträge zur nordischen Philologie, 12 (Basel: Helbig & Lichtenhahn Verlag, 1983)

Grant, Edward, *The Foundation of Modern Science in the Middle Ages: Their Religious, Institutional, and Intellectual Contexts* (Cambridge: Cambridge University Press, 1996)

Green, Dennis H., *The Beginnings of Medieval Romance. Fact and Fiction, 1150–1220* (Cambridge: Cambridge University Press, 2002)

Hannam, James, *God's Philosophers. How the Medieval World Laid the Foundations of Modern Science* (London: Icon Books, 2009)

Harris, Richard L., 'The Lion-Knight Legend in Iceland and the Valþjófsstaðir Door', *Viator*, 1 (1970), 125–45

Haug, Walter, *Literaturtheorie im deutschen Mittelalter. Von den Anfängen bis zum Ende des 13. Jahrhunderts* (Darmstadt: Wissenschaftliche Buchgesellschaft, 1992)

Heinzle, Joachim, 'Konstanten der Nibelungenrezeption in Mittelalter und Neuzeit', in *Die Rezeption des Nibelungenliedes. 3. Pöchlarner Heldenliedgespräch*, ed. by Klaus Zatloukal (Vienna: Fassbaender Verlag, 1995), pp. 81–108

Heizmann, Wilhelm and Tibor F. Pézsa, 'Reale und nominelle Identität in der Konráðs saga keisarasonar', *Journal of English and Germanic Philology*, 94 (1995), 176–89

Hünemörder, Christian, 'Antike und mittelalterliche Enzyklopädien und die Popularisierung naturkundlichen Wissens', *Sudhoffs Archiv*, 65 (1981), 339–65

Journal of Literature and Science, 1 (2008)

Kalinke, Marianne, *Bridal-Quest Romance in Medieval Iceland*, Islandica, 46 (Ithaca: Cornell University Press, 1990)

—, 'Foreign Influence on Old Norse-Icelandic Literature' in *Medieval Scandinavia*, ed. by Phillip Pulsiano (London: Garland, 1993), pp. 451–54

—, 'The Foreign Language Requirement in Medieval Icelandic Romance', *The Modern Language Review*, 78 (1983), 850–61

—, 'Norse Romance', in *Old Norse Icelandic Literature. A Critical Guide*, ed. by Carol J. Clover and John Lindow (Ithaca: Cornell University Press, 1985), pp. 316–63

Keiser, George, 'Lapidaries', in *Medieval Science, Technology, and Medicine. An Encyclopedia*, ed. by Thomas Glick, Steven J. Livesey, and Faith Wallis (New York: Routledge, 2005), pp. 306–07

Kistler, John M., *War Elephants* (Lincoln, NE: University of Nebraska Press, 2007)

Kitson, Peter, 'Lapidary Traditions in Anglo-Saxon England: Part I, the Background; the Old English Lapidary', *Anglo-Saxon England*, 7 (1978), 9–60

Klinkert, Thomas, *Epistemologische Fiktionen. Zur Interferenz von Literatur und Wissenschaft seit der Aufklärung* (Berlin: De Gruyter, 2010)

Knapp, Fritz Peter, *Historie und Fiktion in der mittelalterlichen Gattungspoetik*, vol. 1: *Sieben Studien und ein Nachwort* (Heidelberg: Winter, 1997)

—, *Historie und Fiktion in der mittelalterlichen Gattungspoetik*, vol. 2: *Zehn neue Studien und ein Vorwort* (Heidelberg: Winter, 2005)

—, 'Subjektivität des Erzählers und Fiktionalität der Erzählung bei Wolfram von Eschenback und anderen Autoren des 12. und 13. Jahrhunderts', *Wolfram-Studien*, 17 (2002), 10–29

Krueger, Roberta L., 'Introduction' in *The Cambridge Companion to Medieval Romance*, ed. by Roberta L. Krueger (Cambridge: Cambridge University Press, 2000), pp. 1–9

LaGrandeur, Kevin, *Androids and Intelligent Networks in Early Modern Literature and Culture. Artificial Slaves* (New York: Routledge, 2013)

Literatur und Wissen. Theroetisch-methodische Zugänge, ed. by Tilmann Köppe (Berlin: De Gruyter, 2010)

Medieval Science Fiction, ed. by Carl Kears and James Paz (London: King's College, 2016)

O'Connor, Ralph, 'History or Fiction? Truth-Claims and Defensive Narrators in Icelandic Romance-Sagas', *Medieval Scandinavia*, 15 (2005), 101–69

On Literature and Science. Essays, Reflections, Provocations, ed. by Philip Coleman (Dublin: Four Courts 2007)

Pethes, Nicolas, 'Literatur- und Wissenschaftsgeschichte. Ein Forschungsbericht', *Internationales Archiv für Sozialgeschichte der deutschen Literatur*, 28 (2003), 181–231

van Ruymbeke, Christine, *Science and Poetry in Medieval Persia. The Botany of Nizami's 'Khamsa'* (Cambridge: Cambridge University Press 2007)

Schreck, Florian, *Science in Medieval Fiction. The Reception of Learned Writing on Natural History in Old Icelandic Romance 1300–1550* (Bergen: University of Bergen, 2018)

Sleigh, Charlotte, *Literature and Science* (Basingstoke: Palgrave Macmillan, 2011)

Wegmann, Milène, *Naturwahrnehmung im Mittelalter im Spiegel der lateinischen Historiographie des 12. und 13. Jahrhunderts* (Bern: Peter Lang, 2005)

Willis, Martin, *Literature and Science: A Reader's Guide to Essential Criticism* (London: Palgrave, 2015)

Worstbrock, Franz-Josef, 'Wiedererzählen und Übersetzen', in *Mittelalter und frühe Neuzeit. Übergänge, Umbrüche und Neuansätze*, ed. by Walter Haug (Berlin: De Gruyter, 1999), pp. 128–42

BRAD KIRKLAND

Continental Ironmongers, Whalers, Smugglers, and Craftsmen

Immigration and Trade Routes and their Influence on the London Armourers' Industry

London's fourteenth-century armourers were essential to the English war efforts, and yet their craft remains one of the least-understood of the period. Firstly, this is because the armourers lacked a formal guild structure in the fourteenth century, and in the absence of well-organized or well-preserved guild records, historians interested in armour have understandably focussed on the more accessible records of the industry from later centuries, or upon their continental counterparts. Secondly, late-medieval English armour was primarily not made of metal,[1] or was made out of poor-quality metal, and these pieces rarely survive to the present (particularly in the English climate). Thirdly, the armour that we know was produced in England in the fourteenth century fails to impress when compared to the fine-quality armour manufactured in Germany, Spain, and Italy. Many scholars of armour have been mystified as to how a market for English armour existed at all in the face of such superior products, concluding that English armourers' products 'seem not to have come up to the standard of fine-quality armour produced on the continent',[2] that

---

1 This is perhaps best illustrated by the depiction of how the great folk and the common folk will fight at the end of days depicted in the Holkham Picture Bible (BL, MS Additional 47682, fol. 40ʳ). The work dates between 1337 and 1340, and depicts '*le grant pouple*' and '*le commoune genz*' each fighting among themselves. The great have metal helms and various metal armour, mail, embossed shields, and horses. The common people have cloth armour, bucklers, metal or leather helms, and bows. While artistic sources must be used very carefully in historical analysis, Michelle Brown convincingly argues that it was painted by a Londoner based on London landmarks creeping into biblical tales, such as the thirteenth-century spire of St Paul's Cathedral taking the place of the Temple of Jerusalem in fol. 19ᵛ. The armour depicted among the common men closely matches the sorts which appear in purchase orders from the same period. See: Brown, 'The Historical Context', p. 17. For a good example of the kinds of purchase orders that appear in the first half of the fourteenth century, see: LBE, p. 170. For a list of the abbreviations used in this chapter see Works Cited.
2 Pfaffenbichler, *Medieval Craftsmen*, pp. 23–24.

Brad Kirkland • recently completed his PhD on the production of armour in medieval England at the University of York. brad.kirkland@alumni.york.ac.uk.

Medieval Science in the North: Travelling Wisdom, 1000–1500, ed. by Christian Etheridge and Michele Campopiano, KSS 2 (Turnhout: Brepols, 2021), pp. 201-218

'local products were evidently thought to be inadequate for noblemen',[3] and that it was 'strange that the [English] craftsmen did not attempt to improve their work when examples of foreign skill were imported in great quantities'.[4] This last criticism is particularly perplexing, but the questions that it poses are important to consider: what was the relationship between English and Continental armourers? Why did English armourers continue to produce such 'inadequate' wares when immigration and import competition should have brought the English craft up to Continental standards? And finally, how did craft knowledge of arms and armour travel between the Continent and England?

Without guild records with which to define the armourers' craft in the fourteenth century, these questions must rely upon prosopography to piece together a picture of the industry from the perspectives of many individuals who contributed to the industry. Examining over a thousand documents and records preserved in London's Letter Books, the Plea and Memoranda Rolls, the subsidy rolls of 1292, 1319, and 1332, the Mayor's Court Rolls, the Possessory Assizes, the Assizes of Nuisance, and Cases of Trespass, alongside royal government records including the Feet of Fines, the Patent Rolls, the Close Rolls, Rymer's *Foedera*, the Fine Rolls, exchequers rolls, customs records, Statutes of the Realm, and Parliament Rolls as well as documents in the National Archives, London Metropolitan Archives, and the Guildhall Library relating to the armourers from 1270 to 1400, I was able to provisionally identify three-hundred and eleven armourers.[5] Of these craftspeople and traders, eighty-seven (or twenty-eight per cent) could be either identified as immigrants to London, or bore surnames which pointed towards their country or city of origin.

Examining the armourers as a collection of individuals helps to identify the types of craft networks that were at work in their early organizations. Instead of a singular industry, the armourers of London were working within a collection of interdependent crafts and trades: linen-armourers, haubergers (mail-makers), heaumers (helmet and plate-armour manufacturers), furbishers (armour and weapons repairers), and specialists in leather, baleen, and other materials. Alongside these craftsmen were specialist traders, importers, exporters (when it was legal to do so), smugglers (when it was not), and raw materials specialists who all depended upon one another for their mutual livelihoods. By the time the armourers were given their charter in 1453,[6] many of these specialist industries had been absorbed by the others, leaving only the 'armourers' who still exist as the Armourers' Company today. The roles that immigrant armourers and the continental armour trade played upon the formation of London's late-medieval armourers' crafts have never been investigated before, but this chapter will examine how immigrant armourers contributed to the development of the armourers' crafts, and how international trade was a prerequisite for the development of the industry. It will also demonstrate that the popularity of continental armour in England was

---

3   Williams, *The Knight and the Blast Furnace*, pp. 53–59.
4   Ffoulkes, *The Armourer and his Craft*, p. 13.
5   Kirkland, 'Now Thrive the Armourers'.
6   *CPR 1452–1461*, p. 105.

not only the result of continental skills but was also affected by the different raw materials available to both markets. Finally, it will examine how the trade routes and expertise that enter London contributed to the illegal armour export industry back to the continent, particularly towards the end of the century.

The Immigrant and Import Communities of London's Armourers

Identifying an individual as an immigrant in the records can be more challenging than identifying an individual's craft alone.[7] Many armourers and immigrant armourers had to be identified by surname analysis, combined with nominal record linkage methods.[8] Surnames only came into common use in England sometime between the end of the twelfth and the middle of the thirteenth centuries, but did not fully stabilize until much later.[9] At the beginning of the fourteenth century, surnames of craft workers most often represented either the person's occupation, or the person's place of origin, if an immigrant, but rarely both in the same record.

Of the eighty-seven possible immigrants from among the armourers' population I identified, only seventeen were continental immigrants. The majority of these came from France, Bruges, and Germany. This is probably because maritime trade routes doubled as immigration routes, especially for those who imported armour into London, such as William de Wolde of Germany, Walter de Mateu of Condom, and Hugh le Armurer of Bruges. As stated above, armour of continental manufacture was often preferred by those who could afford it: Hugh le Armurer of Bruges provided armour for Edward II in 1312,[10] while the German Walter de Mateu was made a yeoman of Edward II and sent to France, Gascony, Navarre, and Spain to find horses and armour for the king's use in 1317.[11] Clearly, their knowledge of continental markets and practices was highly valued.

One very important individual to the early London armourers' networks was Herman le Heaumer, normally found in the London records as Manekyn le Heaumer or Manekyn le Armurer. He and his brother Peter were the sons of Lady Gertrude of Cologne, and so it is presumed that they were either immigrants or first-generation Londoners.[12] Manekyn led a tumultuous life, but the many records relating to him

---

7   The England's Immigrants project is a useful source that was being assembled at the time my research was being carried out and will provide a good opportunity for future work in the area. See <https://www.englandsimmigrants.com/>.
8   For a discussion of these methods, see: Ekwall, *Variation in Surnames in Medieval London*; Ekwall, *Two Early London Subsidy Rolls*, pp. 38–71; Ekwall, *The Concise Oxford Dictionary of English Place-Names*; Fransson, *Middle English Surnames of Occupation*, p. 39; Reaney and Wilson, *A Dictionary of English Surnames*, pp. xiii–lv; Kirkland, 'Now Thrive the Armourers'.
9   Mawer, 'Some Unworked Sources for English Lexicography'.
10 CCR 1307–1313, p. 426.
11 CPR 1313–1317, p. 643; CPR 1317–1321, p. 60.
12 Schofield and others, 'Medieval Buildings and Property Development in the Area of Cheapside', pp. 63–64.

provide an uncommon insight into the early workings of the armourers' networks. He was certainly established in London by 1286, when his apprentice Clement Passemer (literally Clement 'From Beyond the Sea') was enrolled in the City records.[13] By 1299 he had evidently risen to considerable prominence: Manekyn was one of those elected by the Commonalty of London to prosecute the confirmation of Edward I's confirmation of the charter of London.[14] In October 1305, Manekyn and his brother both enjoyed the protection of the Earl of Lincoln, who petitioned the crown to pardon for them for 'all felonies which [Manekyn and his brother Peter] are said to have committed, and all trespasses against the peace whereof they are indicted before the king',[15] which was granted as they were serving in war. These felonies included being 'common thieves, homicides, and robbers' and 'beating men against the peace' including the heaumer Richard Deveneys, who feared to complain of Manekyn 'because [Manekyn and his brother Peter] are common maintainers and supporters of ill-doers and those who make assaults in the City'.[16] This apparently had little effect on Manekyn's station, as later that month he was entrusted with the army's wages to keep in his house (although he failed to actually deliver those wages when they were due).[17]

Despite whatever legal troubles Manekyn faced at the beginning of the century, he had the great favour of the future Edward II. The close relationship evidently was maintained throughout both of their lives: in 1318 Edward II wrote a letter to the City of London personally thanking Manekyn for his services in both finding and leading footsoldiers[18] out of London for the king's expeditions in Scotland (Manekyn's apprentice, known only as 'Mannekynnesmanthearmurer' was killed in this campaign).[19] The king asked in his letter that Londoners grant le Heaumer 'the greatest grace and favour[...] in matters of business that shall concern [him]'.[20] A shrewd businessman, Manekyn was almost certainly also involved in outfitting those soldiers with the helmets from which he took his surname, and records from 1320 attest to Manekyn's broad expertise across multiple crafts beyond the helmetry alone. On 24 June that year, he was one of six armourers charged with appraising the value of nearly a thousand spearheads involved in a debt alongside two haubergers, a furbisher, and two cutlers.[21] Of these occupations, only the cutlers and the furbisher could claim specific craft expertise in this appraisal, but evidently le Heaumer's and the other armourers' actual businesses were varied enough to give them this expertise.

13 *LBD*, p. 143.
14 *LBC*, p. 38.
15 *CPR 1301–1307*, pp. 379, 392.
16 *Calendar of London Trailbaston Trials Under Commissions of 1305 and 1306*, ed. by Pugh, no. 161; *LBC*, pp. 69, 70.
17 *CCR 1302–1307*, p. 302.
18 Manekyn co-led 200 foot-soldiers in that campaign as a centurion alongside Roger atte Water. See: *LBE*, p. 93.
19 *CPR 1317–1321*, p. 286; *Mem*, p. 128.
20 *Mem*, p. 128.
21 *LBE*, p. 132.

This occupational pluralism was a defining feature of the king's armourers, many of whom were immigrants, such as John de Coloigne (presumably an immigrant of Cologne, by his surname). Coloigne was Edward III's armourer, and one of the most important linen-armourers in London in the first half of the fourteenth century. As the king's armourer, Coloigne was not making armour by himself. He ran a large armour workshop, storehouse, and purchasing operation which employed multiple craftsmen, buyers, and specialists from London's smaller armourers' crafts. These included a king's heaumer and a king's hauberger who worked with or under him.[22] As a purveyor of various armours, Coloigne needed to be very familiar with the networks of craftsmen and traders in London, and also presumably the import networks that supplied those markets. As an immigrant, Coloigne was likely chosen for his knowledge of this continental market, but his position ensured that the types of armour he commissioned or imported influenced the fashions of the English armour market. Just as armour was imported by those who could afford the superior continental products, the crown at the beginning of the fourteenth century imported continental expertise, and so they brought with them the methods that were in use on the continent. Furthermore, immigrants in positions of authority like le Heaumer and de Coloigne helped to shape the culture of the early fourteenth-century armourers' socioindustrial organizations, such as the multi-specialist armourers' mistery of 1322,[23] the heaumers' ordinances of 1347,[24] and the linen-armourers' and tailors' fraternity.

The armourers' regulations of 1322 formed the earliest armourers' organization in the London records, and of the twenty-seven assentors to the rules agreed upon in the document, twelve bore surnames suggesting immigration from outside London: the armourer William de Segrave (Leicestershire); the King's Armourer Thomas de Copham (a town in modern Surrey); the armourer William de Lanshulle (Suffolk); the armourer Richard de Kent (Kent); the armourer Robert de Skeltone (North Yorkshire); the armourer, tailor, and warden of the tailors and linen armourers, Elias de Wodeberghe (probably Somerset); the armourer William de Staunford (Norfolk); the armourer, furbisher, and tailor, John de Wyght (Isle of Wight); the armourer William de Lyndeseie (Lincolnshire); the armourer and soldier John de Kestevene (Lincolnshire); and the armourer Roger de Blakenhale (Cheshire).

While all of these persons are of interest, Elyas de Wodeberghe was unique. He was the only armourer named 'Elyas' in the fourteenth century — a name which often appeared in the records appended with the description 'the Jew'. Wodeberghe (also called 'Wordebern', 'Wodebern', and 'Wodebere') first appears in the London records in 1310, admitted to the freedom of the City as a tailor,[25] but his inclusion as an assentor of the armourers' regulations indicates that by 1322 his primary business was in linen-armour. However, he was clearly a prominent figure among

---

22 TNA SC 8/247/12310 demonstrates the level of interaction and cooperation required among these three aspects of the craft among the king's armourers.
23 See: *Mem*, pp. 145–46.
24 *Mem*, pp. 237–38.
25 *LBD*, p. 57.

both the armourers and the tailors, as in 1328, he was elected as one of the governors of the Taylors and Linen Armourers mistery alongside Roger Sauvage, John Tavy (sometimes recorded as 'Tany'), and Henry de Horepol (all of whom were likewise assentors or supervisors of the armourers' regulations).[26] De Wodeberghe would have travelled approximately 150 miles to come to London if he originated from the Woodborough in modern Somerset, although it is also possible that he emigrated from the village of Woodborough in modern Nottinghamshire (approximately 135 miles north of London). The Woodborough in modern Nottinghamshire is doubtful, as it was a very small village at this time, and although within sixty miles of the Port of Boston, would still have had few opportunities for a linen-armourer to practise his craft. Eilert Ekwall associated the 'Elya le Armurer' of Castle Baynard Ward in the 1319 Subsidy Roll with Elias de Wodeberghe,[27] but he believed that Elias hailed from Woodbeer, Dorset (approximately 200 miles travel).[28] This would place Elias near the busy port of Exeter, while Woodborough, Somerset would place him near the port of Bristol.[29]

Another interesting example of how armourers outside London interacted with the London market can be found in the life of William de Glendale (Northumberland) and his children. De Glendale first appeared in the London records married to Agnes, the daughter of the wealthy London armourer Peter Nayer.[30] Nayer died in 1346, and Agnes and William were married by 1349. Between 1346 and 1349, Agnes' mother, all three of her brothers, and one sister died, with her brother Nicholas leaving a nuncupative will.[31] As a result, much of Nayer's wealth and properties passed to his surviving daughters. This is of particular interest as it shows one path that immigrants might have taken to gain access to the very complicated local armourers' networks. If de Glendale had not known the Nayers before Peter's death, it is probable that he and Agnes met while the surviving family restructured their father's armour-manufacturing operation. Marrying into the London armourers' community was certainly a successful strategy, and by combining the Nayer's local industrial connections with de Glendale's wider connections, the new household quickly rose to prominence: by 1363, de Glendale was the king's armourer and granted loans of as much as £500.[32]

After de Glendale's death in 1368, Agnes's financial and social position was strong enough to make her a good match for Roger de la Chaumbre, the escheator of Northamptonshire.[33] Agnes's daughter Joan (by de Glendale) would later marry Sir

---

26 *LBE*, p. 234; *Mem*, pp. 145–46.
27 He is certainly correct: in 1338, 'Elyas de Wodebere', along with three other armourers sat on a jury for the theft of 20s. worth of goods stolen from Castle Baynard Ward. See: *LBF*, p. 249.
28 Ekwall, *Two Early London Subsidy Rolls*, p. 342.
29 For an examination of the activities at the Port of Exeter, see: *The Local Customs Accounts of the Port of Exeter: 1266–1321*, ed. by Kowaleski.
30 *CWPH*, I, 535; *LBF*, p. 207.
31 Suggesting that the family (with her, and her sisters Isolda and Leticia excepted) died of the plague. See: *LBF*, p. 207 n. 15.
32 TNA C 241/144/69; TNA C 131/84/21; TNA C 131/84/22.
33 *CCR 1381–1385*, p. 27.

John de la Chambre (possibly Roger de la Chaumbre's son by a previous marriage),[34] but they clearly maintained a connection to the London armour industry even after the family moved to Oxfordshire in 1404. Joan and Sir John rented out two of de Glendale's armourers' shops in Fridaystreet to the armourer William Langford from 1404 through 1411, and in 1402 they rented out a brewery called 'Le Horshed' which had previously been owned by a mail-maker Reginald le Hauberger, his daughter Alice la Haubergere and her family,[35] and would later be owned by Richard Person, who was the apprentice and heir of the armourer and Sheriff of London, Simon de Wynchecombe (probably himself an immigrant to London from Winchcombe, Gloucestershire).[36] These properties link four of the most prominent families from the armourers' networks during the thirteenth and fourteenth centuries, and their administration from outside of London points to just how complex the inter-city armourers' networks could have been.

Of the survey of eighty-seven immigrant armourers found in the records, sixty-nine could be traced to other English cities before their settlement in London. However, while local immigration helped to replenish London's post-Black Death population, it also points towards a growing trade and intellectual network with the continent. The largest immigrant populations came from modern-day Lincolnshire, London's hinterlands and surrounding cities, Norfolk, and North Yorkshire. These populations are particularly interesting, because modern scholarship has suggested that London was the 'only important centre' for armour in the country.[37] Assuming that some of these immigrants acquired their crafts locally before immigrating to London, larger numbers of armourers coming from the same area suggest that the armour trade in England was more complex than previously assumed, but this can only be confirmed by a thorough investigation into the local records of each of these areas. York and its hinterlands may exemplify these secondary centres of armour manufacture: York was home to a vibrant armourers' industry which included female armourers, and may have had its own guild by 1376, based on the earliest records of the York Corpus Christi pageant.[38] Furthermore, London armourers (such as Manekyn le Heaumer) commonly travelled between York and London in order to help supply soldiers bound for Scotland, bringing with them their skills in the same way that continental and regional immigrants may have brought their techniques to London.

When hubs of armourer emigration are compared with ports where the raw materials necessary for armour manufacture were imported, and where armour was

---

34 *CPMR*, II, 167–68; *CPMR*, III, 272.
35 The use of the feminine *la haubergere* is extremely uncommon (making her literally, Alice the Mail-makeress), and so it is very likely that she actually practised her family's craft, like the daughter of the York armourer Adam Hecche, to whom he left 'all the instruments of [his] craft of mailwork'. See: Swanson, *Medieval Artisans*, p. 115 and Goldberg, *Women, Work, and Life Cycle*, p. 128.
36 *CPMR*, III, 272; HGL, 'St Mary Colechurch 105/26'; *DCoAD*, IV, A 6869.
37 Pfaffenbichler, *Medieval Craftsmen*, p. 23.
38 Although it is not known if the armourers actually had their play at this time, as the first records of craft associations with the pageant dates from 1415. See: Pfaffenbichler, *Medieval Craftsmen*, p. 23 and Mill, 'The York Plays of the Dying', pp. 866–76.

illegally exported to the continent, a pattern emerges. Many of London's English immigrant armourers bore surnames identifying them with those trading ports directly, such as the several 'atte Hulls' and 'de Yernemouths' (Yarmouth). As port cities offered better access to imported raw materials as well as a profitable (if illegal) overseas market, it is no surprise that several of London's most successful immigrant armourers maintained properties or business connections in ports in Lincolnshire, Norfolk, Suffolk, Essex, and Kent. One such example was the abovementioned very-wealthy armourer and probable-Gloucestershire immigrant, Simon de Wynchecombe. De Wynchecombe's will is a particularly interesting window into his personal life: he ran a prosperous armour manufacturing and retail business, was politically involved (he was the Sheriff of London during the Mayoralty Riots), and had an apprentice he cared for and to whom he left multiple items of armour (some of which were identified as being 'of London make' and others not (i.e., likely imported through one of the English port cities in which de Wynchecombe owned properties and conducted business)). He was well-liked by his peers and was deeply devoted to his first wife, for whose memory he erected two altars in two separate churches in London.[39] His second marriage to the daughter of a wealthy Ipswich (Suffolk) burgess was clearly more for her property in the port than any shared affection between them, as his will granted her only 'such share of his goods as of [...] custom [...] she ought to have, and no more'.[40] However, the match afforded him some of the wealth that contributed to his success, as well as several important properties in Ipswich.[41] As Nicholas Amor argued, the trade in Ipswich's ports primarily flowed from London in the forms of finished goods and wool, but the port also would have brought in raw materials useful to de Wynchecombe's London businesses.[42]

Port cities were vital to the English armour industry, as they provided English industries with access to Scandinavian baleen and iron. Baleen was a cheap substitute for iron plates in armour manufacture, and was described as being 'fitting for footsoldiers' in 1324, when mail and plate armour was determined to be too expensive.[43] However, hunting the whales that provided baleen for industry required fleets of large ships working together, and as impressment of English ships locked the English merchant fleets in ports, the whalebone used in England must have been imported, most likely from Scandinavia.[44]

---

39 London, Consistory Court, MS 9171 fols 431–35 identifies bequests to construct altars to Saint Anne and Saint George. The Hustings Roll summary omits Saint George. It is impossible to know whether the original Hustings Will included this bequest for an altar to St George as the original document, CLA/023/CP/01/128/17, has gone missing from the London Metropolitan Archives. See: *CWPH*, II, 324, 340–42.
40 *CWPH*, II, 340–42.
41 See: Amor, *Late Medieval Ipswich*, p. 87; *CCR 1399–1402*, pp. 133–34; *CCR 1354–1360*, p. 401; *CWPH*, II, 341–42.
42 Amor, *Late Medieval Ipswich*, pp. 47, 87.
43 *CPR 1324–1337*, p. 622.
44 See: *CPR 1324–1327*, p. 29; de Smet, 'Evidence of Whaling in the North Sea', pp. 304, 307, and Moffat, Spriggs and O'Connor, 'The Use of Baleen', p. 207.

Iron for Horseshoes: The Metal Armour Industry in England in the Fourteenth Century

Continental skill was of particular importance to the establishment of England's fledgling metal-armour industry in the first half of the fourteenth century: in 1317, Edward II sent David le Hope (possibly of Derbyshire), to Paris so that he might 'learn the method of making swords for battle'.[45] Evidently, the swords that were produced in London were not to the king's liking. The same year, Edward sent the abovementioned Walter de Mateu abroad to procure horses and armour for his use.[46] Similar records appear throughout the first half of the fourteenth century: in 1337, the Florentine merchants of the Bardi were requested by the crown to import armour for 100 men,[47] and it is almost certainly these same men who were tasked with importing the earliest known cannon used in England recorded in the City's inventory of munitions of war in 1339.[48] In 1350, the king sent Daniel van Mulneham, who was a sergeant of the London mail-maker, Giles de Colonia (Cologne) to Flanders to cross the channel with 200 florins on the crown's business (presumably to procure arms and armour). Similarly, in 1312 the crown paid £20 16s. 2d. to the armour merchants Hugh le Armurer (of Bruges) and Lorettus of Dynaunt (Dinant, Belgium), for armour they imported only five days after the crown seized all armour and warhorses found in London.[49]

It was not only royal envoys who sought armour from the continent. Local archival records from English ports and court records relating to piracy claims are particularly useful in examining the import industry in the fourteenth century. In 1317, Aymer de Valence, 2nd Earl of Pembroke, commissioned a ship to bring him £60 worth of helmets, haubergeons, and other armour from Bordeaux, but the cog (a type of single-masted vessel) was taken by force by pirates off the coast of Sandwich.[50] It is worth noting that the armour here represented nearly half the value of everything aboard the ship, implying that Pembroke was far more interested in the import of his armour than he was in the wine and wheat that was also lost to piracy. Similarly, in 1342, *la Rede Cogge* was arrested by a sergeant of the king of France in Normandy and its owner lost a considerable sum of wine and salt, along with '24 pairs of plates, 9 bassinets with their aventails, 12 pairs gauntlets of plate and 13 actons' worth £17 8s.[51] The English were also not above seizing ships on the high seas: in 1315, John de Butetourt was ordered to take sufficient force out of Yarmouth to meet thirteen great cogs as they emerged from Sluys on their way to Scotland in order to seize the armour on board.[52] Frank Rexroth has argued that ensuring that the army

---

45 Stapleton, 'A Brief Summary', p. 343. Charles Ffoulkes misidentified this as occurring in 1322. See Ffoulkes, *The Armourer and his Craft*, p. 57.
46 *CPR 1313–1317*, p. 643; *CPR 1317–1321*, p. 60.
47 *CCR 1337–1339*, p. 63.
48 *Mem*, pp. 204–05.
49 *CCR 1307–1313*, pp. 426, 428; *LBD*, p. 290.
50 *CCR 1313–1318*, pp. 563–64; *CCR 1318–1323*, p. 192.
51 *CCR 1341–1343*, p. 435.
52 *CCR 1313–1318*, pp. 218–19.

was properly equipped was 'the greatest problem for [...] Londoners', above even 'ensur[ing] the internal and external security of the city', because the latter was a prerequisite to the former.[53] In the absence of strong armour industry in England at the beginning of the century, importing armour, attracting immigrant craftsmen, and ensuring the supply of necessary raw materials became the greatest economic and military problem of the day.

It has long been assumed that the crown's taste for arms and armour of continental manufacture was solely due to the skills possessed by continental craftsmen. While the skills of continental craftsmen were certainly sought after, and many continental immigrants found prestige working in London, the skill found on the continent was only one factor in what made continental armour superior to that manufactured in England. Continental skill enriched the London armour craft both directly in the form of continental masters moving to the City, and indirectly as imported goods and continental immigrants improved the armour industries of English port cities, whose armourers in turn immigrated to London themselves. But those who sought bespoke armour still looked outside of England, because English iron was of comparatively poor quality to that found on the continent, and the infrastructure for working that iron was rudimentary by comparison. As a result, regardless of the skill that might be found in England, continental armour would always be preferable to those who could afford it because it was made out of superior raw materials to those which could be procured in England for the same price. Furthermore, these materials shortages meant that English armourers were better suited to create cost-effective armour of the sort which the crown called 'fitting for footsoldiers',[54] rather than the much more-expensive bespoke armour that relied upon superior imported raw materials.

English iron mines and bog iron operations were very rich, and the ironworks that developed in Durham and the Forest of Dean greatly contributed to the iron supply market.[55] However, as H. R. Schubert argued, 'English iron was only used to a limited context [...] mainly [...] nails, horseshoes, wedges, spades, pickaxes, and the smaller types of bolt used as missiles for crossbows [...] Iron [...] which had to be particularly strong, was imported'.[56] This superior iron primarily came from Normandy, Germany, Spain, and Sweden in the form of 'osmunds'. Jane Geddes (and others) have suggested that the term 'osmunds' referred to small lumps of iron containing a low-phosphorus content, derived from bog ore, and manufactured in Sweden,[57] while S. H. Rigby also argued that it also referred to 'any similar iron sold in small bars or rods, by the barrel or sack, and not by weight'.[58] Schubert further defined osmunds by the process that was used to make them by melting

---

53 Rexroth, *Deviance and Power*, trans. by Selwyn, p. 31.
54 *CPR 1324–1327*, p. 29.
55 Schubert, *History of the British Iron and Steel Industry*, pp. 94, 98.
56 Schubert, *History of the British Iron and Steel Industry* , pp. 109–10.
57 See: Geddes, 'Iron', p. 168; Williams, *The Knight and the Blast Furnace*, p. 883 and Keene, 'Metalworking in Medieval London', p. 98.
58 *The Overseas Trade of Boston*, ed. by Rigby, p. 274.

pig-iron in a hearth and catching the melting iron on the end of a spinning staff which formed the iron into a ball as it cooled.[59] The size of an osmund also clearly varied: they tend to appear in shipping manifests measured by the barrel (as balls cannot be bundled easily), but while Rigby suggested that they were sold in small bars or rods, clearly some osmunds were very large. By the end of the fourteenth century, the London armourer Stephen atte Fryth (Berkshire) was forging with 'great pieces of iron called "Osmund" […] into "brestplates", "quysers", "jambers" and other pieces of armour'.[60]

Records from the Port of Boston show shipments arriving in Boston carrying iron of various types. For example, on 16 August 1390, a single ship belonging to Gerardus Westvale landed carrying twenty-two barrels of osmund iron, forty-six bundles of botolf iron,[61] and nineteen bundles of frewold iron.[62] Another ship arriving 20 August 1390 carried seventeen bundles and one last of frewold iron,[63] fourteen bundles of Hungarian iron, sixteen barrels of osmunds, seven barrels of landiron,[64] one last of botolf iron;[65] while a third ship arriving on 28 August 1390 carried 700 hundredweights of Spanish iron.[66] Even if armourers exclusively used osmunds, the ships that brought those osmunds to England and ultimately London were part of a complex iron-trading network spanning the entirety of the continent. That network supplied the raw materials necessary for the fledgling bespoke metal-armour industry; however, the iron trade also traced the paths that continental immigrants would follow into London, and the later armour-smuggling routes that would export English-made products back to the continent.

Creating the kind of infrastructure that could take advantage of England's native iron supply was a long process, but it was expedited in the fourteenth century by three factors: the demands of the Hundred Years' War, the larger armies of the 'Infantry Revolution',[67] and the shortages of man-power to supply that market resulting from the Black Death. As Schubert argued, it was 'absolutely necessary to off-set the labour shortage by improving methods at every stage from raw material to finished product'.[68] The size and complexity of metal armour was limited by the size of the furnaces used to smelt the iron. As Alan Williams and

---

59 Schubert, *History of the British Iron and Steel Industry*, pp. 300–01.
60 *LAoN*, p. 617.
61 Iron originating from the Bytów region in northern Poland.
62 Iron likely originating from Eastern Europe. See: Childs, 'England's Iron Trade', p. 34 and *The Overseas Trade of Boston*, ed. by Rigby, pp. 96–97.
63 A measure of capacity which varied by the commodity. Rigby states that a 'last' of iron equated to twelve barrels worth. See *The Overseas Trade of Boston*, ed. by Rigby, p. 261.
64 Iron originating from Germany.
65 *The Overseas Trade of Boston*, ed. by Rigby, p. 99.
66 *The Overseas Trade of Boston*, ed. by Rigby, p. 99.
67 The concept of an 'infantry revolution' has been introduced and excellently debated by Rogers, Curry, and Stone. See: Rogers, 'The Military Revolution'; Rogers, 'The Military Revolutions of the Hundred Years War'; Curry, 'Medieval Warfare', pp. 99–101 and Stone, 'Technology, Society, and the Infantry Revolution'.
68 Schubert, *History of the British Iron and Steel Industry*, p. 145.

H. R. Schubert have both shown, bloom weights and furnace complexity both increased considerably throughout the fourteenth century.[69] However, it is only towards the end of the fourteenth century that bloom weights of greater than ten kilograms became common.[70]

These heavy blooms were required for armour manufacture, as material wastage between the bloom and the finished product could be as high as seventy-five per cent, and a single plate for a large piece of armour might weigh two and a half kilograms.[71] The continental iron industry had a much more advanced network of ironworks, some of which were even powered by water. Evidence for the use of water-powered ironworks in England is scarce: Schubert has argued that water-powered ironworks might have existed in England as early as 1200, but this argument is based on the idea that water-powered ironworks began to become regular on the continent from that period and an assumption that 'it is unlikely that the monastic orders […] would have failed to introduce such a useful device when they settled in England'.[72] The earliest strong evidence for an English water-powered ironworks comes from the fourteenth century (compared to 1228 in Italy).[73] Water power increased the output of ironworks as well as the maximum possible size of the blooms (as they could operate larger bellows and hammers at no cost).[74]

It is not surprising then that most of the earliest mentions of English-made metal armour are made from either small plates attached to leather or fabric bases, or are made from wire knitted into mail: imported iron was expensive, and the higher slag content of English iron could make larger plates brittle, but small plates or wires could be made from smaller blooms manufactured in England, or from smaller pieces of imported iron to save on materials costs.[75] Furthermore, the practice of overlapping small plates or mail which defined the English armour industry for much of the fourteenth century allowed English armour-manufacturers to make the best use of the native iron: the gaps between overlapping plates and interlocking mail rings could act as crack-stoppers.[76]

---

69 Williams, *The Knight and the Blast Furnace*, pp. 877–79 and Schubert, *History of the British Iron and Steel Industry* , pp. 133–43.
70 Williams, *The Knight and the Blast Furnace*, pp. 877–79. Schubert has shown that by the end of the fourteenth century, very large blooms were being produced in England, with blooms found at Byrkeknott from 1408 weighing as much as eighty-eight kilograms. See: Schubert, *History of the British Iron and Steel Industry*, p. 137.
71 See: Williams, *The Knight and the Blast Furnace*, pp. 877–79. Adam Thiele has shown experimentally that one to two kilo blooms can be easily processed with small furnaces and technology available at the time, but the blooms his experiments produced were brittle and would have been unsuitable for arms and armour. However, his method may be where some of the plates used for wiredrawing mail or lamellar plate armour could have come from. See: Thiele, 'Smelting Experiments', pp. 2, 6.
72 Schubert, *History of the British Iron and Steel Industry*, p. 133.
73 Geddes, 'Iron', p. 172 and Cortese, 'Medieval Ironworking', p. 56.
74 According to Geddes, the highest cost in bloomeries was for the bellows-workers, amounting to twenty per cent of the total expenses at Tudeley. See: Geddes, 'Iron', p. 171.
75 The ductility of Swedish osmund iron made it particularly suited to wire-drawing. See: Schubert, *History of the British Iron and Steel Industry*, p. 111.
76 Williams, *The Knight and the Blast Furnace*, p. 879.

This type of armour only functioned when it was combined with other materials. The growth of the English armour industry therefore relied upon the growth of multiple industries working together: the linen-armourers, the *haubergers* (mail-makers), the *heaumers* (helmet-makers), the plate-armourers, alongside other materials specialists such as leatherworkers, baleen-armourers, *furbishers* (armour-repairers), alongside the armour retail market and the various raw materials trades which supplied the industry. These industries took advantage of the expertise that was imported from the continent but, even with the greatest technical skill, competing with continental armour using English iron was impossible. The comparative scarcity of high-quality iron in England ensured that the English armour industry became one which specialized in producing armour 'fitting for footsoldiers'. This armour rarely survives to the present, does not impress when compared to the finest craftsmanship of the continent, and would not have interested those with the ability to import the superior continental products. However, this sort of armour was ideal for the large armies of footsoldiers that England employed in the fourteenth century.

London's Illegal Armour Export Market

Even as the London's armourers' network was enriched by continental expertise, they could not meet all of the demand for armour in England. While modern scholars might deride the quality of English armour, the consumer revolution ensured a growing market for armour among less-affluent consumers, and the larger armies fielded by the English crown made the armour industry one of London's most important (if overlooked) industries. The crown could not risk the products of this industry falling into the hands of its enemies, nor could it allow armour to be exported, as this would reduce the local armour supply and therefore drive up prices because of the increased demand.[77] As a result, the wartime importation of armour and raw materials from the continent was encouraged, while its export was strictly prohibited.[78] At times of greatest need, the crown even ordered that armourers' homes and places of business be searched so that any armour found there could be sold by the king's yeomen 'at a reasonable price'.[79] These practices reduced profit in the local market, and drove armourers to increase their prices, lower their quality, or attempt to export their wares illegally. In the fourteenth century, prohibitions against exporting armour or references to those prohibitions in response to cases of armour exportation appear thirty-four times in the Close Rolls, Patent Rolls, Plea Rolls, and Letter Books, indicating that the export market for London's armour was very large, and that it

---

77 The crown regularly claimed that 'seeing the need of those about to set out [to war] [...], [armourers] now strive to sell [...] armour [...] at an excessive price'. *CCR 1354–1360*, p. 134; *LBH*, pp. 60, 269; *CCR 1385–1389*, pp. 262, 606; *CPR 1385–1389*, p. 261.
78 For statutes prohibiting iron export due to scarcity, see: *SotR*, I, 345.
79 Any armourers who were found to have concealed or attempted to eloign their wares were to forfeit their goods entirely. See: *CCR 1354–1360*, p. 134.

remained lucrative despite increasing royal penalties.[80] The retail armourers were further prohibited from raising their prices on nine occasions between 1350 and 1390;[81] and on nine occasions between the enrolment of the heaumers' ordinances and the aftermath of the 1383–1384 mayoralty riots, armourers were forcibly moved to the Tower of London to work their craft for the king for low or no wages, on penalty of indefinite imprisonment.[82]

The growth of the illegal export market is hardly surprising: as Mark Ormrod commented in his examination of the wool trade, smuggling 'was so endemic in medieval England that the Crown could hope only to reduce rather than eliminate this practice'.[83] As the practice was illegal, records of this practice only appear when exporters were caught: the Lombard John de Plesancia was charged in 1339 with buying arms in the Conduit and exporting them illegally to the king's enemies,[84] and commissions were established in 1341 and 1342 to seize English ships leaving from Lincolnshire, Essex, Norfolk, and Suffolk laden with armour to be sold in Scotland and Norway.[85] This is particularly important because as discussed above, the majority of iron used by London armourers came from the North Sea trade,[86] and the baleen occasionally used as a cheap replacement for metal plates certainly came from Scandinavian whalers.[87] Therefore, the simultaneous import of raw materials from Scandinavian countries, and illegal export of completed armour made by the London industry using those materials is an important area for future research.

---

80 See: *CCR 1307–1313*, pp. 44, 225, 522; *CCR 1313–1318*, p. 218; *CCR 1318–1323*, pp. 369, 694; *CCR 1323–1327*, p. 545; *CCR, 1327–1330*, p. 403; *CCR 1330–1333*, pp. 289; *CCR 1333–1337*, pp. 671, 675, 731; *CPMR i*, p. 102; *CPR 1340–1343*, p. 212; *CCR 1341–1343*, pp. 351, 496; *CCR 1349–1354*, p. 134; *CCR 1354–1360*, p. 62; *LBG*, p. 109; *CCR 1360–1364*, pp. 127, 405; *CCR 1364–1368*, p. 370; *CCR 1369–1374*, p. 114; *CCR 1374–1377*, p. 358; *LBH*, p. 27; *CCR 1396–1399*, p. 510; *CCR 1327–1330*, p. 403; *CCR 1330–1333*, p. 289 ; *CCR 1333–1337*, pp. 671, 675, 731; *CPMR i*, p. 102; *CPR 1340–1343*, p. 212; *CCR 1341–1343*, pp. 351, 496; *CCR 1349–1354*, p. 134; *CCR 1354–1360*, p. 62; *LBG*, p. 109; *CCR 1360–1364*,pp. 127, 405; *CCR 1364–1368*, pp. 370, 370; *CCR 1369–1374*, pp. 114, 387, 568; *CCR 1374–1377*, p. 358; *CCR 1377–1381*, pp. 17, 424; *CCR 1381–1385*, p. 421 *LBH*, p. 27.

81 *CCR 1354–1360*, p. 134; 'Edward III: June 1369', in *PROME*, <http://www.british-history.ac.uk/no-series/parliament-rolls-medieval/june-1369>; *LBH*, pp. 69, 160, 269, 288; *CPR 1385–1389*, p. 261; *CCR 1385–1389*, pp. 261–62, 606.

82 Knoop and Jones's study of masons' impressment suggests that the practice was used far more than it was recorded. See: Knoop and Jones, 'The Impressment of Masons', p. 58; *CPR 1354–1358*, p. 11; *CPR 1358–1361*, pp. 221–22, p. 323, p. 422; *CPR 1361–1364*, p. 282; *CPR 1367–1370*, pp. 240–41, 300; *CPR 1381–1385*, pp. 230, 574; TNA SC 8/247/12310.

83 Ormrod, 'The English Crown and the Customs', p. 31.

84 *CPMR*, I, 102.

85 *CPR 1340–1343*, p. 212; *CCR 1341–1343*, pp. 351; 496. The records of the King's Remembrancer in E 122 at The National Archives also contain some detailed accounts of arms intended for illegal export (and thus forfeit by the various statutes discussed above). TNA E 122/89/144 provides some particulars of seizures of arms illegally shipped from London in 1367/8, but illegal shipping of armour originally made in London is likely represented in many of the southern ports. However, this is a very large record series, and while it is organized by port, a complete survey of these documents as they relate to the London armour industry goes beyond the scope of this chapter.

86 Geddes, 'Iron', 168; Williams, *The Knight and the Blast Furnace*, p. 883.

87 *CPR 1324–1327*, p. 29.

Despite the great need for well-equipped armies, many of the crown's policies relating to the London armour market stunted its legal industrial activities. These short-sighted policies were intended to create a surplus of armour, and so keep costs down. However, those policies taught the armourers that it was not wise to develop a surplus of product within the king's grasp, or risk its seizure. Without a booming wartime local market to sell their products in, London's illegal armour export industry grew steadily through the latter half of the fourteenth century, as evidenced by the increasing penalties against this practice. In this period, the penalty for being caught smuggling armour out of England increased from seizure of the contraband goods and recording of the offender's name in 1360 to indefinite imprisonment in 1367, 'certifying the king from time to time of the names of those arrested'; and by 1369, to imprisonment of all buyers, sellers, and crew of any ships without the possibility of mainprisal, and the additional forfeiture of all of the offenders' lands and chattels.[88] Finally, one of Richard II's last orders relating to the armourers increased this punishment to 'forfeiture of life and limb' in 1399.[89] While the crown was trying to prevent potential enemies from benefitting from his country's armour industry, his wartime policies of impressment, purveyance, and seizure were the most effective encouragements for the armourers' civil disobedience and the growth of the illegal armour export industry.

The relationship between English and continental armourers was more complex than a simple 'inferior/superior' dichotomy. The English armour industry was certainly slower to develop than on the continent, but lacking the high-quality materials and ready trade enjoyed by continental armourers, the English industry developed in a unique position. England's ready access to Scandinavian baleen, local leather, and the largest wool and linen market in Europe encouraged the development of the linen-armour industry and armour that could be applied to linen armour to reinforce it. The transmission of craft knowledge from the continent to London is inherently tied to the development of London's multi-specialist network of interdependent armourers' crafts. The import of fine armour from Germany, Italy, Bruges, Normandy, France, and Spain all contributed to the development of a market for the skills of armourers from those areas. Under the aegis of the royal armouries, those talented continental armourers helped to develop London's prestige armour market and influenced the organization of London's nascent armourers' industry. However, infrastructural and raw materials limitations on the English industry created a niche for lower quality armour produced *en masse* to serve the larger consumer market. These challenges combined with the importation of labourers from London's eastern ports along the raw materials trade routes helped to develop the fourteenth-century English style of armour. While that armour does not attract the praise of modern historians, its lower costs and wider market had direct implications for the English war effort, and as the illegal export market has shown, that armour was very popular among Norwegian consumers. To understand fully the trade networks of armour and the influences between continental and English armourers, a much wider, multi-city and multi-national study is required.

---

88   *CCR 1360–1364*, p. 405; *CCR 1364–1368*, p. 370; *CCR 1369–1374*, p. 114.
89   *CCR 1396–1399*, p. 510.

Works Cited

Manuscripts and Archival Sources

London, British Library, MS Additional 47682
London, Consistory Court, MS 9171 fols 431–35
The National Archives (TNA), Kew, C 131: Chancery, Extents for Debts, Series I
—, C 241: Chancery, Certificates of Statute Merchant and Statute Staple
—, E 122: Exchequer, King's Remembrancer, Particulars of Customs Accounts
—, SC 8: Special Collections, Ancient Petitions

Abbreviations

CPMR: *Calendar of Plea and Memoranda Rolls*
CPR: *Calendar of Patent Rolls*
CCR: *Calendar of Close Rolls*
CWPH: *Calendar of Wills Proved and Enrolled in the Court of Husting*
DCoAD: *A Descriptive Catalogue of Ancient Deeds*
HGL: Derek Keene and Vanessa Harding, *Historical Gazeteer of London Before the Great Fire Cheapside; Parishes of All Hallows Honey Lane, St Martin Pomary, St Mary Le Bow, St Mary Colechurch and St Pancras Soper Lane* (London: Centre for Metropolitan History, 1987) <http://www.british-history.ac.uk/no-series/london-gazetteer-pre-fire>
LAoN: *London Assize of Nuisance: 1301–1431*, ed. by Helena Chew and William Kellaway. vol. X of *London Record Society Publications* (London: London Record Society, 1973)
LBA, LBB, LBC …: *Calendar of Letter-Books of the City of London: A-L*
Mem: *Memorials of London and London Life in the XIIIth, XIVth, and XVth Centuries*, ed. by Henry Thomas Riley (London: Longmans Green, and Co., 1868)
PROME: *Parliament Rolls of Medieval England*, ed. by Chris Given-Wilson (Woodbridge: Boydell, 2005)
SotR: *The Statutes of the Realm*

Primary Sources

Calendar of London Trailbaston Trials Under Commissions of 1305 and 1306, ed. by Ralph Pugh (London: Her Majesty's Stationary Office, 1975)
Ekwall, Eilert, *Two Early London Subsidy Rolls* (Lund: Gleerup, 1951)
The Local Customs Accounts of the Port of Exeter: 1266–1321, ed. by Maryanne Kowaleski (Exeter: Devon and Cornwall Record Society, 1993)
The Overseas Trade of Boston in the Reign of Richard II, ed. by S. H. Rigby (Woodbridge: Boydell, 2005)
Stapleton, Thomas. 'A Brief Summary of the Wardrobe Accounts of the 10th, 11th, and 14th years of King Edward the Second', *Archaeologia or Miscellaneous Tracts Relating to Antiquity Published by the Society of Antiquaries of London*, 26 (1836), 318–45

Secondary Studies

Amor, Nicholas R., *Late Medieval Ipswich* (Woodbridge: Boydell, 2011)
Brown, Michelle, 'The Historical Context', in *Holkham Picture Bible: A Facsimile*, ed. by Michelle Brown (London: The British Library, 2007), pp. 1–24
Childs, W. R., 'England's Iron Trade in the Fifteenth Century', *The Economic History Review*, New Series, 34.1 (Feb. 1981), 25–47
Cortese, Maria Elena, 'Medieval Ironworking on Mount Amiata', in *Prehistoric and Medieval Direct Iron Smelting in Scandinavia and Europe*, ed. by Lars Christian Norbach (Aarhus: Aarhus University Press, 1999), pp. 55–62
Curry, Anne, 'Medieval Warfare. England and her Continental Neighbours, Eleventh to the Fourteenth Centuries', *Journal of Medieval History*, 24.1 (1998), 81–102
Ekwall, Eilert, *The Concise Oxford Dictionary of English Place-Names*, 4th edn (Oxford: Clarendon Press, 1960)
—, *Variation in Surnames in Medieval London* (Lund: Gleerup, 1945)
Ffoulkes, Charles, *The Armourer and his Craft from the XIth to the XVIth Century* (New York: Dover Publications, 1988)
Geddes, Jane, 'Iron', in *English Medieval Industries*, ed. by John Blair and Nigel Ramsay (London: The Hambledon Press, 1991), pp. 161–88
Goldberg, P. J. P., *Women, Work, and Life Cycle in a Medieval Economy: Women in York and Yorkshire, C.1300–1520* (Oxford: Clarendon Press, 1992)
Fransson, Gustav, *Middle English Surnames of Occupation: 1100–1350* (Lund: Kraus Reprint Ltd., 1967)
Keene, Derek, 'Metalworking in Medieval London: An Historical Survey', *Historical Metallurgy*, 30.2 (1996), 95–102
Kirkland, Brad. "'Now Thrive the Armourers': The Development of the Armourers' Crafts and the Forging of Fourteenth-century London' (Unpubl. PhD thesis, University of York, 2015)
Knoop, Douglas, and G. P. Jones, 'The Impressment of Masons in the Middle Ages', *The Economic History Review*, 8.1 (Nov. 1937), 57–67
Mawer, Allen, 'Some Unworked Sources for English Lexicography', in *A Grammatical Miscellany Offered to Otto Jespersen on his Seventieth Birthday*, ed. by Niels Bøgholm and others (Copenhagen: Levin & Munksgaard, 1930), pp. 11–16
Mill, Anna J., 'The York Plays of the Dying, Assumption, and Coronation of Our Lady', *PMLA*, 65.5 (Sep. 1950), 866–76
Moffat, Ralph, James Spriggs, and Sonia O'Connor, 'The Use of Baleen for Arms, Armour and Heraldic Crests in Medieval Britain', *The Antiquities Journal*, 88 (2008), 207–15
Ormrod, W. M., 'The English Crown and the Customs, 1349–63', *The Economic History Review*, New Series, 40.1 (Feb. 1987), 27–40
Pfaffenbichler, Matthias, *Medieval Craftsmen: Armourers* (London: British Museum Press, 1992)
Reaney, P. H., and R. M. Wilson, *A Dictionary of English Surnames* (London: Routledge, 1991)
Rexroth, Frank, *Deviance and Power in Late Medieval London*, trans. by Pamela Selwyn (Cambridge: Cambridge University Press, 2007)

Rogers, Clifford J., 'The Military Revolution in History and Historiography', in *The Military Revolution Debate: Readings on the Military Transformations of Early Modern Europe*, ed. by Clifford J. Rogers (Boulder: Westview Press, 1995), pp. 1–12

—, 'The Military Revolutions of the Hundred Years War', in *The Military Revolution Debate: Readings on the Military Transformations of Early Modern Europe*, ed. by Clifford J. Rogers (Boulder: Westview Press, 1995), pp. 55–94

Schofield, John and others, 'Medieval Buildings and Property Development in the Area of Cheapside', *Transactions of the London and Middlesex Archaeological Society*, 41 (1990), 39–234

Schubert, H. R., *History of the British Iron and Steel Industry* (London: Routledge, 1957)

Stone, John, 'Technology, Society, and the Infantry Revolution of the Fourteenth Century', *Military History*, 68.2 (April 2004), 361–80

de Smet, W. M. A., 'Evidence of Whaling in the North Sea and English Channel during the Middle Ages', in *Mammals in the Seas*, vol. 3: *General Papers and Large Cetaceans: Selected Papers of the Scientific Consultation on the Conservation and Management of Marine Mammals and Their Environment*, ed. by Joanna Clark (Rome: Food and Agriculture Organization of the United Nations, 1981), pp. 301–09

Swanson, Heather, *Medieval Artisans: An Urban Class in Late Medieval England* (New York: Blackwell, 1989)

Thiele, Adam, 'Smelting Experiments in the Early Medieval Fajszi-Type Bloomer and the Metallurgy of Iron Bloom', *Periodica Polytechnica – Mechanical Engineering*, 54.2 (2010), 1–6

Williams, Alan, *The Knight and the Blast Furnace* (Leiden: Brill, 2003)

General Index

Page numbers in italics refer to maps, images, and plates

19-year cycle: 30–41

Abbasid Caliphate: 15
Abbotsley: 51
Absalon, archbishop of Lund: 98–100
Abū Isḥāq Ibrāhīm al-Zarqālī (Arzachel):
   8, 19, 36, 118, 120, 126, 129, 133
   Canones in motibus celestium corporum:
      8, 118, 120, 129, 133
account books: 113, 114
Adam Marsh: 77
Adam of Exeter: 77
Adelard of Bath: 17, 33
   Quaestiones naturales: 17
Æfintýr: 161, 164, 165, 169, 171, 173, 175
Africa: 12, 16, 164 n. 24
Al-Andalus: 33, 54, 120
al-Battānī: 118
   De scientia astrorum: 118
al-Bitrūjī: 16
al-Farghānī (Alfraganus): 36, 125
al-Khwārizmī: 33
al-Kindī (Alkindus): 47, 51
   De aspectibus: 51
al-Ma'mūn, caliph: 15
Albert Behaim, papal legate: 83
Albertus Magnus, saint: 17, 118, 119, 125,
   132, 144, 149, 152, 153, 172 n. 51
   De animalibus: 118, 132
   De mineralibus et lapidibus: 119
   Physicorum libri cum commentario: 132
Alchemy: 82, 152, 153, 156
Alexander the Great: 78–81, 84, 89, 90,
   163, 185
Alexander Neckam: 39, 40, 195 n. 174
   De naturis rerum: 195 n. 75
Alexander of Tralles: 116, 119, 134

   Practica: 116, 119, 134
Alexander de Villa Dei: 144
Alexandria: 195
Alfanus, bishop of Salerno: 16
   De natura hominis: 16
Alfonso VI, king of Castile: 16
Alfonso X, king of Castile: 120
Andrew Sunesen, archbishop of Lund: 97,
   99–103, 106
   Hexameron: 99
   Law of Scania: 99
   On the Seven Sacraments of the Church: 99
Anglo-Norman language: 46
annals: 141, 151, 168
Anonymous Bohemus: 148
Apocalypse of Saint John: 69
Apollonius of Perga: 53
Apollonius of Tyana: 79
Arabia: 13, 159, 160, 162–67
   Arabs: 30, 34, 37, 82, 83, 163, 166, 167
   Arabic language: 12, 13, 16, 17, 19, 27, 28,
      47, 48, 51, 52, 63, 77–84, 92, 93, 97, 131
      n. 57, 182, 188
   Arabic science: 15–17, 19, 27, 28, 31–37,
      46, 52, 53, 63, 77–80, 82, 84, 92, 93, 97,
      131 n. 57, 166, 167, 182, 188
Aragón: 32
Archimedes: 53
Aristotle: 7, 16, 20, 21, 45, 47, 48, 50–55, 62,
   78–84, 89–91, 93, 97, 104, 105, 115, 117–22,
   124–27, 131, 132, 144
   De anima: 53 n. 36, 118, 119, 124, 125, 131
   De animalibus (On Animals): 16, 51
   De caelo et mundo (On the Heavens): 51,
      62, 77, 117–19, 122, 131
   De generatione et corruptione: 117–19,
      122, 131

De longitude et breviate vite: 117
De memoria et reminiscentia: 7, 117, 118, 127, 131
De motu animalium: 118, 131
De sensu et sensato: 117, 118, 131
De somnis et vigiliis: 117, 131
Metaphysica: 117–20, 122, 131
Meteora (Meteorology): 51, 52, 54, 122
Organon: 97
Physica (Physics): 47, 117, 119, 120, 125, 132
Posterior Analytics: 47, 48, 51
Parva Naturalia: 117, 119
Sophistical Refutations: 104
arithmetic: 21, 47, 91, 92, 117, 119, 121, 122
Arnö: 146
Articella: 116, 124, 134
Asia: 12, 84, 164 n. 24
astrolabe: 27, 32
astrology: 16, 18, 21, 27–29, 35, 41, 78, 80, 85, 93, 123, 149
astronomical clocks: 125
astronomy: 11, 13, 15, 16, 18–21, 27–41, 47, 78, 80, 82, 85, 87, 88, 90, 92, 106, 115, 117–23, 125, 126, 133, 141, 149–52, 156, 181, 190
atoms: 33–37
Augustine, bishop of Hippo, saint: 85, 97, 144
Augustinian order: 100
Augustinus de Dacia: 146, 147
Rotulus pugillaris: 146
Austria: 119, 132 n. 58
Aymer de Valence, 2nd Earl of Pembroke: 209

Babylon: 190–92, 194
Baghdad: 13
Baltic Sea: 8, 142, 145, 155
Bartholomeus Anglicus: 118–20, 122, 124, 135, 191, 193, 194
De proprietatibus rerum: 118–120, 122, 124, 135, 191, 193, 194
Bayt al-Ḥikma (House of Wisdom): 15
Bede, Venerable, saint: 14, 15, 52

De natura rerum: 15, 52 n. 26
Benedictine order: 16, 116
Bergen: 22, 168, 170, 172
Bero Magni de Ludosia: 121, 122
Disputata super libros de anima: 121
Bessarion, cardinal: 18
bestiary: 116, 117, 183
Bible: 87, 97, 141, 142, 143, 156, 191, 201 n. 1
Birger Gregersson, archbishop of Uppsala: 146
Birgerus Magni, bishop of Västerås: 123
Birgittine order: 21, 124
Black Death: 207, 211
Boethius: 119, 121
De arithmetica: 119, 121
Bohemia: 142, 148
Bologna, University of: 21, 106, 117–19, 123, 134 n. 63, 168–73, 175
books: 22, 97, 98, 106, 107, 116, 120–26, 143, 145–47, 152, 164, 173, 177, 190, 191
booksellers: 118
Bordeaux: 209
Boston: 206, 211
botany: 15, 186, 191
Brandr Jónsson, bishop of Hólar: 163
Brescia: 177
Bristol: 206
British Isles: 14, 15
Bruges: 98, 203, 209, 215
Brussels: 68
Byzantium: 23

Calcidius: 15, 17, 116
Commentary on Timaeus: 15, 17, 116
calendars: 15, 19, 30–41, 50, 114, 116, 133, 133 n. 61, 147
Caliphate of Cordoba: 16
Cambridge, University of: 21, 119, 120
canon law: 21, 100, 101, 122, 142
Carmelite order: 118
Carolingian: 15
Carolus Erlandi, canon of Uppsala: 122
Castilian language: 163
cathedral schools, 13, 107, 117
catoptrics: 53, 54, 56

celestial bodies: 81, 82, 85, 190
celestial influences: 89
celestial motions: 29, 30, 40
Central Europe: 68
Chaldaeans: 30, 34–36, 39, 40
Charles V, king of France: 18
Chartres Cathedral School: 17, 20
Cheshire: 205
chiromancy: 122
chivalric literature: 13, 14, 174, 183
Chrétien de Troyes: 182, 187 n. 36
   Cligès: 182
   Yvain, ou le Chevalier au Lion: 182, 187 n. 36
Christ: 40, 46, 69, 154, 155, 177 n. 67
Cistercian order: 142
Claudius Ptolemy: 7, 20, 33, 36, 37, 40, 47, 53, 56, 62–67, 69, 90, 125
   Almagest: 18, 37, 40, 90
   Optics: 63
Cologne: 205, 209
Cologne, University of: 107, 108, 204, 206, 210
computus: 15, 19, 27–41, 115, 116, 122, 123, 133
computus naturalis: 30–32, 35, 36, 41
computus vulgaris: 29 n. 11, 31, 35
conjunctions: 30–36, 39
Conrad of Strasbourg: 40
Constantine the African: 16, 124
   De coitu: 124
Constantinople: 13, 97, 189–91
Copenhagen, University of: 21, 107
Copham: 205
Corpus Aristotelicum: 97
Cosimo de Medici: 145
cosmology: 15, 18, 47, 77, 80, 81, 85, 92, 93
Council of Lyon: 83
Council of Nicaea: 36, 40
Cracow, University of: 149
Cunestabulus: 36–38
curriculum: 20, 21, 98, 115, 117, 125
Cyprus: 65
Czech translation: 148

Dante Alighieri: 83
David le Hope: 209

De morali principis institutione: 147
deferents: 30
demagnification: 7, 57–59, 61
Denmark: 13, 21–23, 97–107, 111, 120 n. 32, 142
   Danes: 21, 97, 98, 100, 105 n. 18, 106, 107, 145
   Danish: 20, 22, 99, 101, 102 n. 12, 105, 106, 121, 144, 145, 150, 151, 156
   Danish language: 142, 143 n. 4, 145
Derbyshire: 209
Dhū al-Qarnain: 84
Dialogus creaturarum: 123
Diarium Bibliothecae Sorbonae: 118
Dies Aegyptiaci: 116
Dinant: 209
dioptrics: 53, 56
disputation: 104, 105, 119
Dominic, saint: 143, 177
Dominican order (Order of Preachers): 8, 13, 14, 17, 19–22, 100, 119, 122, 123, 141–56, 161, 163, 168–70, 174, 177
   Dominican province of Dacia: 13, 20–22, 122, 142–51, 154
Dominus Castri Goet: 83
Don Juan Manuel: 163
   El Conde Lucanor: 163
Dublin: 22
Durham: 210
Dutch congregation: 145, 150

earth: 33, 69 n. 80, 80, 87–89, 91, 155
Easter: 15, 27, 28, 116, 154
Eastern Europe: 211 n. 62
eccentric: 30, 33
eclipse: 30–32, 36, 86 n. 61, 133 n. 60, 151
Edward I, king of England: 204
Edward II, king of England: 203, 204, 209
Edward III, king of England: 205, 214 n. 81
Egypt: 53, 65, 67
Eilífr korti Árnason: archbishop of Nidaros: 170, 175, 177
Einar Haflidason: 169, 170
Eiríks saga víðfǫrla: 195, 196
Elyas de Wodeberghe: 205, 206

encyclopaedias: 14, 15, 39, 52, 78, 116, 135, 182, 186, 187, 191, 194, 195
engineering: 167
England: 12–14, 19–23, 28, 32, 50, 77, 105, 105, 106, 116, 119, 135, 144 n. 20, 171, 183, 201–03, 207, 209–15
 English: 13, 22, 23, 33, 48, 50, 77, 81, 93, 112, 117–20, 133 n. 61, 135 n. 64, 148, 187, 201, 202, 205, 207–15
Epicycles: 30, 33
Epistola Alexandri ad Aristotelem: 194
equator: 149
equinoxes: 37, 149
Erfurt, University of: 21, 106, 123, 124
Eric VII, king of Denmark: 18
Essex: 208, 214
Ethiopia: 13, 159, 167
Euclid: 51, 53, 56, 62, 119, 120, 122, 125
 Catoptrica (Catoptrics): 51, 57
 Optics: 53, 78
Eugenius of Sicily: 63
Europe: 12, 15–17, 19, 21, 23, 31, 34 n. 27, 47, 48, 54, 55 n. 42, 62, 69, 81, 83, 106, 115, 117, 142, 147, 149, 163, 164, 167, 182, 186–88, 215
Exeter: 206

Faculty of arts: 17, 18, 21, 97, 115, 117, 125
Faculty of medicine: 17, 21, 117
Fibonacci: 17, 18
 Liber Abaci: 17, 18
Finland: 111, 113–15, 125, 126
Finnish: 13, 112, 113, 114, 115, 125
Flanders: 14, 209
Florence: 145, 209
Forest of Dean: 210
Fornsvenska legendariet: 148
France: 13, 14, 18, 20, 32, 50, 118, 122, 131, 132 n. 59, 141, 145, 159, 161, 165–68, 170, 176, 177, 183, 203, 209, 215
 French: 99, 100, 247, 149 n. 40, 159–61, 165, 166 n. 31, 172 n. 51, 178, 187, 188
Francesco Petrarca: 171
 Rerum memorandum libri: 171
Franciscan order: 17, 19, 22, 46, 50, 77, 84, 85, 97, 105, 120, 123, 124, 142–144, 146, 154

Franciscan province of Dacia: 123
Frederick II, Holy Roman Emperor: 16, 18, 83
Friar Gregorius of Stockholm: 147–49
Friar Guali: 177

Galen: 20, 116
 Tegni: 116
Gascony: 203
Gedeon Gedeonis: 146
gemstones: 183, 187, 190, 194
geography: 15, 87, 141, 149, 156, 191
geomancy: 82, 91
geometry: 11, 17, 21, 45, 47–56, 59, 88, 91, 92, 117, 119, 121, 126
Georg Peuerbach: 18, 19
Gerard of Cremona: 16, 79, 120
Gerard of Cremona/Montpellier: 122, 144 n. 19
Gerbert of Aurillac: 16
Gerland: 31–36, 41
Germany: 22, 116, 132 n. 58, 133 n. 61, 145, 183, 201, 203, 210, 211 n. 64, 215
 German: 18, 106, 107, 116, 119, 125, 142, 147, 170, 172 n. 51, 187
 Middle High German language: 82 n. 39, 183, 185
 Middle Low German language: 145, 160, 164 n. 25, 170
Gilbert Foliot, bishop of Hereford and London: 28
Giles of Rome: 132
 Super de anima: 132
Giovanni Dondi: 18
Gloucestershire: 207, 208
Golden Numbers: 39, 40
Goliards: 174, 178
Gotland: 125
Great Malvern: 20, 32
grammar: 98, 144
Greece: 70. 84, 183
Greek: 12, 14–20, 37, 46–48, 51, 62, 69 n. 80, 77, 78, 80, 82–84, 87, 91, 97, 131 n. 57, 188
Greenland: 150

Gregorian calendar: 41
Gregory IX, pope: 144
Gregory XIII, pope: 41
Greifswald: 22, 150
Greifswald, University of: 21, 123, 150
Guido, bishop of Tripoli: 82
Guillelmus de Saliceto: 119, 134
    Summa conservationis et curationis: 119, 134
Gunner, bishop of Viborg: 20, 97, 102–05, 107
Gustav I Vasa, king of Sweden: 112, 113 n. 6, 122
Gustav II Adolf, king of Sweden: 124

Hákon IV Hákonarson, king of Norway: 187
Hārūn ar-Rashīd, caliph: 15
Hektors saga: 183, 188 n. 44
Helsingborg: 22, 146
Helsinki: 112, 113
Hemming, canon of Uppsala: 117, 120, 121, 126
Henrici de Ansbeke: 118 n. 28, 131
Henry of Schwerin, count: 101, 102
herbal: 15, 116
Hereford Cathedral: 20, 28, 49, 50
Hermann of Reichenau: 31
hermetic sciences: 16, 20, 78–81, 83, 85
Hildegard of Bingen: 18
Hipparchus: 36, 37
Hippocrates: 116, 124, 134
    Aphorisms: 116, 124
    Prognostics: 116, 124
Hólar Cathedral: 21, 163
Holy Roman Empire: 13, 14, 16, 18, 21, 119
Honorius III, pope: 101–02
Horace: 184, 185
    Ars poetica: 184, 185
Hugh Foliot, archdeacon of Hereford: 50
Hugh le Armurer: 203, 209
Hugo of Sanctalla: 79
Humbertus de Romanis: 143, 163
    De dono timoris: 163
Hungaria: 151

Iberian: 32, 47, 82
Ibn al-Baṭrīq: 78
    Sirr al-Asrār: 78–82, 84
Ibn al-Haytham (Alhazen): 51–54, 57, 59, 62, 63, 67, 69
    De Aspectibus (Book of Optics): 53, 63, 78
    Kitāb al-Manāẓir: 53, 78
    Perspectiva: 53
Ibn Rushd (Averroes): 8, 19, 47, 54, 82, 84, 118, 130, 133
    De substantia orbis: 8, 118, 130, 133
Ibn Sīnā (Avicenna): 19, 47, 52–54, 82, 84, 118, 125, 134
    Canon medicinae: 118, 134
Iceland: 13, 14, 21, 161, 163, 168–70, 174 n. 58, 178, 187, 188
    Icelanders: 174, 178, 188
    Icelandic: 13, 160, 163, 167 n. 34, 169, 173, 186–88
    Old Icelandic language: 159–61, 167, 173, 181, 182, 185–88, 191, 194, 195
In Defence of Ockham: 105
India: 16, 163
Ingeborg, queen of France: 99
Innocent IV, pope: 83
Iohannes Hispalensis (John of Seville): 17, 82
Ipswich: 208
Iran: 53
Iraq: 15, 16
Ireland: 15, 116
Isidore, bishop of Seville, saint: 14, 15, 51, 52, 184, 191, 194, 195 n. 74
    De natura rerum: 52 n. 26
    Etymologiae: 14, 184, 191, 193, 194, 195 n. 74
Islam: 70, 163, 167 n. 34
    Islamic: 15, 16, 19, 23, 33, 47, 52, 59, 62, 67, 78, 80
Israel Erlandi: 122, 148
Italy: 16, 18, 67, 119, 134, 142, 145, 168, 183, 201, 212, 215

Jacobus de Cessolis: 144
Jacobus de Voragine: 147, 148, 177 n. 67, 178 n. 69

Legenda aurea: 147, 148, 156
Jacobus Petri de Röd, canon of Turku: 118
Jean Buridan: 18, 122
    Meteorologica: 122
Jews: 28 n. 7, 30, 34, 37, 92, 205
    Jewish: 15, 19, 23, 28 n. 7, 32–34, 37, 47 n. 7, 83, 91
    Hebrew language: 13, 84, 188
Johann Snell: 123
Johannes Campanus of Novara: 121
Johannes de Muris: 119
    Arithmetica speculativa: 119
Johannes de Sacrobosco: 40, 118–21, 125
    Algorismus, 121
    Computus: 121
    De sphaera: 118, 119, 121, 125
Johannes Garisdale: 124
    Termini naturales: 124
Johannitus (Ḥunayn ibn Isḥāq): 116, 124
    Isagoge: 116, 124
John de Coloigne. 205
John Duns Scotus: 105, 120
    De anima: 120
John Pecham: 119
    Perspectiva communis: 119
John of Salisbury: 82, 84
    Metalogicon: 82
John Scotus Eriugena: 15
Jón Halldórsson, bishop of Skálholt: 14, 19, 21, 22, 161, 163, 165, 168–71, 173–78
    Af meistara Pero ok hans leikum: 160, 169
    Clári saga keisarsonar: 13, 14, 160–70, 174
Jóns þáttr: 169, 170, 173, 175, 177
Julius Caesar, Roman emperor: 193
    De bello gallico: 193
Jupiter: 133 n. 60
Justin: 98
    Breviarium Historiarum Philippicarum Pompei Trogi: 98

Kamāl al-Dīn al-Fārisī: 52, 69
Kanutus Johannis: 123
Kent: 205, 208
Konrad Gruter: 18
    De machinis et rebus mechanicis: 18

Konráðs saga keisarasonar: 6, 13, 181, 187–96
Kungälv: 123

Lambrecht: 185
Lapidary: 116, 183, 194, 195
Lárentíus Kálfsson, bishop of Hólar: 170, 175
Latin: 12, 13, 15–17, 19, 27, 28, 33–41, 47, 51, 55 n. 42, 63, 78–83, 91, 97, 99–103, 116, 120, 122, 124, 134 n. 62, 143–47, 160–64, 168, 170, 172, 174, 175
Latin Europe: 16, 27, 31, 33, 47
Laurentius Magni: 147 n. 32
Laurentius Nicolai: 124
Law of Reciprocity: 55
Leicester: 50
Leicestershire: 205
Leipzig: 149, 150
Leipzig, University of: 21, 118, 119, 126, 132 n. 58
lens: 7, 53, 57–62
Leo of Naples: 81
Leonardo da Vinci: 19
Letter Books: 202, 213
Liber quadrantis et ipse quadrans: 121
liberal arts: 98, 99, 103, 106, 117, 152, 164
libraries: 12, 15, 20–22, 28 n. 4, 97, 106, 112, 117, 120–26, 143–48
light rays: 7, 49, 51, 53–62, 66, 68
Lincoln Cathedral: 20, 46, 49, 50, 77
Lincolnshire: 204, 205, 207, 208, 214
Linköping Cathedral: 117
Lödöse: 121
logic: 20, 97, 97, 104, 105, 122, 124, 125, 144
Lolland: 145
London: 8, 28, 105, 201–16
longitude: 30, 33
Lotharingia: 31, 32
Lübeck: 22, 145, 147 n. 32
lunar: 28, 30–32, 34, 35, 37–39, 41, 133 n. 61
lunations: 33, 34, 36–38, 40
Lund: 22, 100, 123, 146, 149, 150, 151 n. 49
Lund Cathedral: 20, 97–99, 106, 125
Lutheran Reformation: 106, 107, 111, 112, 122, 143

Macer: 116, 124
   De viribus herbarum: 116, 124
Macrobius: 15
   Commentarii in Somnium Scipionis: 15
magnets: 195, 196
magnification: 7, 57–59, 61
Magnús VI Hákonarson, king of Norway: 177
Manekyn le Heaumer: 203, 204, 207
manuscripts: 11, 12, 22, 28, 40, 57 n. 49, 59, 63, 78, 83, 84, 97, 105, 106, 111–35, 143–47, 160, 162, 173, 186–88
manuscript fragments: 13, 20, 21, 111–35
mappa mundi: 120
Marbod of Rennes: 116, 124
   Liber lapidum: 116, 124
Mariu Saga: 177 n. 67
Marseille: 35
Martin of Dacia: 106
Martinus Polonus: 148, 149
   Chronicon pontificum et imperatorum: 148, 149
mathematics: 15, 16, 18, 19, 27, 29, 30, 40, 41, 47, 51, 53, 78, 85–93, 121
Mathesis: 86, 87
Mathias Ripensis: 153–55
medicine: 14, 17–21, 78, 115, 117, 122, 134, 141, 145, 150, 152, 156
Mediterranean Sea: 18
Mercury, planet: 133 n. 60
meridian: 28, 35 n. 36, 50, 133 n. 60
Merton College, University of Oxford: 18, 57
Mesopotamia: 16
Michael Scot: 16, 82, 83, 131 n. 57
Middle East: 15–17
mineralogy: 183, 186, 191, 194–96
mirrors, use in optics: 51, 52, 54, 56, 62, 66
monasticism: 15, 20, 22, 98 n. 4, 99, 113, 142, 148, 156, 212
Montecassino Abbey: 16
moon: 30–41, 133 n. 60, 151
Mǫttuls saga: 160
music: 15, 18, 47
natural history: 18, 181, 186, 188, 191, 193, 196

natural phenomena: 47, 48
natural philosophy: 12, 20, 21, 47, 50, 77, 117, 126, 156, 188, 196

natural sciences: 15, 16, 18, 141, 149, 152, 156
Navarre: 203
Neoplatonism: 17, 20, 79
Nicolaus Copernicus: 19
Nicolaus de Dacia: 149–51, 156
   Libri tres anaglypharum: 149
Nicholas, bishop of Schleswig: 102
Nicholas Oresme: 18, 119
   Latitudines formarum: 119
Nicholas Trevet: 49
Nicolaus Lagonis: 146
Nidaros Cathedral: 21, 168
Norfolk: 205, 207, 208, 214
Normandy: 209, 210, 215
North Yorkshire: 205, 207
Northern Europe: 12, 23, 30, 56, 68, 141, 148–50, 182, 193
Northumberland: 206
Norway: 21, 22, 111, 142, 168, 170, 175, 177, 187, 214
   Norwegian: 21, 169, 170, 188, 215
Nottinghamshire: 206
Nyköping: 22, 120

Olaus Johannis: 125, 126
Olaus Petri: 122
Olaus Pauli: 122, 123
Old Norse language: 81, 177 n. 67, 184, 185, 191, 194
Old Provençal language: 164
optics: 14, 17, 19, 45, 48–56, 63, 69, 77, 119
Oxford: 22, 46, 50, 77, 119, 120
Oxford, University of: 21, 47

Pachomius, saint: 36, 133 n. 61
Padua, University of: 18
Paris: 50, 97–99, 106, 160, 172 n. 51, 209
Paris, University of: 13, 18, 20, 21, 47, 50, 77, 97, 102, 103, 105, 106, 117–21, 126, 131 n. 56, 146, 149, 161, 163, 165 n. 27, 168–70, 172 n. 51, 173, 175–78

Paul II, pope: 121
Persian: 79, 80, 93
perspectiva: 54
Perus, master: 13, 19, 159–78
Peter Nayer: 206
Peter Pauli: 120
Peter Philomena de Dacia: 106, 107
Peter Sunesen, bishop of Roskilde: 99
Petrus Alberti: 107
Petrus Alfonsi: 32, 33
Petrus Astronomus: 125
Petrus de Slaulosia: 145
Petrus Hispanus: 144
   Summulae logicales: 145
Petrus Laale: 145
   Parabolae: 145
Petrus Olai: 125
Philaretus: 116
   On Pulses: 116
Philip II, Augustus, king of France: 99
Philip IV, the Fair, king of France: 161
Philip of Tripoli: 78, 82
Philosophia: 11
physics: 20, 45, 48, 51, 54, 117, 141, 147, 154, 156
physician: 134 n. 62, 182
Physiologus: 116, 122
planets: 18, 81, 87, 89, 90, 120
Plato: 11, 16, 17, 20, 51, 55, 79, 116, 117, 122, 144
   Timaeus: 17, 51, 122
Pliny the Elder: 15, 135, 145, 191, 193–96
   Naturalis historia: 15, 135, 145, 191, 193–96
Poland: 142, 211 n. 61
Polonia: 142
Pomerania: 142
Portugal: 183
Porus, king of India: 164
printing: 19, 22, 23, 114, 123, 124, 145, 146, 147 n. 34
Prussia: 142
Pseudo-Apuleius: 116
   Herbarium: 116
Pseudo-Aristotle: 7, 19, 20, 78, 115, 118, 128, 132

   De coloribus: 118, 132
   De lineis insecabilibus: 118, 132
   De physionomia: 118, 132
   Lapidary of Aristotle: 79
   Liber de causis: 19, 20, 78, 118, 132
   Secretum secretorum: 118, 128, 132
Pseudo-Bede: 124
Pseudo-Burley: 135
   De vita et moribus philosophorum: 135
Pseudo-Euclid: 51

quadrant: 121
quadrivium: 15, 47
quaestio: 105

rainbow: 20, 45–55, 62, 69, 151
Randers: 123
Ranerius, patriarch of Antioch: 83
Rastede Abbey: 116
Raymond of Marseille: 35
reflections in optics: 45, 51–54, 56, 62, 63, 66, 69, 81
refraction: 7, 45, 46, 48, 53, 56–58, 62–69
Regiomontanus: 18, 19
renaissance of the twelfth century: 27
Rhazes (Abū Bakr Muhammad Zakariyyā ar-Rāzī): 134
   Liber medicinalis Almansaris: 134
rhetoric: 100, 102, 189
Ribe: 22, 145, 154
Riddarasaga: 159, 161, 169
Robert Grosseteste: 7, 13, 17, 19–22, 28 n. 5, 41, 45–51, 54–57, 59–63, 69, 77, 84, 85, 93, 123
   Château d'amour: 46, 69
   De iride (On the Rainbow): 45, 47–51, 54–57, 59, 61, 62, 69
   Hexaemeron: 69, 83
   Notes on Physics: 123
   On Comets: 50
   On the Generation of Sounds: 50
Roger Bacon: 13, 17, 19–22, 48 n. 9, 57, 77, 78, 81, 83–93
   De multiplicatione specierum: 87
   Opus Maius: 57, 83–88, 93

Opus tertium: 86
Secretum secretorum: 19, 20, 77–78, 82–93
Roger of Hereford: 13, 19, 27–41
    Compotus: 27–41
Roman: 20, 65, 100
Roman numerals: 32, 33
Romantic Literature: 79, 81, 159, 160, 170, 174, 181–89, 191, 194, 195, 196
Rome: 105, 149, 150
Roskilde: 151
Roskilde Cathedral: 20, 98, 107
Russia: 111, 113

Sabean: 17
Saint Victor Abbey: 20
Salerno, schools of: 16, 119, 134 n. 62
Sami: 111
Sandwich: 14, 209
Sassanian: 80
Saxo Grammaticus: 98 n. 4
Saxony (Saxland): 159, 161, 163, 164–67, 189
    Saxon: 159
Saxonia: 142, 147 n. 32
Scandinavia: 12–14, 20–23, 112, 117, 123, 125, 126, 141, 146, 183, 188, 208
    Scandinavian: 21–23, 117, 118, 121, 141, 145, 148, 208, 214, 215
Schleswig: 20
schools: 12, 13, 16, 20, 47, 50, 106, 107, 117, 141, 142, 151, 156, 163, 170, 175, 176
Scotland: 13, 204, 207, 209, 214
    Scottish: 107, 166, 167
Sendibréf Alexanders: 194
Sicily: 16, 83
sidereal: 31, 38
Siena Cathedral: 67, 68
Sigtuna: 22, 122, 123, 143, 147, 148
Sigurðar saga þǫgla: 182, 183, 188 n. 44, 191 n. 61
Silesia: 142
Simon de Wynchecombe: 207, 208
Siward, bishop of Uppsala: 116, 117
Sixtus IV, pope: 125
Skálholt Cathedral: 21, 22, 161, 168–70, 173, 175, 177, 178

Skänninge: 22, 148, 151
Skara Cathedral: 20, 121, 122, 147 n. 33
Sluys: 209
Snell's Law: 7, 53, 64–67
solar: 31, 36–38, 87, 151
solstices: 37
Somerset: 205, 206
Sophistics: 103–05
Sorbonne College, University of Paris: 118
South English Legendary: 148
Spain: 16, 183, 201, 203, 210, 215
    Spanish: 211
stars: 29, 55, 85, 87, 89–91, 154, 165
Stephen Langton: 99
Stephanus de Borbone: 147, 154, 163
    Tractatus de diversis materiis praedicabilibus: 163
Stjórn: 191, 193
Stockholm: 22, 112–14, 123–25, 147–49
Strasbourg: 123
Studia: 17, 22, 118
studium: 169, 178
studium generale: 119, 123
studium particulare: 119
subalternation: 45, 46, 48, 49, 54, 55
Suffolk: 49, 205, 208, 214
Sun: 30–32, 37–41, 51, 52, 55, 69 n. 80, 78, 87, 133 n. 60, 149, 151
sunlight: 51, 52
Suno Karoli, canon of Linköping: 117, 118
Surrey: 205
Sweden: 19–23, 111–26, 134 n. 61, 142, 143, 148, 149, 210, 221 n. 75
    Swedish: 13, 21, 107, 112–16, 119, 125, 134 n. 63, 148
    Old Swedish language: 122, 124, 148
synodic month: 30–33, 35, 40
syzygy: 33, 35

talismans: 16, 79, 81, 90
Tallinn: 22, 143, 150
technology: 12, 14, 15, 17–19, 67, 125, 167, 196, 212, 213
telescopes: 61, 62
terminus paschalis: 36

Teutonia: 142
textbooks: 30, 40, 97, 101
Thābit ibn Qurra: 17, 36
The Netherlands: 183
    Dutchman: 107
Theresa, queen of Portugal: 82
Theodoric of Freiburg: 69
theology: 47 n. 6, 50, 62, 77, 86, 93, 97–99, 103, 106, 107, 114, 115, 121, 122, 124, 141–56, 182
Theophilus (Presbyter): 68
Theophilus: 116, 134
    De urinis: 116, 134
Theorica planetarum: 120, 121
Thomas Aquinas, saint: 133, 144, 146, 147, 152, 153
    Commentarius ad libros de caelo et mundo: 133
    Expositio in Posterium: 133
    Summa theologiae: 144, 146, 153
Thomas Bradwardine: 119
    Proportiones breves: 119
Thomas Docking: 77
Thomas of Cantimpré: 194
    Liber de natura rerum: 194
Thuo of Viborg: 106
Tideman of Närke: 119–20
    Termini naturales: 120
Toledan Tables: 33–35, 38, 120, 121, 126
Toledo: 16, 35, 133
translations: 12–17, 19, 27, 33, 47, 51, 55 n. 42, 62, 63, 77–70, 82, 83, 93, 97 n. 2, 99, 131 n. 57, 145, 148, 160, 161, 163 n. 22, 182, 187, 188, 191, 194
tropical year: 37
Turku: 113
Turku Cathedral: 20, 111, 118

universe: 80, 92
universities: 12–14, 17–22, 40, 47, 83, 105–07, 115–22, 126, 132 n. 58, 156, 178, 182, 186
Uppsala Cathedral: 20, 111, 116, 117, 120–22, 125, 146
Uppsala, University of: 21, 107, 123–26, 143 n. 6, 146

Vadstena Abbey: 21, 124, 125, 143 n. 6, 146, 151 n. 53
Valdemar II, king of Denmark: 101, 102
Valerius Maximus: 98
    Facta et dicta mirabilia: 98
Vallentuna Missal: 116
Västerås: 112
Västerås Cathedral: 20, 123
Venus: 133 n. 60
Viborg Cathedral: 20, 105
Vienna, University of: 20, 21, 119, 121, 126, 132 n. 58
Vincent of Beauvais: 147, 191, 194
    Speculum historiale: 147, 191
    Speculum naturale: 147, 191, 194
Virgin Mary, saint: 46, 154, 177 n. 67
Visby: 22, 59 n. 54, 146

Wadi Natrun: 65
Walcher of Malvern: 32–34, 41
Walter Burley: 104, 135 n. 64
    De Puritate Artis Logicae: 104
Walter of Châtillon: 81, 163
    Alexandreis: 81, 163
Western Europe: 164
William de Glendale: 206, 207
William de Vere, bishop of Hereford: 49
William of Auvergne, bishop of Paris: 50
William of Conches: 117, 122
    Glosae super Platonem: 117, 122
William of Heytesbury: 120
    Termini physicales: 120
William of Moerbeke: 131
William of Ockham: 105–07
Witelo: 7, 46, 48, 62–68
    Perspectiva: 46, 63, 66, 67

Yarmouth (Great): 208, 209
York: 28 n. 28, 207
Yūḥannā ibn al-Biṭrīq: 52

Zealand: 100
zodiac: 32, 34, 37, 38, 133 n. 61
zoology: 13, 183, 186, 190–94

Knowledge, Science, and Scholarship in the Middle Ages

All volumes in this series are evaluated by an Editorial Board, strictly on academic grounds, based on reports prepared by referees who have been commissioned by virtue of their specialism in the appropriate field. The Board ensures that the screening is done independently and without conflicts of interest. The definitive texts supplied by authors are also subject to review by the Board before being approved for publication. Further, the volumes are copyedited to conform to the publisher's stylebook and to the best international academic standards in the field.

Titles in Series

Faith and Knowledge in Late Medieval and Early Modern Scandinavia, ed. by Karoline Kjesrud and Mikael Males (2020)